CARBON REINFORCED EPOXY SYSTEMS

VOLUME ONE
MATERIALS TECHNOLOGY SERIES

Edited by

CARLOS J. HILADO

a TECHNOMIC® publication
TECHNOMIC Publishing Co., Inc.
265 W. State St., Westport, Conn. 06880

CARBON REINFORCED EPOXY SYSTEMS

VOLUME ONE
MATERIALS TECHNOLOGY SERIES

265 W. State St., Westport, CT 06880

a TECHNOMIC® publication

Printed in U.S.A.
Library of Congress Card No. 74-83231
Standard Book No. 0-87762-149-7

To Connie

INTRODUCTION

The use of materials has served as a measure of the progress of civilization: the Stone Age, the Bronze Age, the Iron Age. To the proponents of modern synthetic materials, this is the Age of Plastics. The complexity of modern civilization is so great that we are in the age of metals, the age of plastics, and the age of other types of materials as well, because all types of materials have found areas of usefulness in our life and work.

Because of the abundant flow of information about a wide variety of materials, information relevant to a particular group or type of materials tends to be scattered among various publications and different issues of the same publication. Compilations of articles on specific types of materials are a valuable literature resource, and the Materials Technology series of books embodies this concept.

The Materials Technology series consists of compilations of articles pertinent to specific types of materials, drawn from journals published by Technomic Publishing Company.

Composite materials seek to utilize the best features of widely differing types of materials, and fiber reinforced composites are based on the use of high strength and high stiffness materials. Carbon reinforced epoxy systems represent an important class of composite materials.

The twenty-seven articles in this volume are reprinted from the Journal of Composite Materials, with the hope that it will prove to be a valuable literature resource in this field of knowledge.

CONTENTS

CONTENTS (continued)

PUBLISHER'S NOTE

Strain Rate Effect on the Ultimate Tensile Stress of Fiber Epoxy Strands*

T. T. Chiao and R. L. Moore, *Lawrence Radiation Laboratory, University of California, Livermore, California 94550*

(Received October 24, 1970)

The strain rate effect on composite performance has been the subject of many arguments. This note points out some of the problems involved in measuring strain rate effect and presents our tensile data on some simple unidirectional composites.

EXPERIMENTAL

For this study we used only one spool of single end glass (SCG 150-1/0-1Z-HTS 901, 204 filaments). The spool was kept in a container with a drying agent when not in use. A flexible epoxy formulation in 40% acetone solution (Dow DER 332/Union Carbide ERL-4206/Celanese Epi-Cure 855,† 70 parts/30/40), was used in the program. This low viscosity resin, which has higher elongation to break than the fiber, and a torsional shear strength of 3.8 Ksi, can be cured at room temperature; however, in order to save time, we evaporated off the acetone and cured the specimens at 350°F for one hour. The properties of the cast pure resin will be determined in connection with our work on stress-rupture of composites.

A complete description of our specimen fabrication and test method for fiber strands was reported previously [1]. Briefly, a strand is cut from the spool of fiber, allowed to untwist by holding one end up, dipped in a disposable trough containing epoxy-acetone solution, and cured vertically under constant low tension in an oven. Through the use of clamps and an alignment fixture, the specimen is then fixed to a 10 in. gage length. The clamps (cushioned with the combination of pressure-sensitive aluminum foil, 0.015-in. Neoprene sheet, and household cement) are held by four 6-32 screws, each having 50 to 75 in.-oz. torque.

We tested the specimen-clamp system in an Instron tensile test machine, model TTDM, with a load capacity of 10,000 kg. At the higher crosshead speeds, i.e., 2-20 in./min, the conventional load-weighing system is completely inadequate, particularly for fibers of low failure strain.

---

* Work performed under the auspices of the U.S. Atomic Energy Commission.

† Reference to a company or a product name does not imply approval or recommendation of the product by the University of California or the U.S. Atomic Energy Commission to the exclusion of others that may be suitable.

Table 1. Strain Rate Effect on the Ultimate Tensile Stress of Fiber/Epoxy Strands.

Material	Strain Rate (in./in.-min)	Av. Tensile Strength (psi)	95% Conf. Limits	Std. Dev. (psi)	Coeff. of Var. (%)	No. of Specimens
S-Glass/Epoxy[a]	1.97×10^{-4}	497,700	±15,400	21,500	4.3	10
	1.97×10^{-3}	522,900	±18,000	25,000	4.8	10
	1.97×10^{-2}	555,100	±15,000	21,000	3.8	10
	1.97×10^{-1}	607,500	±11,400	15,900	2.6	10
	1.97	610,300	±11,800	16,500	2.7	10
Graphite/Epoxy[b]	1.97×10^{-4}	225,700	± 5,800	8,200	3.7	10
	1.97×10^{-3}	224,000	± 7,400	10,300	4.6	10
	1.97×10^{-2}	218,800	± 5,200	11,100	5.1	20
	1.97×10^{-1}	231,800	± 5,600	11,900	5.1	20
	1.97	237,100	± 5,800	10,500	4.4	15
Boron[c]	1.97×10^{-4}	469,000	± 8,100	11,000	2.4	10
	1.97×10^{-3}	466,800	±17,700	25,800	5.6	10
	1.97×10^{-2}	464,600	±33,900	47,200	10.3	10
	1.97×10^{-1}	466,800	±80,000	104,000	22.4	10
	1.97	469,000	±39,100	82,600	17.7	20
Beryllium	5×10^{-4}	155,700	± 400	600	0.3	10
	1.97×10^{-3}	159,000	± 600	800	0.6	10
Wire[d]	5×10^{-3}	160,600	± 800	1,200	0.8	10
	1.97×10^{-2}	163,900	± 500	700	0.4	10
	1.97×10^{-1}	166,300	± 700	800	0.6	10
	1.97	168,700	± 1,000	1,200	0.7	8
Steel Wire[e]	1.97×10^{-4}	529,200	± 3,600	3,100	0.5	5
	1.97×10^{-3}	527,100	± 300	200	--	5
	1.97×10^{-2}	527,100	± 300	200	--	5
	1.97×10^{-1}	527,100	± 4,700	3,600	0.7	5
	1.97	533,300	± 6,700	5,200	1.0	5

(a) Owens Corning Corp., Hts-901, cross sectioned area, $A = 2.14 \times 10^{-5}$ in.^{2}

(b) Hitco, treated HMG 50 fiber, $A = 6.39 \times 10^{-5}$ in.^{2}

(c) Texaco Experiment Inc., 4-mil Borofil, $A = 1.356 \times 10^{-3}$ in.^{2}

(d) The Beryllium Corp., 4-mil wire, $A = 1.233 \times 10^{-3}$ in.^{2}

(e) Crucible Steel Co., 5-mil wire, $A = 1.935 \times 10^{-3}$ in.^{2}

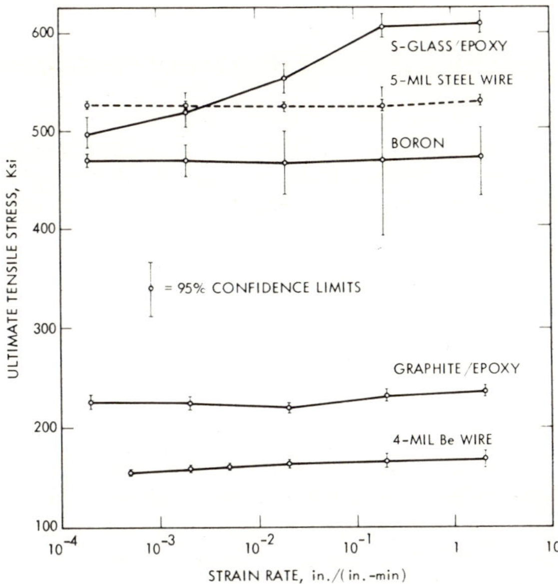

Figure 1. *Strain rate effect on the ultimate tensile stress of fiber/epoxy strands.*

Instead of the commonly-used bridge amplifier-recorder system, we excited the load cell with a regulated D.C. power supply never exceeding 4 volts. The load cell output was amplified about 100 times in a D.C. amplifier and recorded on an oscillograph.

RESULTS AND DISCUSSION

Table 1 summarizes the effect of strain rate on the ultimate failure tensile stress of several common fiber/epoxy strands. The same data is plotted in Figure 1. What we call strain rate in this study is really the crosshead speed of the tensile machine divided by the gage length of the specimen, 10.0 ± 0.02 inches. This definition, of course, is subject to debate. However, considering the relatively low breaking loads of the strands vs. the machine capacity, and the impracticality of using strain gages on so many strands, we feel that it is justifiable for comparative purposes.

For mono-filament fibers, such as boron (B), beryllium (Be) and steel, we wish to point out that there was no epoxy coating on the specimens. Our prior work (Ref. 1) verified that matrix had no effect on single-filament B and Be fibers.

Figure 1 shows that the strain rate does influence the ultimate failure stress of the S-glass/epoxy strand. For all practical purposes, however, we can ignore the rate effect on the B, Be, steel, and graphite/epoxy strands, at least within the limited speed range of most standard tensile machines. Our curve on S-glass/epoxy strands shows the same trend, but with less scatter, than the results on E-glass/epoxy unidirectional laminates obtained by Ishai et al. [2]. It also agrees well with the very limited data of Pirzadeh and Kennedy on S-glass/epoxy strands [3]. On advanced fibers, our results do support Rauch's comments in his recent book [4]: "The cross-head speed used in testing filaments is not generally critical because most of the new, advanced fibers are not particularly strain-rate sensitive in the range normally used for testing (approximately 0.0014 to 0.1 min^{-1}). It is important, however, not to apply the load so rapidly that the response of the recorder becomes a limiting factor." However, he reported no data.

Further examining the S-glass/epoxy data from an engineering point of view, we find that a designer can ignore the strain rate effect and make no more than 10% error for each ten-fold change in rate.

SOME PROBLEMS

In measuring strain rate effect, we found that the speed of response of the standard recording instruments was too slow to indicate the true readings. This matter has too often been overlooked by many laboratories. Actually, we were forced to study the rate effect of composites ourselves after two independent contracting laboratories reported conflicting data—one showed that the ultimate tensile stress of S-glass/epoxy strands decreased with strain rate; the other indicated that there was no effect. We believe that they both had instrumentation problems.

Another problem in studying composites is that the fabrication of the test specimens often is not under control. This is particularly true of specimens of complex form, such as laminates or filament wound articles. We therefore selected the simple fiber/epoxy strand to avoid this problem. Hopefully, the simple specimen can approach a uniform state of stress during testing.

REFERENCES

1. T. T. Chiao and R. L. Moore, "A Tensile Test Method for Fibers," *J. Comp. Mat.*, Vol. 4 (1970), p. 118.
2. O. Ishai, A. E. Moehlenpah, and A. Preis, "Temperature and Time Effects on Yield and Failure of Unidirectional Glass Epoxy Composites," Monsanto/Washington University, ONR/ARPA Association, AD 861189 (July 1969).
3. N. Pirzadeh and P. B. Kennedy, *Symposium on Standards for Filament Wound Reinforced Plastics,* ASTM STP-327 (1962), p. 174.
4. H. W. Rauch, *Ceramic Fibers and Fibrous Composite Materials,* Academic Press, New York (1968), p. 34.

Tensile Strengths of Notched Composites*

P. W. R. BEAUMONT AND D. C. PHILLIPS*

Materials Department
School of Engineering and Applied Science
University of California
Los Angeles, California 90024

(Received November 1, 1971)

ABSTRACT

The tensile strengths of several different aligned fiber-epoxy resin composites containing cracks perpendicular to the fibers have been measured as a function of crack size. The fibers were surface treated and surface untreated, high and low modulus, carbon; and S-glass. Thus, the effect of varied fiber-matrix bond strengths were observed. Surface treated carbon fiber composites were notch-sensitive but the surface untreated carbon fiber composites and the glass fiber composites were notch-insensitive. This effect is due to differences in the shear strengths of the materials. The tensile strengths of the notch-sensitive materials are discussed in terms of various failure criteria, and it is shown that as the relative sharpness of cracks is increased, strengths tend towards the prediction of linear elastic fracture mechanics. The fracture surface energies of all these materials were measured by a work of fracture technique. The works of fracture of the carbon fiber composites are best explained by a model in which shear stress at the interface is maintained during pull-out. The works of fracture of the glass fiber composite are best explained by a debonding model.

INTRODUCTION

AN IMPORTANT FEATURE feature of fibre composites is that they can be fabricated so as to have considerable resistance to crack propagation perpendicular to the fibres. This property makes them candidates for applications where high structural reliability is required. The relative ease with which cracks can propagate in solids can be compared by means of the fracture surface energy (γ). This is defined, empirically, as the minimum energy required to create unit area of fracture surface, microscopic surface irregularities being ignored. It varies from a few Jm^{-2} in brittle

*Present address: Atomic Energy Research Establishment, Harwell, Didcot, Berks, England.

materials to values in excess of 10^5 Jm^{-2} for some tough materials, and thus provides a quantitative basis to intuitive ideas of toughness. As well as being a comparative material parameter γ can sometimes be used as a design parameter to predict, by means of linear elastic fracture mechanics (LEFM)**, the strength of a structural component under conditions where failure occurs by propagation of the pre-existing flaws which often occur in-service. An important question in composite technology at the present time is whether such a LEFM approach can be used in the design of composite structures, or whether other failure criteria are more applicable. This note describes the results of some experiments which have been carried out in an attempt to answer this question for crack propagation perpendicular to the fibers in composites consisting of aligned carbon and glass fibers in polymer matrices.

EXPERIMENTAL

Materials

The four materials studied are described in Table 1. All contained fibre aligned in a single direction. The carbon fibre was manufactured from PAN by the R.A.E. process and some was surface treated by proprietary processes to increase the fibre-matrix interfacial bond strength. The two type I (high modulus) carbon fibre composites were fabricated in this laboratory by the preimpregnation technique [1, 2].

Table 1. The Composite Materials

			Fibre Properties		
Fibre	*Matrix*	*Fibre Volume Fraction*	*Diameter (m)*	*Tensile Strength (GNm^{-2})*	*Young's Modulus (GNm^{-2})*
R.A.E. Type I Carbon fibre, Surface treated	Epoxy 828/MNA/ BDMA	0.40	8×10^{-6}	1.58	360
R.A.E. Type I Carbon fibre, Surface untreated	Epoxy 828/MNA/ BDMA	0.40	8×10^{-6}	1.58	360
Whittaker-Morgan MODMOR II (Surface treated)	Proprietary Epoxy Resin	0.64	8×10^{-6}	2.41	240
S-glass	Proprietary Epoxy Resin	0.76	10×10^{-6}	3.44	85.5

**In LEFM terminology $\gamma \equiv G_c/2$

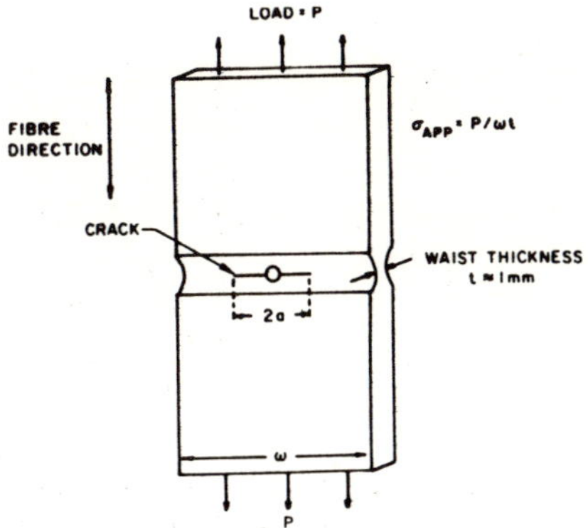

Figure 1. *A center-notched plate.*

Briefly, aligned carbon fibres were coated with a 1 $^{wt}/_{o}$ solution of epoxy resin in acetone to form an impregnated fibre mat. Excess resin was removed by rolling the material and the mat allowed to dry in air at room temperature. The resulting 'pre-preg' was cut into 150 mm squares which were laid up into a laminate and hot-pressed at 130°C for 16 hours. The material containing type II (high strength) carbon fibres was supplied in the form of a plate by the Whittaker-Morgan Corporation, and was their Modmor II product. The S-glass composite was fabricated by hot-pressing a laminate made from 'Scotchply' S-glass fibre/epoxy resin sheet in a manner similar to the carbon fibre composites.

Tensile Testing of Notched Plates

Figure 1 is a diagram of a centre-notched plate. The anisotropy of fibre composites encourages crack propagation in directions parallel to the fibres through splitting due either to shear stresses or tensile stress components perpendicular to the fibres. In order to inhibit this mode of failure the plates were grooved on each side to produce a narrow section along the desired crack propagation direction. Most of the plates were 25.4 mm wide but some testing was carried out with plates of greater width to determine the effect of specimen dimensions. All the plates were approximately 1 mm thick at the narrowest section. The artificial crack was introduced by cutting the plate with a jeweler's saw, and in some of the specimens this was sharpened with a scalpel. Crack lengths were measured with a travelling microscope.

Tensile testing was carried out on a floor model Instron using a 5000 Kg load cell and a cross-head speed of 1.27 mm. min^{-1}. The specimens were clamped into grips with serrated surfaces [2] and aligned by universal joints. A disadvantage of this apparatus was that at small notch-sizes, where fracture loads were high, the specimens tended to pull out of the grips before fracturing. Therefore, the tensile strengths of unnotched composites were obtained from tests on specimens of smaller width.

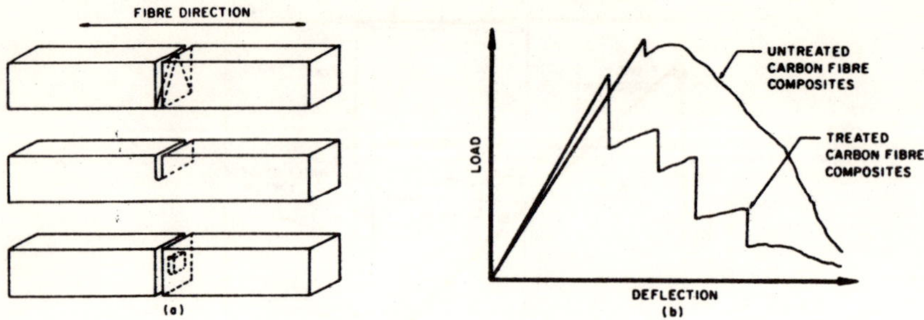

Figure 2. (a) Work of fracture specimens; (b) Load vs. deflection curve of γ_F specimen.

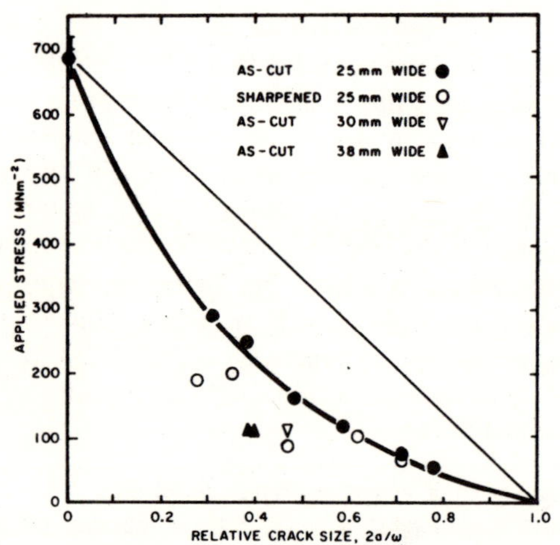

Figure 3. Surface treated, type I carbon fiber composite.

Work of Fracture Tests

Works of fracture [3] were measured by the controlled fracture in bending of notched bars. Several different types of notches were employed (Figure 2a) including roof-top [3] edge-notched [4] and circumferentially notched [5] in an attempt to obtain a completely controlled failure. All, however, fractured in the quasi-controlled stepwise manner shown in Figure 2b, although the S-glass and untreated carbon composites were more controlled than the others. Work of fracture was calculated by dividing the integrated load-deflection curve by twice the fracture cross-sectional area. Testing was carried out on a floor model Instron with a 500 Kg load cell at a cross-head speed of 1.27 $mm.min^{-1}$ and a span of 40 mm.

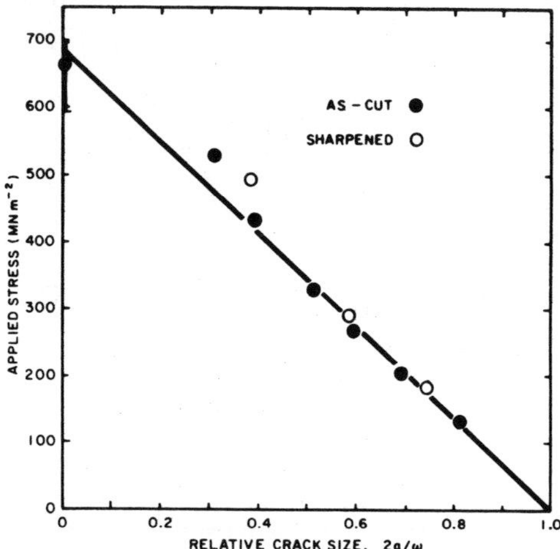

Figure 4. Surface untreated, type I carbon fiber composite.

EXPERIMENTAL RESULTS

Table 2 shows the works of fracture, and the tensile strengths of unnotched specimens. Figures 3 to 6 shows the applied stress at failure σ_{APP} (maximum load divided by gross section in the plane of the crack) plotted against the relative crack length ($2a/w$) for both sharpened and unsharpened cracks. The procedure used for

Table 2. Strength, Work of Fracture and Pull-Out Lengths

Material	*Mean Tensile Strength (GNm^{-2})*	*Mean Work of Fracture (kJm^{-2})*	*Mean Fibre Pull-Out Lengths (m)*
Type I, surface treated, carbon fibre composite	0.69	4.0	2.0×10^{-5}
Type I, surface untreated, carbon fibre composite	0.67	18.7	2.4×10^{-4}
Type II, surface treated, carbon fibre composite	1.24	14.6	5×10^{-5}
S-glass fibre composite	1.45	121	1×10^{-3}

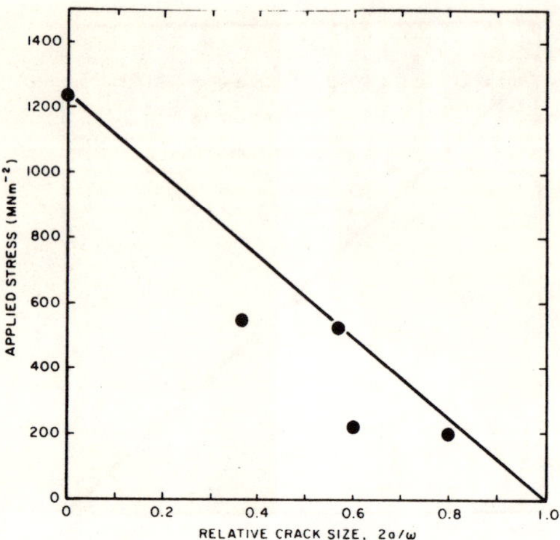

Figure 5. *Surface treated, type II carbon fiber composite.*

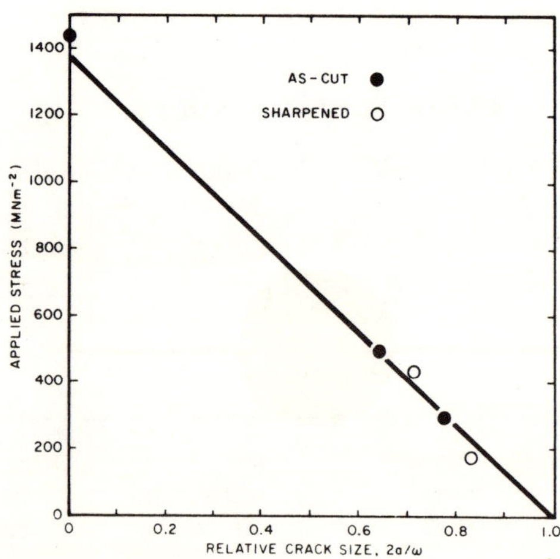

Figure 6. *S-glass fiber composite.*

sharpening the cracks resulted in very variable amounts of sharpness. Figure 7 shows a crack as cut by a jeweler's saw and compares it with a well sharpened crack. The variability in the sharpness of sharpened cracks is reflected in the scatter of points in Figure 3 as compared with the consistency of data obtained from as-cut cracks.

While some of the tensile specimens were being loaded the behavior of the crack was recorded by filming it. During loading of the S-glass tensile specimens, considerable crack opening displacement was observed taking place prior to failure, as shown in Figure 8. Rather less occurred with the untreated carbon fibre composite and none was observed with the two treated carbon fibre composites.

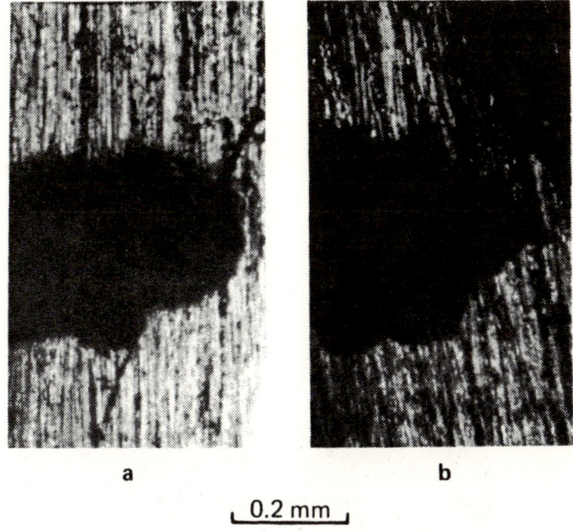

Figure 7. Crack tips (a) as-cut; (b) well-sharpened.

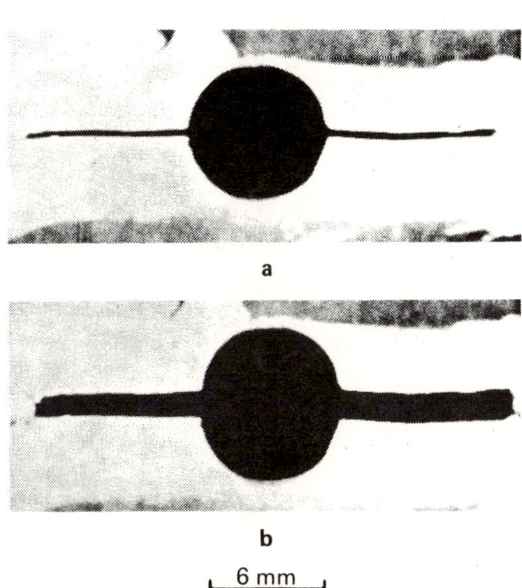

Figure 8. Crack opening displacement of an S-glass fiber composite (a) crack prior to loading; (b) immediately before failure.

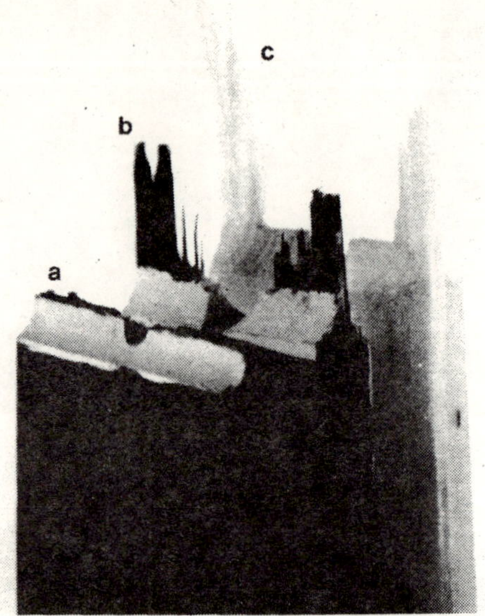

Figure 9. *The fracture surfaces of some of the composites (a) treated carbon fiber; (b) untreated carbon (c) S-glass.*

The fracture surfaces of the S-glass tensile specimens were characterized by the well-known "shaving brush" appearance. Similar features were observed on the untreated carbon fibre specimens where long fibre pull-out lengths were observed with in some cases, pull-out of strips of composite as opposed to just fibres. Very much shorter fibre pull-out lengths were observed on the treated type I carbon fibre specimens. These features are illustrated in Figure 9. The treated type II carbon fibre specimens fell into two categories. The two which lay near the straight line in Figure 5 displayed some pull-out of strips of composite at the tip of the artificial crack. The other two did not. The fracture surfaces of the work of fracture specimens also showed considerable differences and Table 2 shows the mean lengths of fibre protruding from the fracture surfaces of both work of fracture and tensile specimens. These lengths have been estimated approximately from scanning electron microscopic studies.

DISCUSSION

Notch Sensitivity and Insensitivity

The materials fall clearly into two categories: notch insensitive, where the fracture stress on the net cross-section is constant and independent of crack length so that the applied stress varies linearly with crack length; and notch sensitive.

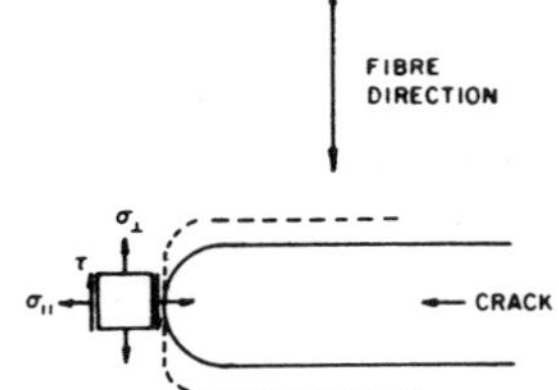

Figure 10. The stress distribution at the crack tip.

Notch insensitivity was associated with long fibre pull-out lengths or pull-out of strips of composite material at the tip of the crack, and with observable crack opening displacements as in Figure 8. When a cracked solid is loaded in tension perpendicular to the crack, the resulting stress distribution may be resolved into tensile stresses perpendicular to the crack ($\sigma_\perp$), tensile stresses parallel to the crack ($\sigma_\parallel$) and shear stresses perpendicular to the crack (τ), (Figure 10). Notch insensitivity could occur due to crack blunting prior to macroscopic tensile fracture from microfracture or deformation on a plane perpendicular to the crack either due to the tensile stress $\sigma_\parallel$ or to the shear stress τ. Kelly [6] has pointed out that in carbon fibre composites the ratio of stresses is such that a pre-cracked plate is more likely to shear parallel to the fibres. The results presented here agree with this. At the tip of a crack, the ratio of $\sigma_\perp : \sigma_\parallel : \tau$ in an isotropic solid is 2:0.4:0.6 [7] and in a highly anisotropic material with the elastic constants of a composite containing 50% of carbon fibres it is 6.3:0.13:0.57 [8]. The only fabrication difference between the treated and untreated type I carbon fibre composites was the strength of the fibre-matrix bond. This difference in bond strength is known to have a considerable effect on the shear strength of carbon fibre composites, the treated carbon fibre composites used in this work having a shear strength of ~80MN m^{-2} and the untreated carbon fibre composites having a shear strength of 23 MN m^{-2}. From Gilliland's results, presented above, it would be necessary for the ratio of $(\sigma_\perp)$ ultimate/$(\sigma_\parallel)$ ultimate to be ~50 for notch insensitivity due to tensile splitting parallel to the fibres, but this ratio is typically ~14. However, $\sigma_\perp/\tau$ in a 40% composite is close to 11, and the ratio of $(\sigma_\perp)$ ultimate/τultimate of the treated and untreated type I composites respectively were 8.5 and 29. Thus the untreated fibre composite would be expected to shear at the crack tip before crack propagation occurs. This is in fact what occurs, the shear failure giving rise to the observed crack opening displacement in the untreated carbon fibre composite, and similarly in the S-glass composite (Figure 10). Where notch-insensitivity occurred the appropriate failure criterion is that where the net stress on the cracked cross-section reaches the unnotched fracture stress.

Notch Sensitivity

All of the treated type I carbon fibre composites and some of the treated type II were notch sensitive. The type II material had a higher volume content of fibre than the type I, was fabricated with a different matrix and by a different process, and was microstructurally less homogeneous. The as-cut treated type I material of width 25.4 mm all fall on a smooth curve. There is much more scatter in the sharpened specimens, probably due to poor reproducibility in the relative sharpnesses of their notches. However, it is clear that sharpening the notches leads to a decrease in strength. There is also an effect due to varying the plate width.

There are several possible distinct approaches to explaining the notch sensitivity of a fibre composite material. The material may be treated as homogeneous but elastically anisotropic. In that case, if the cracks are sharp a LEFM approach might be suitable. Alternatively, since the concept of a sharp crack in a composite is not at all clear, the cracks may be treated as having a finite crack tip radius which is either equal to the radius produced by the cutting technique, or is some material parameter such as the critical fibre transfer length ℓ_c [9]. Appropriate stress concentration factors may then be calculated. Another approach is to consider the material's inhomogeneity. Then, where the fibres have a much greater elastic modulus than the matrix, a shear lag analysis may be used to calculate the stress concentration effect in fibres adjacent to a region of broken fibres or a crack [10, 11].

Figure 11 shows stress intensity factors (K_c) calculated for the notch sensitive, treated type I carbon fibre composites. Sih, Paris, and Irwin [12] have shown that the stress intensity factors calculated from isotropic elasticity theory are applicable to elastically anisotropic materials fracturing in the opening mode in a principal symmetry direction. Accordingly the K_c values have been calculated from the analysis, due to Isida, described by Brown and Srawley [13]. In anisotropically elastic media, K_c is related to the critical strain energy release rate (G_c) by [12]

$$G_c = K_c^2 \left(\frac{a_{11} a_{22}}{2}\right)^{1/2} \left[\left(\frac{a_{22}}{a_{11}}\right)^{1/2} + \frac{2a_{12} + a_{66}}{2a_{11}}\right]^{1/2} \qquad (1)$$

where the a_{ij} are components of the elastic compliance matrix. The elastic compliance of the material has been calculated from the analysis due to Tsai, [14] and G_c thus calculated from the K_c values. Figure 11 also shows these G_c values. If LEFM is applicable, K_c (or G_c) values should be constant over a range of notch sizes. Further, the $G_c/2$ values should be approximately equal to the work of fracture (γ_F). However both these statements need qualifying as stress intensity factors may vary with crack size if there is some effect due to the interaction of the crack stress field with free surfaces, and work of fracture values might differ from $G_c/2$ values as the latter is a measure of the energy released during crack initiation and the former is a measure of the energy released during overall crack propagation. Figure 11 shows that there is a considerable variation of K_c and $G_c/2$ with crack size. However, it can be seen by inspection of the data from the as-cut specimens, that as crack length increases, $G_c/2$ values decrease towards the measured γ_F values, and the K_c values tend towards the theoretical K_c value calculated from the work of fracture and equation 1. Further, at any given relative crack size, sharpening the cracks tends to lower K_c (or G_c) values. Unfortunately the variability in "sharp crack" sharpness

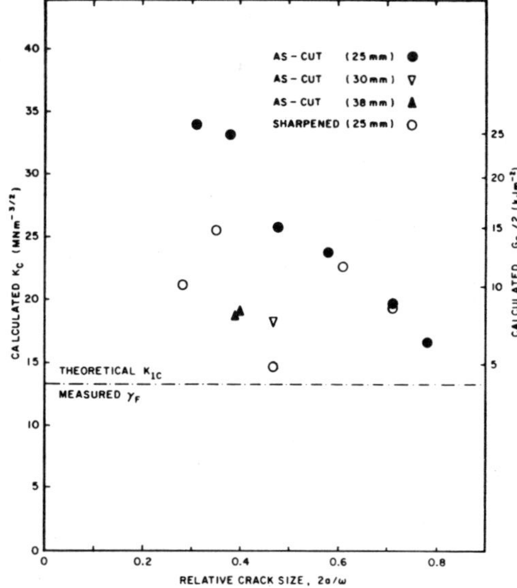

Figure 11. Critical stress intensity and strain energy release rates calculated from the surface treated, type I carbon fiber data.

precludes a more quantitative conclusion. None of the K_c or $G_c/2$ values are lower than the theoretical and it appears that as the relative crack sharpness i.e., the crack length divided by crack tip radius, increases, the data falls closer to the prediction of LEFM. This leads us to the tentative suggestion that "sharp" cracks in this material would tend to obey LEFM.

An alternative approach is to consider the stress concentration at the tip of the notch. This may be carried out either by treating the material as an elastically anisotropic homogeneous material, or by employing stress concentration factors calculated for inhomogeneous materials.

The former approach may be carried out approximately by treating the cracks as elliptical in shape. The stress at the tip of an elliptical crack of length $2a$ in an infinite, isotropic elastic plate under an applied stress σ_{app} is

$$x_{tip} = \sigma_{app}\left(1 + 2\sqrt{\frac{a}{c}}\right) \tag{2}$$

where ρ is the radius of curvature of the crack tip. An expression of this form may be applied to the present data if the finite size of the plate and its anisotropically elastic nature are taken into account. Dixon [15] has shown experimentally that the effect of finite plate size on stress concentration factors in elastically isotropic materials can be taken into account by a correction factor. Thus

$$K_F = K_\infty \left[1 - \left(\frac{2a}{w} \right)^2 \right]^{-\frac{1}{2}} \tag{3}$$

where K_F is the stress concentration factor for a finite plate and K_∞ is that for an infinite plate. The effect of elastic anisotropy in homogeneous materials of similar elastic constants to the fibre composites studied here would be to increase the stress concentration factors. Gilliland has shown that in the particular case of a material with the elastic constants of an aligned composite, containing 50% of RAE carbon fibres in an epoxy resin in which $e/a \rightarrow 0$, the stress concentration factor is increased from the isotropic value of 2 to a value of 6.3. The radius of curvature of an as-cut crack tip was, as shown in Figure 7, typically 5 x 10^{-4} m. Sharpened cracks had smaller radii than this, and the critical transfer length of the fibres was a few fibre diameters so that a radius of curvature based on this material property would be ~ 4 x 10^{-5} m. Thus the smallest calculated stress concentration factor would be that based on the 5 x 10^{-4} m radius. Table 3 shows the stress concentration factors calculated on the basis of an elliptical crack of tip radius 5 x 10^{-4} m in an isotropic plate of width 25.4 mm for three crack sizes. All the theoretical values are much greater than the experimental as-cut crack values, even without taking elastic anisotropy into account. The effect of elastic anisotropy would be to multiply the theoretical stress concentration factors by between two and three and thus to make the theoretical values much larger than even the sharpened cracks experimental values. This suggests that there may have been a crack blunting effect during loading, similar perhaps on a microscopic scale to that shown in Figure 8.

Hedgepeth [10] has used a shear lag analysis to derive expressions for the stress concentration factors applicable to the stresses in continuous fibres adjacent to broken fibres, and thus taken into account inhomogeneity. For a two-dimensional model of an array of fibres the stress concentration factor for r broken fibres is

$$K_r = \frac{4.6.8.\ldots(2r+2)}{3.5.7.\ldots(2r+1)} \tag{4}$$

Table 3. Theoretical and Experimental Stress Concentration Factors for 25.4 mm Wide Plates of Type I Treated Carbon Fibre Composites

Relative Crack Size $\frac{2a}{w}$	*Theoretical Stress Concentration Factors*		*Experimental Stress Concentration Factors*	
	Elliptical Crack (isotropic case)	*Hedgepeth Analysis*	*As-cut Crack*	*Sharpened Crack (approximate)*
0.3	6.9	18.1	2.4	3.5
0.5	9.4	26.5	4.3	7.7
0.7	13.2	36.6	6.9	9.2

The number of "broken" fibres along the crack in a plate may be calculated from the fibre volume fraction and diameter (8 x 10^{-6} m). Equation (4) has been combined with (3) and Table 3 shows the theoretical stress concentration factors for three different crack sizes. All the theoretical values are very much bigger than the experimental values.

Work of Fracture

Several different physical processes have been postulated to account for the energy absorption during fracture of fibre composites. The two simplest models are debonding and fibre pull-out, and controversy surrounds their relative merits. Outwater and Murphy [15] have pointed out that in glass fibre/polymer the composites, the adhesive bond between fibres and matrix is sometimes observed to be destroyed for some distance along the fibres in the vicinity of the macrocrack. This results in the fibres having a debonded length (y) immediately prior to their fracture. The stored elastic strain energy in a debonded length of fibre immediately before fracture is irreversible when the fibre snaps and if this lost energy is equated to the fracture surface energy, the following expression may be obtained.

$$\gamma_D = \frac{V_f \sigma_f^2 y}{4E_f} \tag{5}$$

where V_f is the volume fraction of fibres of strength σ_f and Young's modulus E_f. This, of course, requires that the interfacial shear strength of the glass fibre/polymer bond be zero after debonding. The pull-out model [6, 17] on the other hand requires that interfacial shear stresses be maintained while the fracture faces are being separated so that fibres pull-out against a restraining force. Cottrell [17] and Kelly [6] have shown that, where this applies, the energy dissipated during the pull-out of discontinuous fibres of length equal to the critical transfer length 1_c is given by

$$\gamma_D = \frac{v_f \sigma_f 1_c}{24} \tag{6}$$

This may be extended to the real case of continuous fibre composites with a distribution of strengths by making the approximation that the lengths of fibres protruding from the fracture surface vary between zero and $1_c/2$ with a mean of $1_c/4$. The two models may therefore be compared by using the protruding lengths of fibre to estimate the debonded length or the critical transfer length.

Table 4 shows the theoretical and experimental values for the composites studied. It can be seen that the pull-out model is in better agreement with the observed γ_F values for the carbon fibre composites, while the debonding model more closely agrees with the observed glass fibre composite data.

CONCLUSIONS

(i) Untreated carbon fibre and S-glass fibre composites are notch insensitive because of low shear strengths. Treated carbon fibre composites are notch sensitive because of higher shear strengths.

(ii) Linear elastic fracture mechanics, combined with work of fracture values, underestimated the strengths of all the notch sensitive pre-cracked plates. As the relative crack sharpness (crack length/tip radius) increased, the experimental strengths tended towards the predictions of LEFM.

(iii) Theoretical stress concentration factors underestimated the strength of the notch-sensitive composites.

(iv) The work of fracture of carbon fibre composites is more closely approximated by a model in which the interfacial shear stress is maintained during pull-out, then by a debonding model. The reverse is true for the S-glass fibre composites.

ACKNOWLEDGMENTS

We are grateful to Professors A. S. Tetelman and G. H. Sines for several useful discussions. We are also grateful to Whittaker Morgan Inc. for giving us the type II carbon fibre composite.

Table 4. Theoretical and Experimental Works of Fracture

Material	*Experimental γ_F (kJ m^{-2})*	*Theoretical γ_F (kJ m^{-2})*	
		Debonding	*Pull-Out*
Type I, surface treated, carbon fibre composite	4.0	0.07	2.2
Type I, surface untreated, carbon fibre composite	18.7	0.70	26.0
Type II, surface treated, carbon fibre composite	14.6	1.1	13.8
S-glass fibre composite	121	105	436

REFERENCES

1. L. N. Phillips, "Forming Processes for Carbon Fibre and Resin," *Composites,* Vol. 1 (Dec. 1969), p. 101.
2. P. W. R. Beaumont, "Fracture and Fatigue of Carbon Fibre-Reinforced Plastics," Ph.D. Thesis, University of Sussex, 1970.
3. H. G. Tattersall and G. Tappin, "The Work of Fracture and its Measurement in Metals, Ceramics and Other Materials," *Journal of Mat'ls. Sci.,* Vol. 1 (1966), p. 296.
4. R. W. Davidge and G. Tappin, "The Effective Surface Energy of Brittle Materials," *J. of Mat'ls. Sci.,* Vol. 3 (1968), p. 165.
5. D. C. Phillips, "The Fracture Energy of Carbon Fibre Reinforced Glass," A.E.R.E. Rept., R-6916, 1971.
6. A. Kelly, "Interface Effects and the Work of Fracture of a Fibrous Composite," *Proceedings of the Roy. Soc.,* Series A, Vol. 319 (1970), p. 95.
7. J. Cook and J. E. Gordon, "A Mechanism for the Control of Crack Propagation in All-brittle Systems," *Proceedings of the Roy. Soc.,* Series A, Vol. 282 (1964), p. 508.
8. J. M. Gilliland, Unpublished work quoted in reference 6.
9. See for example A. Kelly and G. J. Davies, "The Principles of the Fibre Reinforcement of Metals," *Met. Reviews,* Vol. 10 (1965), p. 1.
10. J. M. Hedgepeth, "Stress Concentrations in Filamentary Structures," NASA, TN-D-882, 1961.
11. J. M. Hedgepeth and P. Van Dyke, "Local Stress Concentrations in Imperfect Filamentary Composite Materials," *J. Composite Mat'ls.,* Vol. 1 (1967), p. 294.
12. G. C. Sih, P. C. Paris and G. R. Irwin, "On Cracks in Rectilinearly Anisotropic Bodies," *Int.'l. J. of Fracture Mech.,* Vol. 1 (1965), p. 189.
13. W. F. Brown Jr., and J. E. Srawley, "Plane Strain Crack Toughness Testing of High Strength Metallic Materials," ASTM, STP 410, 1966.
14. S. W. Tsai, "Formulas for the Elastic Properties of Fibre-Reinforced Composites," Monsanto/Washington University ONR/ARPA Association, HPC 68-61, 1968.
15. J. R. Dixon, "Stress Distribution Around a Central Crack in a Plate Loaded in Tension; Effect of Finite Width of Plate," *J. of the Roy. Aeronautical Soc.,* Vol. 64 (1964), p. 141.
16. J. O. Outwater and M. C. Murphy, "The Fracture Energy of Uni-directional Laminates," 24th Annual Tech. Conf., Reinforced Plastics/Composites Div., The Soc. of Plastics Industry, Inc., 1969.
17. A. H. Cottrell, "Strong Solids," *Proceedings of the Roy. Soc.,* Series A, Vol. 282 (1964), p. 2

A Comparative Study of Tensile Fracture Mechanisms

J. V. MULLIN AND V. F. MAZZIO, *Space Sciences Laboratory General Electric Company, P.O. Box 8555, Philadelphia, PA 19101*

(Received January 31, 1972)

The purpose of the work described here is to identify failure mechanisms for some of the more widely used carbon fibers in epoxy under tensile loading as a basis for more reliable prediction of their performance. A similar study was performed on boron/epoxy [1] with interesting results. Besides the obvious design related benefits of such a study there is the added advantage of being able to improve composite performance through the development of means to control the failure process.

The approach is first to encapsulate single fibers and bundles of fibers in the epoxy matrix and observe their failure mechanisms under tensile loading. These observations can then be used to analyze the failure process in more heavily reinforced composites to establish the critical parameters in optimizing performance of a given fiber in a given matrix. The behavior of three kinds of carbon fibers was studied: high strength (HT), high modulus (HM) fibers and low modulus (Type A) fibers. The resin matrix selected was DEN 438, an epoxy novolac capable of being modified to change its crack sensitivity.

Table 1 gives the properties of three fiber types including the strain energy density $\sigma_f^2/2E_f$ released at fracture. Note that the lower modulus fibers have higher strain energy density at failure than the higher modulus fibers since the strengths do not differ appreciably. The DEN 438 resin system modified with PPG 425 has a modulus about half that of the unmodified resin (2×10^5 compared to 4×10^5 psi) but the total elongation is increased by an order of magnitude (25% compared to 2.5%). This extreme variation in elongation results in three times as much toughness for the modified formulation.

Single fibers and tows were encapsulated in each resin formulation to identify their failure mechanisms under tensile loading. All three types of carbon fibers failed by cleavage of the resin at the first fiber fracture in the unmodified resin. A typical failure is shown in Figure 1 with the fracture surface showing clearly how the fracture initiated at the fibers and propagated through the resin. There was no evidence of debonding and no other fiber fractures were found after careful microscopic examination of the entire specimen. In contrast to this behavior a number of cleavage cracks and debonded regions were observed in the modified resin as shown in Figure 2. Note that both the HT and HM fibers generate cleavage cracks in the matrix normal to the fibers, but the cracks are of smaller diameter in the HM fiber specimens. Since the strain energy released by the fiber and matrix is absorbed by

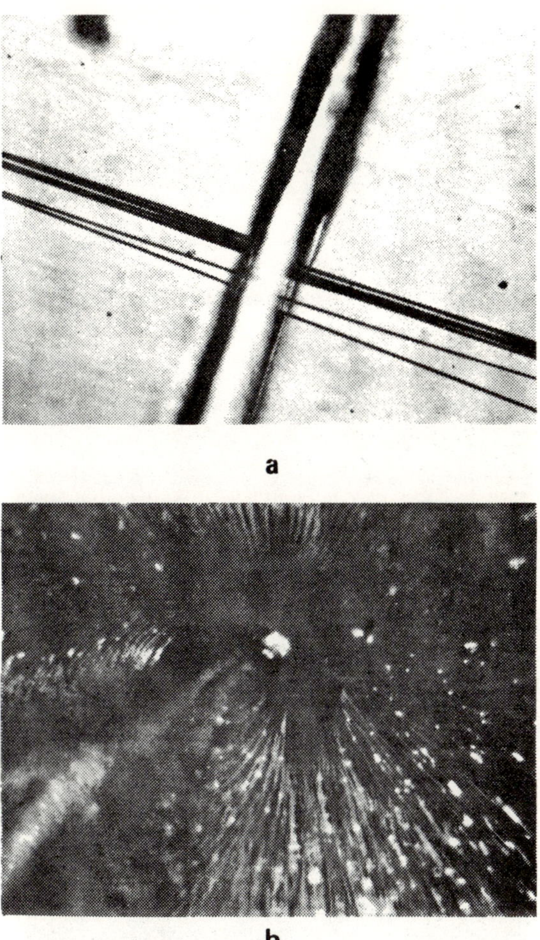

Figure 1. Tensile behavior of HT fibers in unmodified resin where (a) shows single fiber site at ultimate load, and (b) fracture surface showing fibers which initiated fracture. Magnification 28×.

generating new crack surface, this observation is consistent with the Griffith concept of crack stability. Note also that the Type A fibers, although they have high energy densities, do not generate cracks normal to the fibers for the most part, but tend to debond from the matrix when the fibers fail in tension. Energy absorbed in this manner is desirable in that it is achieved without destroying the integrity of surrounding fibers. This phenomenon may explain the fact that Type A fibers often yield composite strengths equivalent to those of HT and HM fibers even though the

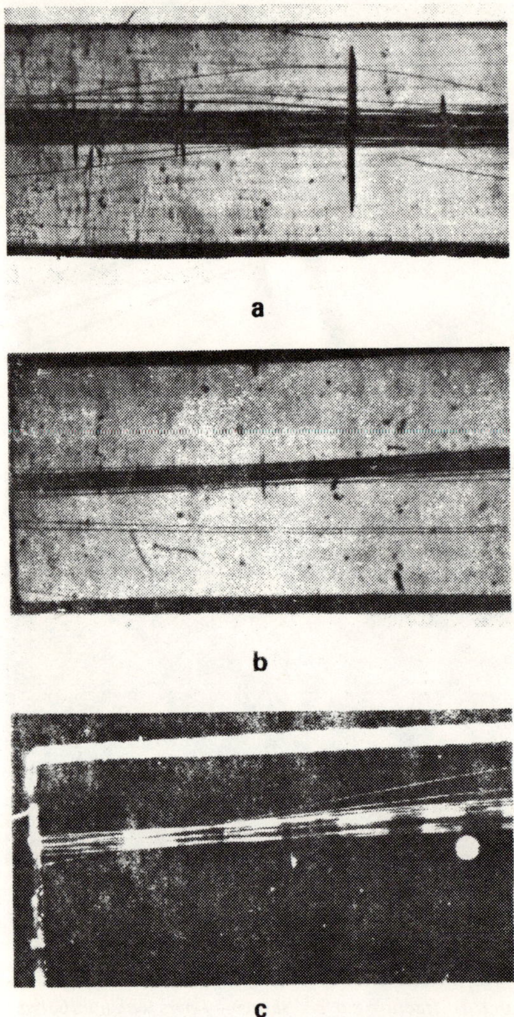

Figure 2. Comparison of tensile failure modes for (a) HT, (b) HM, and (c) Type A carbon fibers in modified resin.

Type A fibers have inherently lower basic properties. Figure 3 shows details of each type of fiber fracture at higher magnification. Note that HT fibers exhibit a very concentrated cleavage fracture site while the HM fibers show somewhat less concentration with some debonding evident. Individual HM fibers initiate local fractures on different planes in the same area. Type A fibers show little or no matrix cracking normal to the broken fibers, but extensive debonding on either side of the fiber fracture site (as evidenced by the bright areas in the lower photo). A more detailed discussion and analysis of the energy considerations for these two failure processes is given in Reference 2.

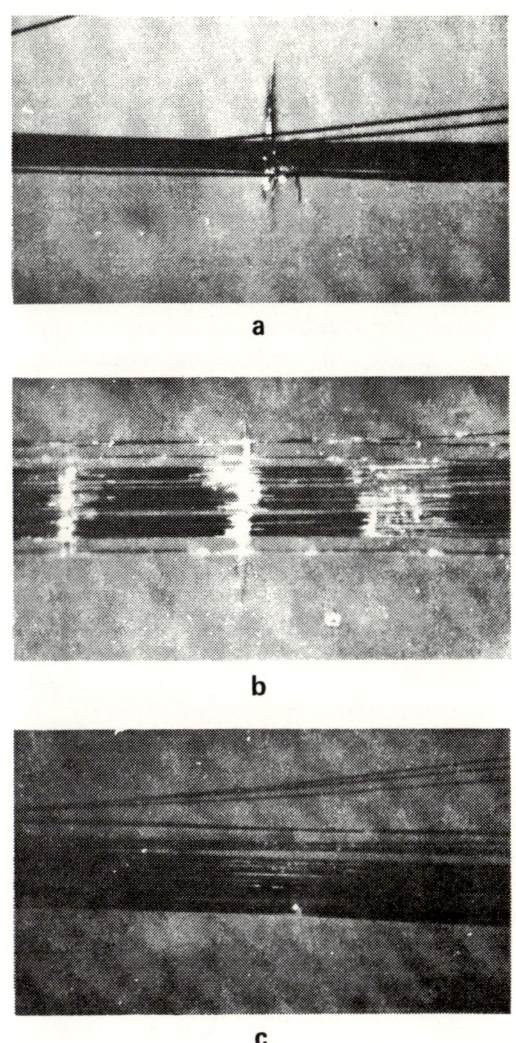

Figure 3. Comparison of tensile failure modes for (a) HT, (b) HM, and (c) Type A fibers in modified resin.

Although both the modified and the unmodified resin can be expected to behave differently in a high volume fraction composite, these observations indicate that there is at least more likelihood of damage to adjacent fibers and matrix when a single fiber fails in a rigid matrix. It is also clear that the fracture mechanisms for HT fibers are quite different from those of the Type A fibers in the modified resin. The next step was to compare these observations with test data from higher volume fraction composites made with the same untreated fibers.

TENSILE TESTS OF STANDARD SPECIMENS

Because of the difference in fundamental failure mechanisms observed between the HT carbon fibers and the Type A fibers these two fibers were selected for comparison of gross failure modes using ASTM test method D-638-38. Since these resin formulations were applied in the laboratory to untreated fibers, wet prepregging was done by hand, and this resulted in lower volume fractions than would have been obtained by commercial prepreg operations. A preweighed batch of fibers was immersed in each resin formulation and placed in a mold. After prestaging in a circulating air oven the impregnated fiber mass was hot pressed for two hours at 180°F and then cured in steps to 350°F and held for another two hours.

To facilitate comparison, the tensile test data was normalized to a 0.35 volume fraction for each type of fiber in both the modified and unmodified matrix and is given in Table 2. Note that the average tensile strength of the Type A fiber composite is slightly higher in the unmodified resin than in the modified resin. However, if the single low value for the unmodified resin is neglected there is a 20 per cent lower strength for Type A fibers in the modified resin formulation. By contrast, the HT fiber composites show significant improvement in strength in the modified compared to the unmodified resin. Looking at the data by resin formulation and comparing fibers we see that in the unmodified resin both fibers give about the same average composite strength (approximately 55 to 59 ksi), while in the modified resin the stronger HT fibers come somewhat closer to reaching their ultimate potential.

The less crack sensitive modified resin seems to be effective in containing the cleavage (Mode I) cracks initiated by the HT fibers and therefore allows more of the high strength fibers to reach their ultimate load carrying capacity before an unstable crack results. But the Type A fibers appear to be generating Mode II cracks at the interface and making the resin tougher cannot contribute much to improved properties unless resin modification also changes the bond strength to an even more optimum level. From this limited data it appears that the bond is less optimum in the modified system however, since the composite strength is reduced. One crude but effective measure of the efficiency with which we develop the potential strength of the fibers in the composite can be obtained by applying the rule-of-mixtures. The fiber efficiency percentages given in Table 2 have been computed by comparing the actual composite strength to the theoretical strength. They suggest the HT fiber composite with unmodified resin was matrix crack limited and by modifying this resin to reduce its crack sensitivity, it was possible to isolate individual fiber failures more effectively, thus the increase in composite strength. Had a fiber coating been applied which would optimize the bond strength it is possible that even higher efficiency could have been obtained in the HT fiber tests. However, there is insufficient data to draw such conclusions at this time.

It should be stated again that the fibers used in this study were untreated and the volume fractions obtained by the wet lay-up fabrication process are somewhat lower than would be obtained with more uniform commercial prepreg. These considerations, together with the limited number of specimens tested, preclude any far-reaching conclusions regarding strength variations in high volume fraction composites. Nonetheless, the strength results obtained are consistent with the results obtained by others [3, 4] for resins similar to the unmodified resin formulation with high fiber loadings. Since the major goal of these tests was to compare single fiber observations to gross failure modes, more emphasis was placed on the nature of the gross failure modes as shown in Figures 4 and 5, for the HT and Type A fiber specimens in the two resin formulations.

In the unmodified resin (Figure 4) note that the Type A fiber specimen shows considerable fracture surface area parallel to or at slight angles to the fiber (and load) axis indicating that the final failure involved considerable interfacial fracture as well as cleavage normal to the fibers. Such failures usually result from subcritical cracks generated at several locations throughout the specimens which ultimately link up in a complex manner. By contrast, the HT fiber specimens (Figure 4b) failed by unstable cleavage on what is nearly a single plane normal to the fiber axes. There is no evidence of fracture on planes parallel to the fiber axes and no fiber pullout. These observations are entirely consistent with the failure modes for individual HT fibers in the unmodified resin shown earlier, while the Type A fiber behavior seems to be less catastrophic than it was in single fiber tests with unmodified resin.

Figure 5 shows a similar comparion in the modified resin for the two types of fibers. The Type A fibers shown in Figure 5a still show evidence of interfacial failure, but the fracture area is much more localized than it was in the unmodified resin. There is much more evidence of fiber pullout in the fracture zone which extends over an axial distance equal to the specimen width. This suggests a change in bond strength and unstable Mode II fracture in that a broken fiber debonds over a region many times greater than the transfer length. Adjacent fibers must carry the additional load over that greater distance and therefore the likelihood of this larger length of overloaded adjacent fiber containing a flaw increases. This is analogous to testing adjacent fibers over longer gage lengths with strength diminishing proportionately.

Finally, the HT fibers in the modified resin in Figure 5b show some evidence of irregular fracture with more than one fracture site apparent in two of the specimens. Although this mode of failure is not grossly different from the cleavage failures of the unmodified resin specimen in Figure 4b, it is of diminished intensity and some degree of fiber pullout is evident in the fracture surface. This shift away from the unstable Mode I (cleavage) failure may explain the increase in tensile strength which results from modification of the matrix to a less crack sensitive state. No attempt was made to change the interfacial behavior by coating the fibers in this study, although related work has been done using this approach. It may be possible to further increase the tensile strength of the HT fiber composite by coating the fibers to obtain a controlled debonding failure condition. However, both matrix modification and fiber coatings may have a negative effect on transverse properties and interlaminar shear strength. The significant point is the two types of fibers behave quite differently and the single fiber tests provide valuable insight into bulk composite behavior.

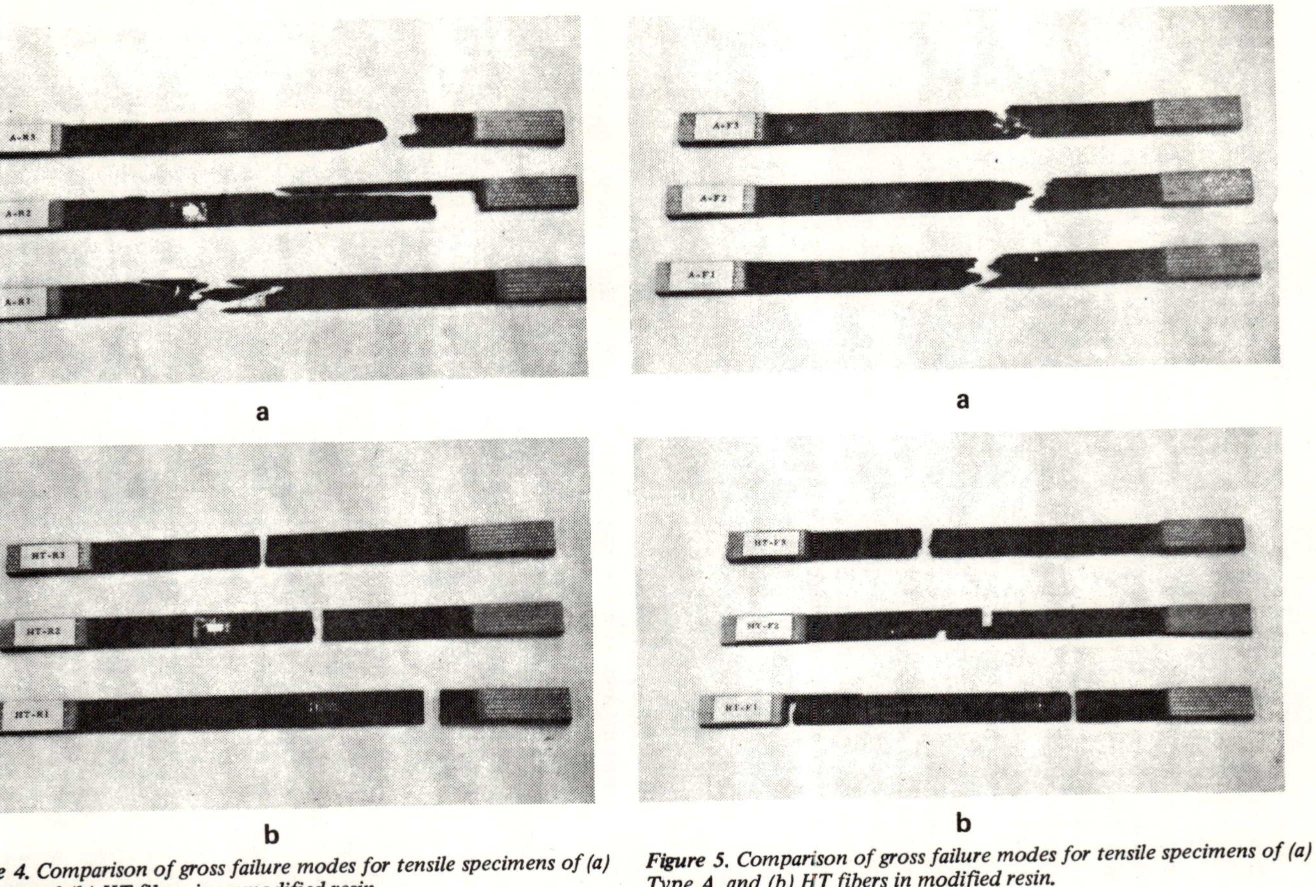

Figure 4. *Comparison of gross failure modes for tensile specimens of (a) Type A, and (b) HT fibers in unmodified resin.*

Figure 5. *Comparison of gross failure modes for tensile specimens of (a) Type A, and (b) HT fibers in modified resin.*

Table 1. Fiber Properties

Fiber Type	*Diameter* $\times 10^{-3}$ *psi*	*Ultimate Strength* $\times 10^3$ *psi*	*Young's Modulus* $\times 10^6$ *psi*	*Strain Energy Density,* μ_f *in lbs./cu. in.*
HM of HM-S	0.275	250-350	55-60	570-1020
HT or HT-S	0.275	300-400	35-50	1285-2000
A	0.275	275-325	28-35	1360-1510

Table 2. Tensile Strength Data for HT and Type A Fibers in Modified and Unmodified Resin

Fiber Type	*Unmodified Resin Tensile Strength, psi*	*Fiber Efficiency %*	*Modified Resin Tensile Strength, psi*	*Fiber Efficiency %*
A	65,500	62.4	54,000	51.4
	62,600	59.6	56,500	53.8
	43,250	45.7	57,900	55.1
	65,300	62.2	52,950	55.6
	Avg. 59,160	Avg. 57.5	Avg. 55,335	Avg. 54.0
HT	63,010	51.4	77,900	63.2
	51,860	42.3	61,260	48.9
	40,000	32.7	86,805	69.5
	76,220	62.2	56,500	45.2
	45,250	36.9	—	—
	Avg. 55,270	Avg. 45.1	Avg. 70,620	Avg. 56.7

Data Normalized to 0.35 V_f

ACKNOWLEDGMENTS

This work was sponsored by NASA under Contract Number NASw-2093. The authors wish to thank B. G. Achhammer and J. J. Gangler for their helpful technical suggestions.

REFERENCES

1. J. V. Mullin, J. M. Berry, and A. Gatti, "Some Fundamental Fracture Mechanisms in Advanced Composites," *J. Composite Materials,* Vol. 2 (1968), p. 82.
2. J. V. Mullin and V. F. Mazzio, "Basic Failure Mechanisms in Advanced Composites," NASA-CR-121621, April 1971.
3. Hercules, Inc., "Advanced Composites Technical Data Handbook," Volume 1, 1971.
4. Fothergill & Harvey, Ltd., "Carboform® and Technical Data Bulletin," 1971.

Free-Edge Delamination of Tensile Coupons

J. M. WHITNEY AND C. E. BROWNING, *Nonmetallic Materials Division Air Force Materials Laboratory, Wright-Patterson Air Force Base, Ohio 45433*

(Received March 4, 1972)

It has been experimentally observed [1, 2, 3], that both the in-plane uniaxial static and fatigue strength of orthotropic symmetrically laminated composites (i.e., in the usual notation $A_{16}=A_{26}=B_{ij}=0$) are a function of the ply stacking sequence. Pagano and Pipes [4] have attributed this phenomenon to the interlaminar stresses induced at the free-edge of a flat coupon. They also outlined a procedure for qualitatively assessing the nature of the free-edge stresses and optimizing laminate stacking sequence for protection against free-edge delamination. It is the purpose of this note to present experimental data on 90°, ± 45° graphite/epoxy laminates of differing stacking sequence and to interpret the results in terms of the procedure described by Pagano and Pipes [4]. The particular laminate orientation was chosen because it is of practical engineering interest and it also produces significant interlaminar normal stresses at the free-edge when subjected to a uniaxial tensile load.

Consider a composite laminate constructed of Hercules' HT-S graphite fiber and Union Carbide's ERL-2256 epoxy resin system. The following unidirectional ply properties are obtained for 53% fiber volume content.

$$E_L = 19.5 \times 10^6 \text{ psi}, \quad E_T = 10^6 \text{ psi}, \quad \nu_{LT} = 0.25, \quad G_{LT} = 0.6 \times 10^6 \text{ psi}$$

Two stacking sequences are experimentally evaluated, $[+45°, -45°, 90°]_s$ and $[90°, +45°, -45°]_s$. In order to qualitatively evaluate the interlaminar stresses generated in a standard tensile coupon, the procedure outlined by Pagano and Pipes [4] is employed.

It has been previously shown [5] that the interlaminar stresses are confined to a zone which is approximately one laminate thickness, h, away from the free-edge. Thus, for a laminate having a width to thickness ratio greater than unity, the stresses along the line $y=0$ (see Figures 1 and 2) can be calculated from lamination theory (LT), [6, 7]. The free body diagrams in Figures 1 and 2 illustrate the force and moment resultants obtained from equilibrium considerations as in [4]. While both stacking arrangements induce significant interlaminar normal stress σ_z laminates having the 45° plies on the outside lead to tensile values of σ_z. This stacking sequence is, therefore, a strong candidate for displaying a delamination failure mode under the uniaxial load $N_x/h = \sigma_0$.

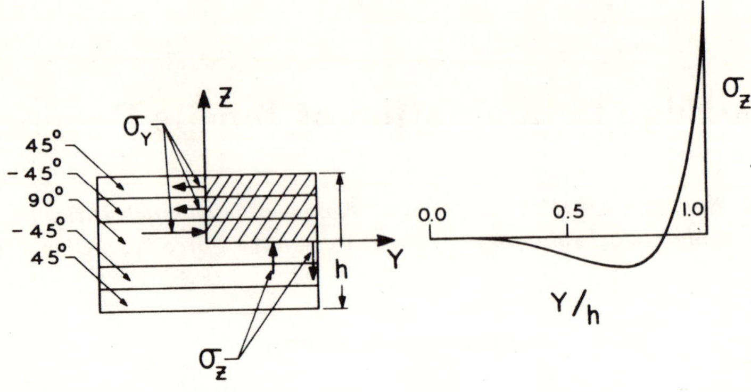

Figure 1. Free body diagrams of $[\pm 45°, 90°]_s$ laminate.

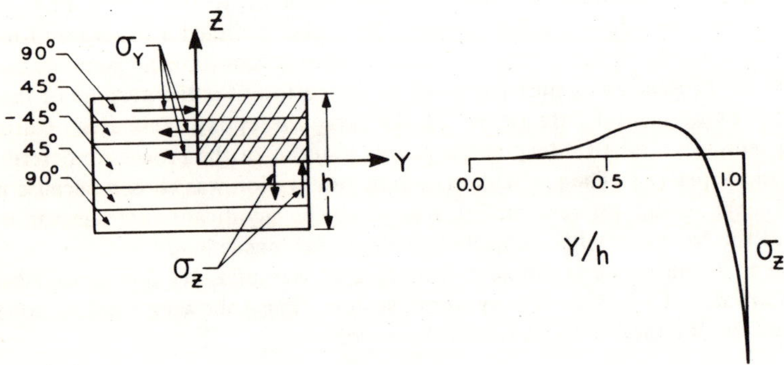

Figure 2. Free body diagram of $[90°, \pm 45°]_s$ laminate.

Experimental results are shown in Table 1 for static and tension fatigue. All specimens of like stacking sequence were cut from the same panel. Modulus was measured by using a strain gage at the center of the specimen. For specimens having 90° outer plies, resin crazing at low strain levels induced very erratic strain gage output. Thus, modulus results are not reported for these specimens. It is interesting to note that *LT* predicts an effective modulus of $E_{11} = 3.26 \times 10^6$ psi which is in close agreement with experimental observation.

Both the static and fatigue specimens having 45° outer plies display significant free-edge delamination. Typical fatigue failures are shown in Figure 3. The top specimen contains 45° outer plies and was subjected to 1000 cycles at 11.2 ksi maximum load and failed statically with a residual strength, S_R, of 15.6 ksi. The edge delamination appeared at approximately N=200. Further opening of this delamination was observed on subsequent cycling. Despite the obvious damage incurred by free-edge delamination, the residual strength was essentially the same as the static strength. The lower specimen in Figure 3 was constructed with 90° outer plies. No

Table 1. Experimental Data*

	Static		Fatigue		
	$\bar{E}_{11}$ (10^6 psi)	σ_{max}(ksi)	N	S(ksi)	S_R(ksi)
$[\pm45°, 90°]_s$	3.2	14.8	–	–	–
$[\pm45°, 90°]_s$	3.0	14.8	–	–	–
$[\pm45°, 90°]_s$	3.4	15.0	–	–	–
$[\pm45°, 90°]_s$	–	–	1000	11.2	15.6
$[\pm45°, 90°]_s$	–	–	1000	10.3	14.7
$[90°, \pm45°]_s$	–	15.9	–	–	–
$[90°, \pm45°]_s$	–	15.4	–	–	–
$[90°, \pm45°]_s$	–	17.8	–	–	–
$[90°, \pm45°]_s$	–	–	187	12.1	–
$[90°, \pm45°]_s$	–	–	690	11.1	–
$[90°, \pm45°]_s$	–	–	1000	9.8	15.9
$[90°, \pm45°]_s$	–	–	1000	10.6	14.6

***All specimens 0.5″ X 0.09″ X 6″**

significant delamination is observed, except in small zone near the cross section of specimen failure. This specimen, however, only survived 187 cycles at 12.1 ksi. The data indicate that the 90° outer ply specimens, in which σ_z is in compression along the free edge, has a higher static strength than the specimens with 45° outer plies.

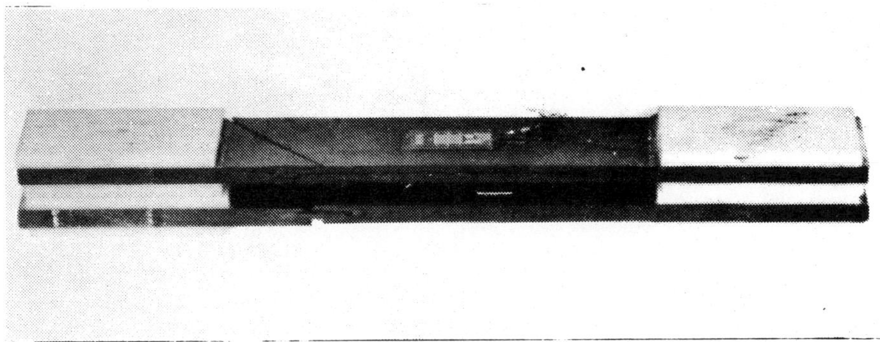

Figure 3. Failed specimens.

It is clear from the data presented that free-edge stresses can lead to significant delamination in a uniaxial tensile coupon. These results also suggest that further experimental work is necessary to assess the effect of stress free edges on design allowables and the interpretation of coupon data. Of particular practical interest is the effect of interlaminar stresses on the response of a laminate containing a hole or cutout.

ACKNOWLEDGMENTS

The authors wish to acknowledge Mr. C. Lovett for composite fabrication and Mr. C. Fowler for his assistance in obtaining the experimental data.

REFERENCES

1. R. L. Foye and D. J. Baker, "Design of Orthotropic Laminates," presented at the *11th Annual AIAA Structures, Structural Dynamics, and Materials Conference,* Denver, Colorado, April 1970.
2. R. G. Spain, "Graphite Fiber-Reinforced Composites," *AFML-TR-66-384,* Air Force Materials Laboratory, Wright-Patterson AFB, Ohio (1967).
3. J. C. Halpin and E. M. Wu, "The Influence of Lamination Geometry on the Strength of Composites," presented at the *Second ASTM Conference, Composite Materials: Testing and Design,* Anaheim, California, April, 1971.
4. N. J. Pagano and R. B. Pipes, "The Influence of Stacking Sequence on Laminate Strength," *J. Composite Materials,* Vol. 5 (1971), p. 50.
5. R. B. Pipes and N. J. Pagano, "Interlaminar Stresses in Composite Laminates Under Uniform Axial Extension," *J. Composite Materials,* Vol. 4 (1970), p. 538.
6. E. Reissner and Y. Stavsky, "Bending and Stretching of Certain Types of Heterogeneous Aeolotropic Elastic Plates," *J. Applied Mechanics,* Vol. 28 (1961), p. 400.
7. J. E. Ashton and J. M. Whitney, *Theory of Laminated Plates,* Technomic, Westport, Connecticut, 1970.

Optimal Experimental Measurements of Anisotropic Failure Tensors

EDWARD M. WU

Washington University
St. Louis, Missouri 63130

(Received September 9, 1972)

ABSTRACT

The resolution of the components of strength tensors derived from experimental data is discussed. It is shown that the variations in the non-interaction components are independent of the tension-compression strength ratios and magnitudes and their relative scatter can be related directly to the scatter in the engineering strength. The variation in the interaction component F_{12}, is dependent on both the magnitude of the biaxial strength and the magnitude of F_{12}; it has no well defined relative scatter and only its absolute resolution can be determined from experimental scatter. Optimal experiments for F_{12} using tubular specimens are discussed. Optimal biaxial stress ratios using compression-internal pressure and tension-torsion tests are discussed and experimentally demonstrated.

INTRODUCTION

IN STRUCTURAL APPLICATIONS using composite materials, it is technologically necessary to develop design methodologies for estimating the load carrying capacity under complex states of stress. One rational approach is to establish a phenomenological failure criterion for anisotropic solids. A phenomenological failure criterion is a mathematical correlation of experimental observations which does not necessarily explain the physics of failure. Because of the technological importance and utility, formulation of anisotropic failure criterion has received international attention and numerous criteria have been proposed.

As pointed out by Tsai and Wu, [1] the majority of these failure criteria are deficient, being either operationally awkward, or limited in capability of correlating wide ranges of complex stresses. In the same article, it was proposed that the failure potential may be expressed in terms of strength tensors in polynomial form, i.e.,

$$f(\sigma_k) = F_i\sigma_i + F_{ij}\sigma_i\sigma_j = 1 \qquad i,j = 1,2,\ldots,6 \tag{1}$$

where F_i and F_{ij} are second and fourth rank strength tensors and where higher order terms are truncated for operational simplicity. The distinguishing features of this failure criterion are: (1) it is mathematically rational and operationally simple; (2) it accounts for strength difference in tension and compression; (3) it acknowledges possible interactions between normal and shear stresses. Graphite reinforced epoxy under off-axis tension, compression and torsion tests has provided significant confirmation of this failure criterion. Equation (1) assumes that, for a path-independent material, the strength tensors F_i, F_{ij} are *intrinsic* material parameters. In order to critically resolve this basic assumption and experimentally verify the existence of the strength tensors, the experimental resolution of the components of the strength tensors computed from experimental data must be understood. We will explore herein the experimental accuracy in the determination of the strength tensors and, in particular, examine the optimal experiments for evaluating F_{12}, the component of the strength tensor which characterizes the interaction of the normal stresses. Determination of normal-shear stress interaction components F_{16} and F_{26}, if they exist, can be inferred from relations developed for F_{12}.

RESOLUTION OF STRENGTH TENSORS

The failure criterion in Equation (1), for a two dimensional composite, contains three components of the 2nd rank strength tensor (F_1, F_2, F_6), and six components of the 4th rank tensor ($F_{11}, F_{12}, F_{16}, F_{22}, F_{26}, F_{66}$). For a typical filament reinforced composite such as graphite reinforced epoxy, most of these strength tensors can be evaluated by simple material tests. If testing is along the principal directions, then F_6, F_{16} and F_{26} are zero from orthotropy. The experimental determination of non-interacting components of the strength tensors is given in Reference [1],

$$F_1 = \frac{1}{X} - \frac{1}{X'}, \quad F_2 = \frac{1}{Y} - \frac{1}{Y'} \tag{2}$$

$$F_{11} = \frac{1}{XX'}, \qquad F_{22} = \frac{1}{YY'}, \quad F_{66} = \frac{1}{SS'} \tag{3}$$

where X, X', Y, Y' are uniaxial tensions and uniaxial compression strengths along longitudinal and transverse directions respectively. S and S' are plus and minus shear strengths in the principal directions. Since components of the strength tensors are calculated from inverses of engineering strengths, their absolute magnitude may be very small. Their significance, as measured by the relative resolutions $|\Delta F_i/F_i|$ and $|\Delta F_{ij}/F_{ij}|$, are relatable to the variability in the original strength measurements $|\Delta X/X|$; e.g.,

$$\left|\frac{\Delta F_1}{F_1}\right| = \psi_1 \left|\frac{\Delta X}{X}\right| \qquad \left|\frac{\Delta F_{11}}{F_{11}}\right| = \psi_{11} \left|\frac{\Delta X}{X}\right|$$

$$\left|\frac{\Delta F_2}{F_2}\right| = \psi_2 \left|\frac{\Delta Y}{Y}\right| \qquad \left|\frac{\Delta F_{22}}{F_{22}}\right| = \psi_{22} \left|\frac{\Delta Y}{Y}\right| \tag{4}$$

$$\left|\frac{\Delta F_{66}}{F_{66}}\right| = \psi_{66} \left|\frac{\Delta S}{S}\right|$$

The magnifying factors ψ_i and ψ_{ij}* are derivable from Equations (2) and (3), as shown in Appendix I,

$$\begin{aligned} \psi_1 &= \psi_2 = -1 \\ \psi_{11} &= \psi_{22} = \psi_{66} = -2 \end{aligned} \tag{5}$$

Equations (4) and (5) imply that the evaluation of F_1, F_2 have the same scatter as the experiments; and those of F_{11}, F_{22}, F_{66} have twice the scatter of the experiments. We will now explore the relation between actual and experimental sensitivity in the case of F_{12}.

The normal stress interaction tensor F_{12} can be determined from many combined states of stress. From the expanded form of Equation (1), it is readily seen that any complex state of stress which has non-zero normal stress components in the longitudinal and transverse direction (σ_1, σ_2) may be used to determine F_{12}. Actually, as depicted in Figure 5 of Reference [1], the sensitivity of F_{12} varies in many of these complex states of stress. Note in particular that a ±3% variation in an uniaxial off-axis tension test strength at 45° will obscure the entire permissible range bounded by the stability condition. This identifies the necessity to understand how experimental scatters are magnified in the computation. Proper experiments must be designed for measuring the interaction of normal stresses and for checking the existence of F_{12}. We propose herein two methods for the critical determination of F_{12}: 1) Optimum ratios for normal biaxial stresses along longitudinal and transverse directions; 2) Optimum ratios for combined normal stress and shear experiments in the off-axis orientation.

Biaxial Normal Stress Experiments

The interaction of normal stresses is characterized by the component F_{12}. It can be determined by a suitable biaxial stress experiment. For this experiment, the

---

*ψ_i, ψ_{ij} are *not* tensors, indical notation is used for the purpose of identification only.

biaxial stress ratio has to be optimized so as to maximize the sensitivity for the determination of F_{12}.

We follow the conventional practice of orienting the 1 and 2 axes along the longitudinal and transverse directions of the composite. For the normal biaxial stresses ($\sigma_6 = 0$), Equation (1) is expanded into:

$$F_{11}\sigma_1{}^2 + 2F_{12}\sigma_1\sigma_2 + F_{22}\sigma_2{}^2 + F_1\sigma_1 + F_2\sigma_2 = 1 \cdot \qquad (1a)$$

Geometrically, the failure surface defined by Equation (1a) is an ellipse in the σ_1,σ_2 plane. If we assume that the components of the strength tensors – F_1,F_2,F_{11},F_{22} – have been measured from previously described material testing experiments, then the axis crossings of the ellipse are known. The remaining constant F_{12} determines the inclination and semi axes of the ellipse. Thus the relative resolution of F_{12} may be interpreted as the accuracy with which the inclination of the ellipse can be measured. This is schematically illustrated in Figure 1 where the biaxial stress ratio B is defined as:

$$B = \frac{\sigma_1}{\sigma_2} \qquad (6)$$

In Figure 1, $\tilde{\sigma}_i$ are failure strength at complex state of stress* and β is defined as the radial intercept between a biaxial ratio B and the failure surface defined by Equation (1a). The relative resolution of F_{12} can be expressed as

$$\left|\frac{\Delta F_{12}}{F_{12}}\right| = \psi_{12}\left|\frac{\Delta\beta}{\beta}\right| = \psi_{12}\left|\frac{\Delta\tilde{\sigma}_2}{\tilde{\sigma}_2}\right| \qquad (7)$$

In order to evaluate the scatter magnification factor ψ_{12}, Equation (1a) may be expressed in terms of the unknowns B, σ_2 and F_{12}:

$$B^2F_{11}\sigma_2{}^2 + 2BF_{12}\sigma_2{}^2 + F_{22}\sigma_2{}^2 + BF_1\sigma_2 + F_2\sigma_2 = 1 \qquad (8)$$

ψ_{12} in Equation (7) can be evaluated by a similar operation as performed in the Appendix.

$$\frac{\Delta F_{12}}{F_{12}} = \frac{1}{F_{12}}\frac{\partial F_{12}}{\partial\sigma_2}\bigg|_{\tilde{\sigma}_2} \Delta\sigma_2 \qquad (9)$$

*In the limiting case, for example, when $\tilde{\sigma}_2 = \tilde{\sigma}_6 = 0$, $\tilde{\sigma}_1 = X$ and $-\tilde{\sigma}_1 = X'$.

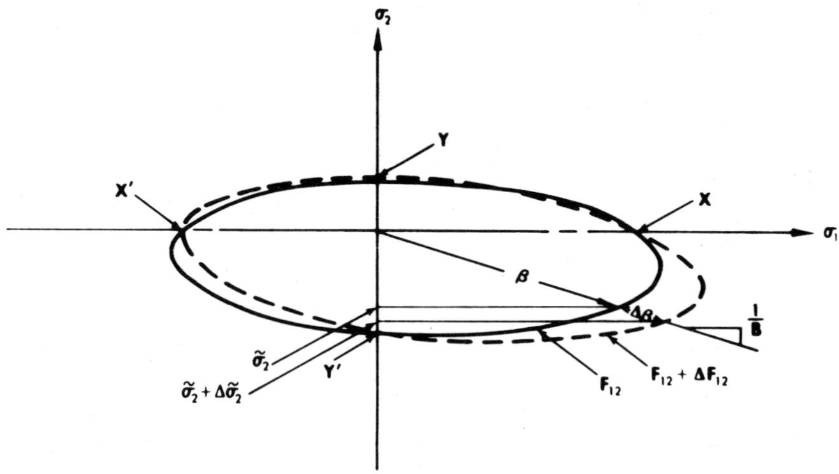

Figure 1. Schematic of failure surface in $\sigma_1 - \sigma_2$ domain with variation of F_{12}.

Differentiating Equation (8) and evaluating for $\sigma_2 = \tilde{\sigma}_2$ (intercept of the failure surface) yields:

$$\left.\frac{\partial F_{12}}{\partial \sigma_2}\right|_{\tilde{\sigma}_2} = -\left(F_{11}B + 2F_{12} + \frac{F_{22}}{B} + \frac{F_1}{2\tilde{\sigma}_2} + \frac{F_2}{2B\tilde{\sigma}_2}\right)\frac{1}{\tilde{\sigma}_2} \tag{10}$$

Thus, by Equation (7),

$$\psi_{12} = \frac{-1}{F_{12}}\left(F_{11}B + 2F_{12} + \frac{F_{22}}{B} + \frac{F_1}{2\tilde{\sigma}_2} + \frac{F_2}{2B\tilde{\sigma}_2}\right) \tag{11}$$

$\tilde{\sigma}_2$ in Equation (11) can be solved from Equation (8)

$$\tilde{\sigma}_2 = \frac{-(F_1B + F_2) \pm [(F_1B + F_2)^2 + 4(F_{11}B^2 + 2F_{12}B + F_{22})]^{1/2}}{2(F_{11}B^2 + 2F_{12}B + F_{22})} \tag{12}$$

Combining Equations (11) and (12), it can be seen that, contrary to the non-interacting tensor components, the relative scatter $\Delta F_{12}/F_{12}$, as characterized by the magnifying factor ψ_{12}, is dependent on the magnitude of the strength measurement $\tilde{\sigma}_2$ and on the magnitude of the strength tensor F_{12} itself.

It is important to note that this conventional interpretation of relative resolution loses physical meaning when $F_{12} \to 0$. Since $F_{12} = 0$ is a physically admissible condition, i.e., no normal stress coupling for failure strength, we have to examine the absolute resolution of F_{12}. The attainable absolute resolution of F_{12} can be computed from the ratio of the absolute scatter ΔF_{12} to the relative scatter of the experimental measurement $\Delta\sigma_2/\sigma_2$. Combining Equations (7) and (11), this ratio becomes:

$$\frac{\Delta F_{12}}{\Delta\sigma_2/\sigma_2} = -\left(F_{11}B + 2F_{12} + \frac{F_{22}}{B} + \frac{F_1}{2\widetilde{\sigma}_2} + \frac{F_2}{2B\widetilde{\sigma}_2}\right) \tag{13}$$

The sensitivity of the interaction tensor F_{12} to the experimentally measurable failure stress $\widetilde{\sigma}_2$ can be observed from graphical representation of Equation (12) which is shown in Figures 2a and 2b.* For example, it can be seen that in the lower Figure 2a for $B = 9$, a small variation in F_{12} would result in a large change in

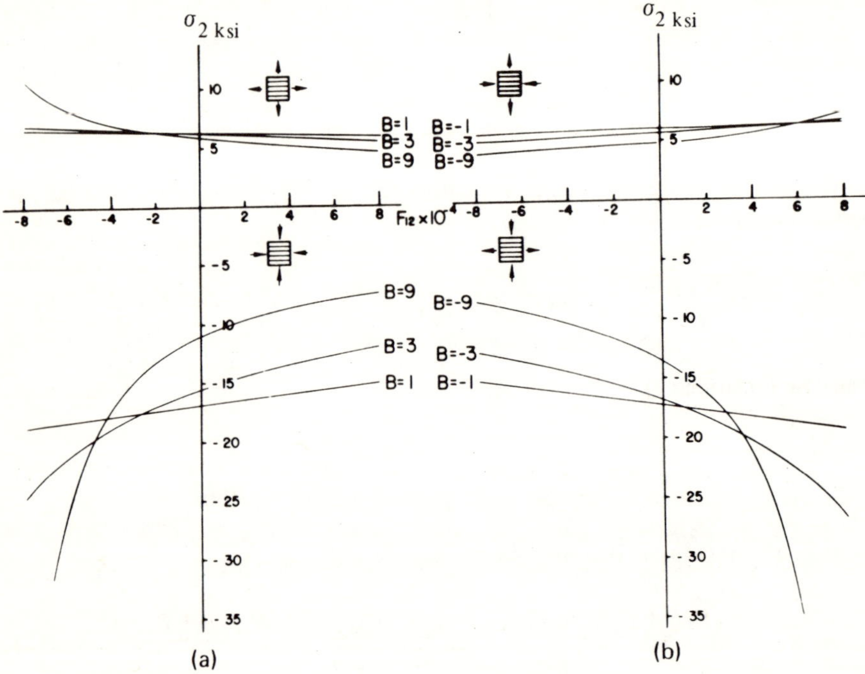

Figure 2. Sensitivity of interaction tensor F_{12} for different biaxial stress ratios $B = \sigma_1/\sigma_2$ [graphical solution of Equation (12)]. σ_1 longitudinal stress, σ_2 transverse stress.

failure stress $\widetilde{\sigma}_2$. The maximum attainable resolution ΔF_{12} for a given experimental scatter $\Delta\sigma_2/\sigma_2$ can be computed from Equation (12) and (13). The simultaneous

---

*Strength tensor components (for stresses in ksi) for graphite-epoxy composite as reported in [1] is used:

$F_1 = -.003 \quad F_{11} = .00007 \quad F_{66} = .01$

$F_2 = \ \ .107 \quad F_{22} = .0098 \quad |F_{12}| \leqslant .0008$ (stability condition)

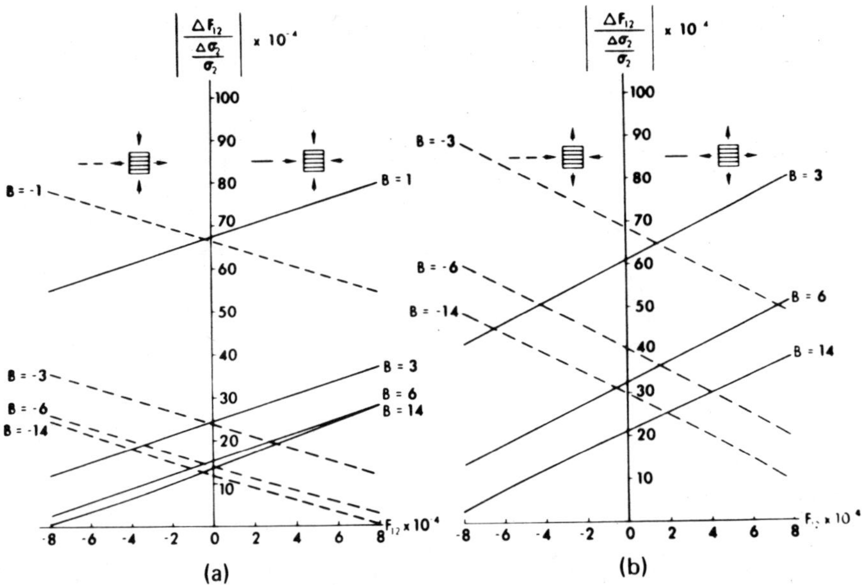

Figure 3. Resolution of interaction tensor F_{12} for different biaxial stress ratios $B = \sigma_1/\sigma_2$ [graphical solution of Equations (12) and (13)]. σ_1 longitudinal stress, σ_2 transverse stress.

solution of these two equations for selected value of B is presented graphically in Figures 3a and 3b. It can be observed that when the biaxial stress ratio B is increased, the resolution of F_{12} increases (or ΔF_{12} decreases). Also, the same biaxial stress ratio B in the 3rd and 4th stress quadrant (Figure 3a) provide significantly higher resolution than those in the 1st and 2nd stress quadrant (Figure 3b). The same observation can also be made from Figures 2a and 2b.

The optimal biaxial ratio B for measurement of F_{12} occurs when ψ_{12} is minimum as seen from Equation (7). The extremum for ψ_{12} is

$$\left.\frac{d\psi_{12}}{dB}\right|_{F_{12}} = 0\,. \tag{14}$$

Equation (14) may be evaluated by the following relation

$$\left.\frac{d\psi_{12}}{dB}\right|_{F_{12}} = \frac{\partial\psi_{12}}{\partial B} + \frac{\partial\psi_{12}}{\partial\sigma_2}\,\frac{d\sigma_2}{dB} = 0 \tag{15}$$

Differentiation of Equations (11) and (12) with respect to B and σ_2 respectively, Equation (15) becomes:

$$\frac{[2\sigma_2(F_{11}B+F_{12})+F_1]}{[2\sigma_2(F_{11}B^2-F_{22})-F_2]} = \frac{-[2\sigma_2(F_{11}B^2+2F_{12}B+F_{22})+F_1B+F_2]}{[B(BF_1+F_2)]} \tag{16}$$

Thus the optimal biaxial ratio B can be expressed in terms of the unknown F_{12} and the constants (F_{11},F_{22},F_1,F_2) by elimination of σ_2 using Equations (16) and (12). It also follows that if F_{12} is not known *a priori*, the optimal ratio B cannot be predetermined. An iterative procedure may be used to determine the optimal ratio B. Initially, an experiment has to be executed at a ratio B for which $\Delta F_{12}/\Delta\sigma_2/\sigma_2$ is small within the range of F_{12} bounded by the stability condition [1]. From this initial experiment, a reasonable estimation of F_{12} may be computed. Based on this F_{12}, an optimal ratio B can be obtained by substituting F_{12} into Equations (12) and (16) and then simultaneously solve for B. Using this ratio B in subsequent experiments leads to a better estimate F_{12}. This process may be repeated to achieve the desired accuracy.

Since solutions to Equation (12) and (16) do not lend themselves to explicit expressions, this iterative procedure is illustrated from graphical solutions, shown in Figure 4a. For each F_{12}, there exist four roots B, one for each stress quadrant. The optimal B can be observed from Figure 4b where the resolution of F_{12} (Equation 13) is plotted for these roots. For a given experimental scatter $\Delta\sigma_2/\sigma_2$, the resolution of F_{12} is high when $\Delta F_{12}/\Delta\sigma_2/\sigma_2$ is small. Thus, tension-compression

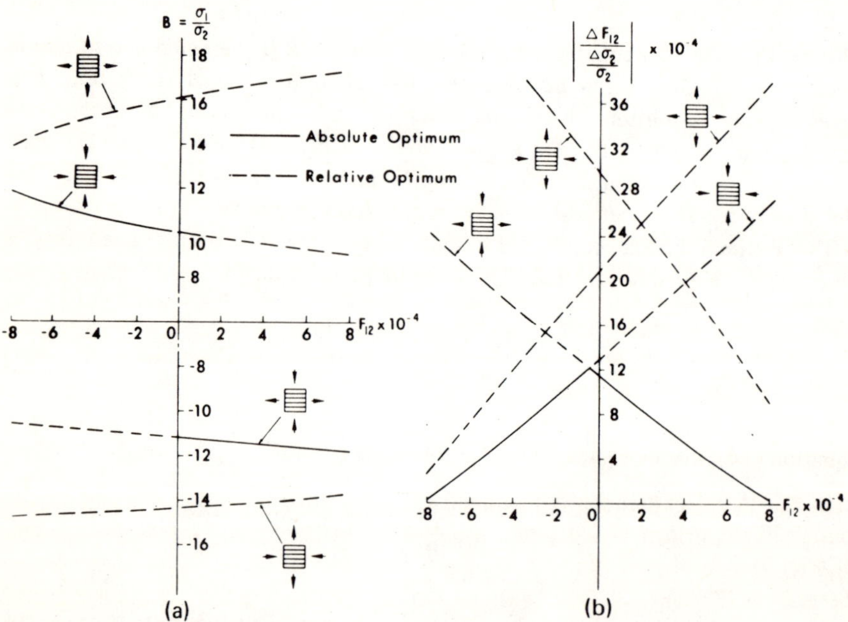

Figure 4. a) Optimal biaxial ratio B for determination of F_{12} [graphical solution of Equations (12) and (16)]. b) Attainable resolution of F_{12} for optimal ratios B.

experiments in the 4th stress quadrant is optimal for $F_{12} > 0$, and compression-compression in the 3rd stress quadrant is optimal for $F_{12} < 0$. For example, for $F_{12} = 6 \times 10^{-4}$, from Figure 4b the biaxial stress in decreasing optimality are 4th, 2nd, 3rd and 1st quadrant; their respective resolutions $\Delta F_{12}/\Delta\sigma_2/\sigma_2$ are 2.8×10^{-4}, 14.6×10^{-4}, 22.8×10^{-4} and 23.8×10^{-4}; and their corresponding biaxial ratios B from Figure 4a are −11.6, −14, 9.3 and 17.1.

At the beginning of an iterative procedure, the sign and the approximate magnitude of F_{12} have to be estimated. For experimentation, we may choose from Figure 3, a suitable biaxial ratio B for which $\Delta F_{12}/\Delta\sigma_2/\sigma_2$ is sufficiently small over the range bounded by the stability limits. For example, Figure 3a indicates that for $B = -3$ in the 4th stress quadrant (tension-compression), the minimum resolution is $\Delta F_{12}/\Delta\sigma_2/\sigma_2 = 35\times 10^{-4}$ at $F_{12} = -8\times 10^{-4}$. Assuming an experimental scatter of $\Delta\sigma_2/\sigma_2 = 0.1$, this ratio provides a meaningful minimum resolution of $\Delta F_{12} = 3.5\times 10^{-4}$. Experiments using this ratio will provide an initial measurement of F_{12}.

The second step of the iteration in measuring F_{12} can be determined from Figure 4a. For example, if the initially estimated F_{12}' is 2×10^{-4} then Figure 4a suggests that the optimal ratio is $B = -11.3$ in tension-compression. Figure 4b indicates that experimental measurements performed at this ratio should provide a resolution of $\Delta F_{12} = 8 \times 10^{-5}$ for a 10% experimental scatter, i.e., $\Delta\sigma_2/\sigma_2 = 0.1$. Alternately, if this 3rd stress quadrant experiment is experimentally not feasible, Figure 4b suggests the 3rd stress quadrant experiment (compression-compression) as an alternative where the relative optimum, again for $F_{12}' = 2\times 10^{-4}$ is $B = +9.7$. In this experimental condition, as seen from Figure 4b the resolution is decreased to $\Delta F_{12} = 16\times 10^{-5}$.

Although there exists an optimal biaxial ratio for each F_{12}, the conventional interpretation of *relative* resolution, as defined in Equation (7), does not have physical meaning. Examination of Equation (7) reveals that when $F_{12} \to 0$, $\Delta F_{12}/F_{12}$ becomes unbounded. Figure 5 illustrates the trend of ψ_{12} for the optimal ratios computed from Equations (12) and (16). Thus, even for the optimal ratios, the relative scatter for F_{12} is unbounded if any finite experimental scatter $\Delta\sigma_2/\sigma_2$ exists. In contrast, as indicated by Equation (4) and Equation (5), the relative scatter is always the same for any magnitude of $F_1, F_{11}, F_2, F_{22}, F_6$ and F_{66}. Hence it is only meaningful to refer to the absolute resolution of F_{12} for a given experimental scatter such as $\Delta\sigma_2/\sigma_2$.

Experimental Considerations

In the initial experiment to estimate F_{12}, sensitivity and feasibility of experimental implementation must be considered together. Since for a given scatter $\Delta\sigma_2/\sigma_2$, the resolution for F_{12} is high when $\Delta F_{12}/\Delta\sigma_2/\sigma_2$ is large, a cursory observation of Figure 3a and 3b shows that experiments in the 3rd and 4th biaxial stress quadrant are far more sensitive than those in the 1st and 2nd stress quadrant.

The practicality of these experiments must also be examined with respect to experimental implementation. For biaxial normal stress materials testing, tubular

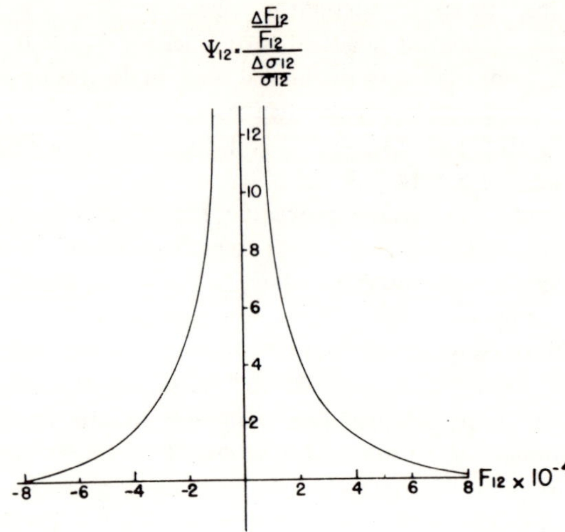

Figure 5. Relative scatter of F_{12} for optimal ratios B.

specimens offer good flexibility. A testing system must apply simultaneous axial force and internal pressure at pre-selected ratios to the tubular specimen. In such an experimental set-up, the biaxial tension experiment (1st stress quadrant) is easiest to perform. Longitudinal compression and transverse tension (2nd stress quadrant) can also be readily performed by using a 0°–tube (with reinforcing fibers oriented along the tube axis) and applying axial compression and internal pressure. The drawback of these simple experiments is the unfortunate lack of sensitivity in the determination of F_{12}.

The biaxial compression experiment (3rd stress quadrant) is most difficult to perform. It requires the experimental implementation of external pressure applied to the tube. One such system has been suggested by Bratt and Kanan [3]. However, their system appears to be applicable only for solid rod samples.

Finally, the longitudinal tension-transverse compression experiment (4th stress quadrant) can be performed using a 90° tube (with fibers oriented along the hoop direction) and applying axial compression and internal pressure. The constraint for this test, in the case of a large bi-axial stress ratio B, is that large internal pressure is required. For example, testing a composite with longitudinal tensile strength of 150 ksi, fabricated into 1 inch I.D. tubes 0.05 inch thick, at $\sigma_1/\sigma_2 = -3$, could require internal pressure close to 7500 psi. Under normal materials laboratory practices, implementation of pressure above 5000 psi is difficult.

Taking these factors into consideration, for a typical graphite epoxy composite, we can suggest that a biaxial longitudinal tension to transverse compression ratio of $-3 < \sigma_1/\sigma_2 < -5$ should provide sensitive initial estimation of F_{12} if the interaction is positive. The optimal ratio from Figure 4a can be used for subsequent experiments. A convenient experiment for determining F_{12} when it is negative is currently not available. Figure 4b suggests that biaxial tension with a large $B = \sigma_1/\sigma_2$ ratio will provide an estimation. From Figure 4b it can be seen that the maximum resolution of F_{12} for 10% experimental scatter is $\Delta F = 1.2 \times 10^{-4}$ for F_{12} close to zero. It follows that F_{12} may be considered zero if it falls within $\pm 0.6 \times 10^{-4}$. These conclusions are based on the strength values measured for graphite epoxy. For other composites, F_1, F_2, F_{11}, F_{22} have to be evaluated, and the optimum ratio and resolution can be arrived at from graphs similar to Figure 3.

Optimum Ratio for Combined Normal and Shear Stress Experiments

In the above optimum biaxial experiments, it is assumed that axial loading and internal pressure experimental equipment for 90° tubular specimens is available. However, there exist other limitations which may make this test impractical. For example, high internal pressure may not be available. Or, for large diameter fibers, (e.g., boron) composite specimen fabrication for a small diameter hoop wound tube is difficult. Under such circumstances, other experimental possibilities have to be explored. It can be easily seen that, for an axial and torsion test of a tubular specimen with reinforcing fibers at a helix angle of θ to the tube axis (see Figure 6), a biaxial state of principal stress can be developed. In order to produce a biaxial stress of σ_1 and $-\sigma_2$ in the principal (fiber) direction, a combined tension and torsion σ_1' and σ_6' must be applied to the tube, where

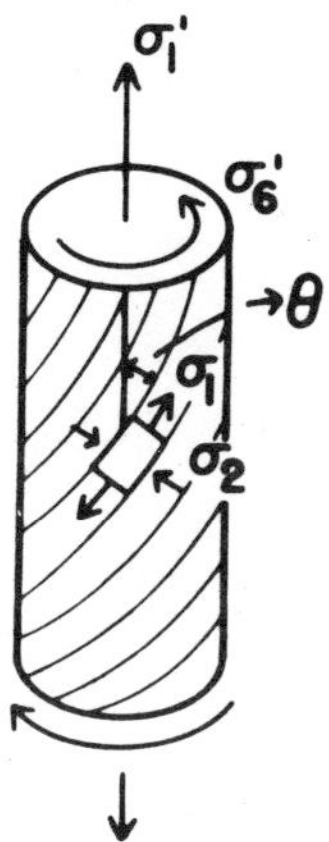

Figure 6. Generation of a biaxial stress ratio from off-axis tension-torsion.

$$\frac{\sigma_1'}{\sigma_6'} = 2 \cot 2\theta \tag{17}$$

The tension-torsion ratio and the required fiber angle θ for a desired optimal ratio B may be derived from the stress transformation equations. This relationship is:

$$\frac{\sigma_1'}{\sigma_6'} = \frac{-(B+1)\, 2 \sin 2\theta}{(1-B) + (1+B)\cos 2\theta} \,. \tag{18}$$

The sensitivity of F_{12} in this tension-torsion test is also seen in the representation of the strength tensor failure criterion Equation (8) in the tension-torsion spaces for various fiber angles. In Figures 7a and 7b, this is illustrated for several values of F_{12} within the stability limits for graphite epoxy at 15° and 30°. The sensitivity of F_{12} in these tests is self-evident. An experimental caution which must be observed is that under high torsion load, buckling stability has to be assured.

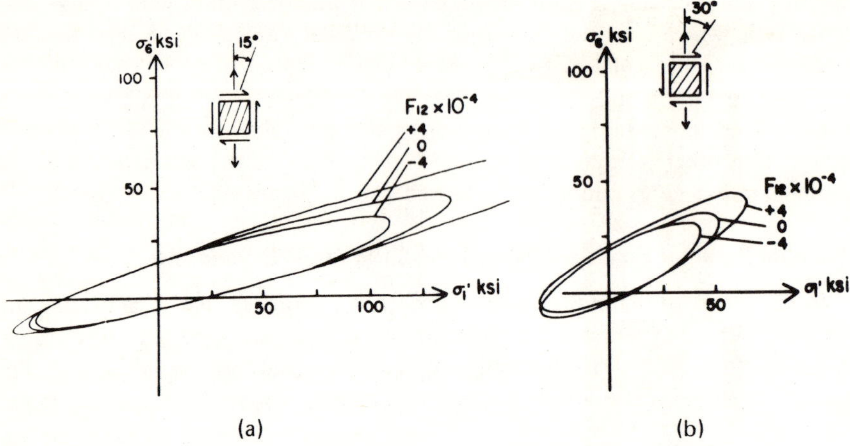

Figure 7. Sensitivity of off-axis tension-torsion strength to interaction tensor F_{12}.

EXPERIMENTAL VERIFICATIONS

Tension-Internal Pressure Tube Test

Initial experiments of longitudinal-tension and transverse compression are performed on graphite-epoxy specimens. Specimen material is the same type used in Reference [1]. The 90° specimen tubes are of 1 inch nominal diameter, have .05 inch wall thickness, and are 5 inches long. The specimen ends are squared with a diamond wheel and are then bonded to an aluminum end-fixture as shown in Figure 8. Shell epoxy 828-Versamid 140 (50/50) is used to bond the outside of the tube to the inside of the end-fixture. The bond length is 1 inch on each end of the tube. Since this epoxy system has a cure shrinkage of approximately 3%, the specimen is under hoop tension on the bond line while in the unloaded state. When subjected to internal pressure and/or compression, the specimen develops hoop strain. This load induced strain minimizes the shrinkage induced strain and produces a relatively uniform state of stress across the length of the tube. Such an end-fixture arrangement proves to be a good compromise between simplicity and minimum grip constraint. While being bonded in the end-plug, the specimen is aligned with a system of *V* blocks. Figure 9 illustrates the bonding operation. When both ends are bonded, Urethane elastomer

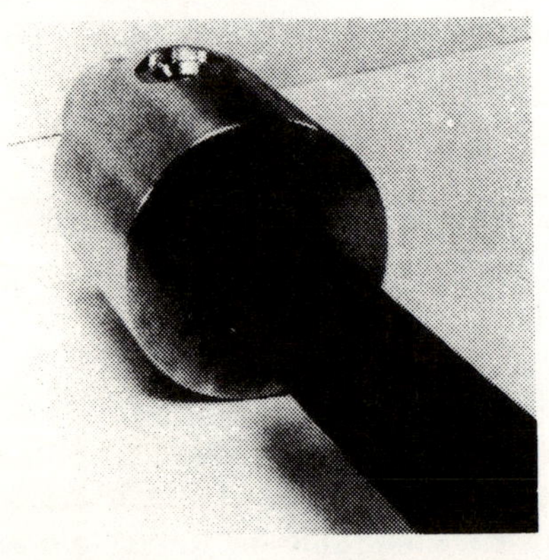

Figure 8. End-fixture for tubular specimen.

Figure 9. Alignment of specimen to end-fixture during bonding.

is centrifugally cast inside the tube for a pressure seal. The elastomer liner is approximately 1/16 inch thick.

Internal pressure is applied to the tube via a combined water-oil system, as shown in Figure 10. Tap water is used to fill the interior of the sample via the quick disconnect *B*. After air is bled off from valve *D*, the sample is disconnected from the filler *E* and connected to the oil line. Oil Cylinder *A* is loaded by a closed-loop hydraulic test machine, which is operated with feedback from the pressure transducer PT. *P* and *G* are high and low range pressure gauges respectively. Both have a shut-off valve *C*. Under this system, closed-loop control can be used to develop internal pressure. 10,000 psi is the operating capacity and 15,000 psi is the burst pressure.

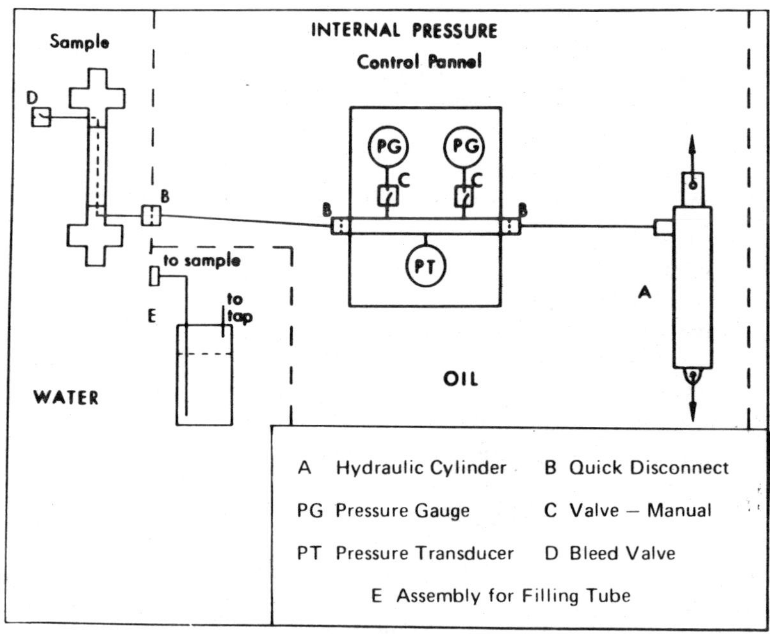

Figure 10. Schematic of internal pressure system for tubular specimen.

Figure 11. Experimental assembly of combined internal pressure-axial-torsion testing of tubular specimen.

The general experimental set-up is shown in Figure 11. The machine applying the axial load is also a closed-loop testing machine. The command signals to the pressure loop and the axial loop can be controlled to produce the desired axial load to internal pressure ratio.

Tension-Torsion Tube Test

Tension-torsion experiments producing large $\sigma_1/-\sigma_2$ ratios are also performed on 30°-tubes and 15°-tubes. These tubes are tested in an axial-torsion testing machine, which is shown on the left of Figure 11. The major difference in experimental technique is the adoption of a different end-fixture attachment. The end plug is bonded to the inside diameter of the tube specimen, thus inducing a compressional hoop strain in curing. In this tension-torsion test, the diameter of the specimen contracts. The load induced strain tends to minimize the curing shrinkage strain for the epoxy bond, and reduces the constraint of the end-fixture.

The tension-to-torsion ratios are computed in accordance with the previous discussion. Command signals, sent to the axial-loop and rotation-loop of the testing machine, are programmed for this ratio.

From results tabulated in Table 1, it can be seen that both internal pressure and tension-torsion tests produce comparable results.

RESULTS AND DISCUSSION

Using the internal pressure-axial compression set-up, two tubular graphite epoxy specimens were tested at a biaxial stress ratio $B = -3$. The 30° off-axis tubular specimen also offers a biaxial stress ratio $B = -3$ in the principal direction when subjected to the tension-torsion ratio as computed from Equation (18). Two

Fiber Orientation	Loading	Measured Biaxial Stress Ratio $\sigma_1/-\sigma_2$	Measured Tension-torsion Ratio σ_1'/σ_6'	Ultimate Stress (ksi)	Computed F_{12} $(\frac{1}{ksi})^2 \times 10^{-4}$	F_{12} (Avg) $(\frac{1}{ksi})^2 \times 10^{-4}$
90°	Axial Compression and Internal Pressure	2.85	–	$\sigma_1 = 62.9$ $\sigma_2 = 22.1$	5.4	
90°	Axial Compression and Internal Pressure	3.01	–	$\sigma_1 = 55.4$ $\sigma_2 = -18.4$	2.0	2.8
30°	Tension Torsion	2.97	1.15	$\sigma_1' = 35.2$ $\sigma_6' = -30.9$	1.5	
30°	Tension Torsion	2.95	1.15	$\sigma_1' = 36.3$ $\sigma_6' = -32.0$	2.2	
90°	Axial Compression and Internal Pressure	6	–	$\sigma_1 = 115.2$ $\sigma_2 = -19.2$	2.6	
90°	Axial Compression and Internal Pressure	6	–	$\sigma_1 = 108.4$ $\sigma_2 = -18.1$	1.9	
15°	Tension Torsion	13.50	3.39	$\sigma_1' = 156.8$ $\sigma_6' = -46.1$	1.6	2.1
15°	Tension Torsion	14.05	3.48	$\sigma_1' = 172.6$ $\sigma_6' = -49.6$	2.4	
15°	Tension Torsion	13.91	3.47	$\sigma_1' = 162.4$ $\sigma_6' = -46.9$	1.9	

specimens were tested in this manner. The result of these tests are tabulated in Table 1 and they were used for the initial estimate of F_{12}. Using this averaged estimate $F_{12} = 2.8 \times 10^{-4}$ additional experiments at the optimal ratio of $B = -11.4$ are suggested by Figure 4a. However, due to the limitation of the internal pressure capacity, we could only execute tests at a compromised ratio of $B = -6$. Two specimens were tested at this ratio. In addition, 15° off-axis tubular specimens offered a biaxial ratio of $B = -14$ as computed from Equation (18). Although neither of these ratios are the optimal ratio, it can be seen from Figure 3a that they provide resolutions not far from that attainable from the optimal ratio. Three 15° specimens were tested at the tension-torsion ratio to provide $B = -14$. These results are also tabulated in Table 1.

From results tabulated in Table 1, it can be seen that both internal pressure and tension-torsion tests produce comparable results. Also tabulated in Table 1 is the predicted variability ΔF_{12} (assuming the strength variability $\Delta\sigma_2/\sigma_2 = 0.1$) obtained from Figure 4a for the appropriate stress ratios B. For comparison, the experimentally measured ΔF_{12} are also listed.

The average F_{12} determined from experiments closed to the optimum condition $(-6 < B < -14)$ for graphite-epoxy composite is $F_{12} = 2.1 \times 10^{-4}$. Although the magnitude of this measured F_{12} appeared to be small, it is by no means negligible. If F_{12} is assumed to be zero, the mean strength, estimated from Equation (1) at a stress ratio of $\sigma_1/-\sigma_2 = 10$, is approximately 15% lower than the mean strength measured and predicted using $F_{12} = 2.1 \times 10^{-4}$.

For general anisotropic materials, in addition to the strength interaction between normal stresses (F_{12}), there exist strength interactions between normal and shear stresses, i.e., F_{16} and F_{26} are not zero. Since F_{16} and F_{26} have a functional structure similar to that of F_{12} in Equation (1), the concept of resolution is similar to that discussed for F_{12} and the determination of optimum stress ratios can be inferred from those relations developed for F_{12} by simple permutations of σ_6 for σ_1 or σ_2.

CONCLUSIONS

In materials strength characterization, significant scatter is inevitably encountered due to material variability. The computation scatter of the components of the strength tensors F_i and F_{ij} evaluated from experimental data was discussed. It was pointed out that, in the cases of the tensor components which do not characterize stress interaction, well defined relative scatter relations exist. These relative scatters are independent of tension-compression or positive-negative shear strength ratios. For F_1 and F_2 the relative scatter is the same as experimental data scatter. For F_{11},F_{22} and F_{66} the relative scatter is twice the scatter of the experiments.

For strength tensor components which characterize stress interactions (F_{12},F_{16},F_{26}), the variation is dependent on the magnitude of the biaxial strength and the magnitude of the tensors themselves. Only the absolute resolutions of these interaction tensors are meaningful. Because of their similarity in functional form, relations developed for F_{12} are applicable for F_{16} and F_{26}.

The resolution of F_{12} is minimum for F_{12} close to zero. For $F_{12}>0$, transverse compression and longitudinal biaxial experiments with biaxial ratio $-6 < B < -14$ provide good resolution. Whereas, for $F_{12}<0$, biaxial compression test with biaxial ratio $6 < B < 14$ provides good resolution. An alternative experiment is suggested where a 90°-tube is not practical and/or internal pressure equipment is not available. This experiment uses an off-axis tubular specimen with helical angle $0 < \theta < 45$ and it is tested in tension-torsion. It was shown that this is equivalent to a transverse compression and longitudinal tension biaxial experiment.

Initial experiments using 90° tubes tested in compression-internal pressure, and 15° and 30° tubes tested in tension-torsion substantiated these findings thus establishing that F_{12} exists for graphite-epoxy composite and it is an intrinsic material property parameter.

APPENDIX I

Relative Resolution of Non-Interacting Strength Tensors

The relative resolution of the non-interacting components of the strength tensors F_1,F_2,F_{11},F_{22} and F_{66} are defined by Equation (4). In Equation (4), the magnifying factors ψ_i and ψ_{ij} may be computed by defining a tensile strength to compressive strength ratio as

$$A = \frac{X}{X'} \tag{A-1}$$

Substitution of Equation (A-1) into Equation (2) yields

$$F_1 = \frac{(1-A)}{X} \tag{A-2}$$

The relative resolution $\Delta F_1/F_1$ can be expressed in terms of the original experimental scatter $\Delta X_1/X_1$ by writing

$$\frac{\Delta F_1}{F_1} = \frac{\partial F_1}{\partial X}\frac{\Delta X}{F_1} \tag{A-3}$$

Using Equation (A-2), Equation (A-2) can be evaluated:

$$\frac{\Delta F_1}{F_1} = \frac{-(1-A)}{X^2}\frac{\Delta X}{F_1} = -\frac{\Delta X}{X} \tag{A-4}$$

Therefore, in Equation (4), $\psi_1 = -1$ and this magnifying factor is independent of the tensile and compressive strength ratio A.

Similarly, F_{11} can be expressed in terms of A in the form:

$$F_{11} = \frac{A}{X^2} \qquad \text{(A-5}$$

and the relative resolution for F_{11} follows from Equation (A-5) is:

$$\frac{\Delta F_{11}}{F_{11}} = \frac{\partial F_{11}}{\partial X}\frac{\Delta X_1}{F_{11}} = -2\frac{\Delta X}{X} \tag{A-6}$$

Thus, in Equation (4), $\psi_{11} = -2$. From the similarity of Equations (2) and (3), the relative resolution for F_2 has the same form as Equation (A-4) and the relative scatter for F_{22}, F_{66} have the same form as Equation (A-6).

Thus, it has been shown that the relative resolution $\Delta F_i/F_i$ and $\Delta F_{ij}/F_{ij}$ of F_1, F_2, F_{11}, F_{22} and F_{66} are constants independent of the magnitude of the ratio of positive to negative strengths measurements and the magnitude of the strength tensors themselves.

ACKNOWLEDGMENT

The author wishes to thank Dr. M. Michno of Washington University for his help in this work and acknowledges the financial support by the Non-metallic Materials Division of the Air Force Materials Laboratory through contract AF-33615-69-C-1404.

NOMENCLATURE

- F_i Strength tensor of the 2nd rank
- F_{ij} Strength tensor of the 4th rank
- X, X' Pure tensile and pure compressive strength along 1-axis
- Y, Y' Pure tensile and pure compressive strength along 2-axis
- S, S' Pure positive and pure negative shear strength in 1-2 plane
- $\tilde{\sigma}_i$ Strength at complex state of stress
- ψ_i, ψ_{ij} Magnification factors of experimental scatter in the computation of F_i and F_{ij}.

REFERENCES

1. S. W. Tsai and E. M. Wu, "A General Theory of Strength for Anisotropic Materials," *J. Composite Materials,* Vol. 5 (January, 1971) p. 58.
2. B. R. Collins and R. L. Crane, "A Graphical Representation of the Failure Surface of a Composite," *J. Composite Materials,* Vol. 5, (July, 1971) p. 408.
3. J. F. Bratt and O Kanan, "Determination of the Yield Condition in the Third Quadrant of the Stress Plane," *J. of Applied Mechanics,* Trans. ASME Vol. 33E (March, 1958) p. 228.

Analysis of the Rail Shear Test-Applications and Limitations

J. M. Whitney

Air Force Materials Laboratory
Wright-Patterson Air Force Base, Ohio

AND

D. L. Stansbarger and H. B. Howell

Northrop Corporation, Aircraft Division
Hawthorne, California

(Received Oct. 15, 1970)

A detailed theoretical and experimental analysis of the rail shear test is presented. A Fourier series solution is obtained for the stresses in an idealized rail shear specimen. The theoretical results are qualitatively verified on fiber reinforced rubber composites. Rail shear data on the modulus and strength of current high-modulus reinforced composites is compared to predicted values obtained from lamination theory. Results show that, despite definite limitations, the rail shear test can be useful in the determination of the in-plane shear properties of a laminated composite.

INTRODUCTION

In the acquisition of design data for fiber reinforced composite materials, the determination of in-plane shear modulus and strength causes particular difficulty. The problem stems from an inability to define a specimen geometry and loading condition which will induce a state of uniform shear. From an applied mechanics point of view a tubular specimen subjected to pure torque provides the most idealized means of measuring the in-plane shear properties of a laminated composite [1]. On the practical side, however, fiber reinforced tubes are expensive to fabricate, require specialized equipment to evaluate, and are often difficult to instrument. Thus, the materials engineer and the structural designer are often forced to use flat coupons for the measurement of in-plane shear properties.

Among the flat coupon type shear tests, the rail shear test is one of the

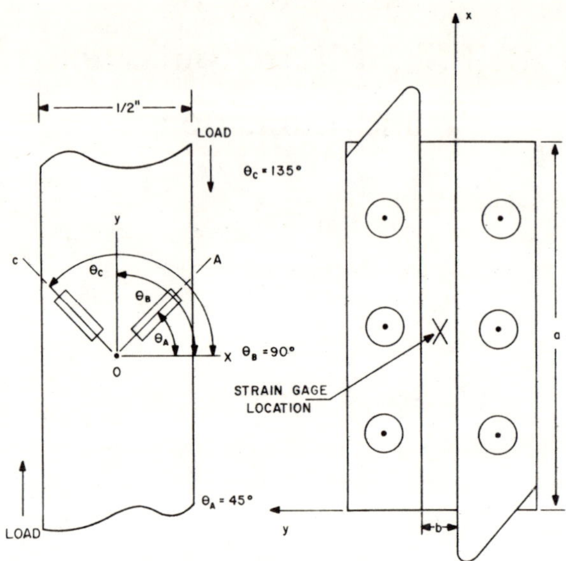

***Figure 1.** Rail shear specimen and orientation of 90°, two element strain rosette.*

most widely used in the aerospace industry. This method consists of clamping the long sides of a rectangular specimen to steel rails, while the remaining sides are free (see Figure 1). Stresses are transmitted to the test specimen by a relative displacement of one rail parallel to the other. From the classical mechanics standpoint this test must be viewed with a great deal of suspicion. In particular, the existence of the free edges causes the stresses in the laminate to differ from pure shear. Although the photoelastic work by Coker [2] on homogeneous glass indicates that a state of uniform shear exists a short distance from the free edges for large length-to-width ratios, the steel rails induce large combined stresses near the specimen corners which may initiate premature failures in an unknown mode. Furthermore, the narrow gage section violates St. Venant's principle. From a practical standpoint, however, the test method is simple, inexpensive, does not require sophisticated apparatus, and is easily adapted to low and high temperature testing. Thus, a study to ascertain the value of the rail shear test is desirable.

In the present paper a detailed theoretical and experimental analysis of the rail shear test is presented. A Fourier series solution is obtained for the stresses in a symmetrically laminated orthotropic plate subjected to a uniform tangential displacement along one edge with the opposite edge clamped and the two adjacent edges stress-free. These are essentially the edge conditions imposed by the rail shear test. The theoretical results are qualitatively verified on rubber reinforced composites. Rail shear data on the modulus and strength of current high-modulus fiber reinforced composites is compared to predicted values obtained from lamination theory.

ANALYSIS

Consider a symmetrically laminated orthotropic rectangular plate lying in the region $0 \leqq x \leqq a$, $0 \leqq y \leqq b$ (see Figure 1). The governing equations for in-plane loading are [3, 4]

$$A_{11} u^0_{,xx} + A_{66} u^0_{,yy} + (A_{12} + A_{66}) v^0_{,xy} = 0 \tag{1}$$

$$(A_{12} + A_{66}) u^0_{,xy} + A_{66} v^0_{,xx} + A_{22} v^0_{,yy} = 0 \tag{2}$$

where partial differentiation is denoted by a comma, u^o and v^o are the mid-plane displacements in the x and y directions, respectively. Also

$$A_{ij} = \int_{-h/2}^{h/2} Q_{ij}^{(k)} \, dz$$

where $Q_{ij}^{(k)}$ are the reduced stiffness coefficients for plane stress of the k th layer and h is the plate thickness. The in-plane constitutive relations in terms of force resultants N_i are

$$\begin{aligned} N_x &= A_{11} u^0_{,x} + A_{12} v^0_{,y} \\ N_y &= A_{12} u^0_{,x} + A_{22} v^0_{,y} \\ N_{xy} &= A_{66} (u^0_{,y} + v^0_{,x}) \end{aligned} \tag{3}$$

It is convenient to consider the free-edge shear problem as a superposition of the following two boundary value problems:

$$\begin{aligned} 1.\ & u^0(x,0) = v^0(x,0) = v^0(x,b) = N_x(0,y) \\ & = N_x(a,y) = 0 \\ & u^0(x,b) = c = \text{constant} \\ & N_{xy}(0,y) = N_{xy}(a,y) = A_{66}\frac{c}{b} \end{aligned} \tag{4}$$

$$\begin{aligned} 2.\ & u^0(x,0) = u^0(x,b) = v^0\,(x,0) = v^0(x,b) \\ & = N_x(0,y) = N_x(a,y) = 0 \end{aligned} \tag{5}$$

$$N_{xy}(0,y) = N_{xy}(a,y) = -\ A_{66}\frac{c}{b} \tag{6}$$

The solution to problem No. 1 leads to a state of pure shear

$$u^0 = \frac{c}{b} y, \qquad v^0 = 0 \qquad (7)$$

It is easily seen that Equation (7) satisfies Equations (1) and (2), and the boundary conditions Equation (4).

Displacement solutions for problem No. 2 are sought in the region $0 < x < a, 0 < y < b$, and are assumed to be of the form

$$u^0 = \sum_{m=1}^{\infty} \sum_{n=0}^{\infty} A_{mn} \sin \frac{(2m-1)\pi x}{a} \sin \frac{2n\pi y}{b} \; (0 < a < a, 0 \leqq y \leqq b) \qquad (8)$$

Figure 2. *Rail shear specimen in test fixture.*

$$v^0 \quad \sum_{m=1}^{\infty} \sum_{n=0}^{\infty} B_{mn} \cos \frac{(2m-1)\pi x}{a} \cos \frac{2n\pi y}{b} \; (0 \leqq x \leqq a, 0 \leqq y \leqq b) \qquad (9)$$

Because of symmetry considerations only odd terms are taken in x and even terms in y. If Equations (8) and (9) are assumed to be term-by-term differentiable, the boundary conditions, Equations (5) and (6) cannot be satisfied. Thus, it is necessary to use a procedure suggested by Green [5] for the bending of isotropic plates. This method has been recently applied to the bending of laminated anisotropic plates [6, 7]. Details are presented in the appendix.

EXPERIMENTAL PROCEDURE

The basic test method and specimen configuration as outlined by Floeter and Boller [8] was used to determine the shear properties of the boron-epoxy and graphite-epoxy composites in Table I with strain measurements being monitored by a two-element rosette strain gage located on the specimen surface as shown in Figure 1. The test specimen configuration is such that an aspect ratio $R = a/b = 12$ results.

The specimens were placed in the test fixtures, as shown in Figure 2, with 80 ft/lbs of torque applied to each of the six high tension bolts to prevent slippage between the specimen and the rails and loaded to failure with strain readings taken with a Datran II automatic strain indicator and recorder.

RESULTS AND COMPARISONS

As previously pointed out, it is assumed in the rail shear test that the free edge effects are minimal for large values of R. This assumption is based on the work of Coker [2], in which the stress distribution in a rail shear specimen of homogeneous isotropic glass was investigated by photoelastic techniques. For high aspect ratios R, a state of uniform shear stress was found a short distance from the free ends. As a result the rail shear test is usually employed with $R \geqq 10$.

The results of the current stress analysis, however, show that the stress distribution in an orthotropic laminate is strongly dependent on various inplane stiffness ratios. For example, if $A_{22}/A_{66} >> 1$, a uniform shear stress can be obtained throughout most of the specimen for as low an aspect ratio as $R = 2$. This is illustrated experimentally in Figure 3 where a 90° nylon reinforced tire-cord specimen is deformed in a rail shear jig with $R = 2$ and $A_{22}/A_{66} = 500$. The same material in a 0° orientation is shown in Figure 4. For this case $A_{22}/A_{66} = 4$ and the shear stress is non-uniform over a large region near the free edges. The wrinkled grid lines indicate that some local buckling may be occurring.

A complete parametric study reveals that with one exception a uniform state of shear can be obtained across the center of the specimen for $R \geqq 10$. The one exception occurs when $A_{12}/A_{11} \simeq A_{12}/A_{22} \simeq 1$, i.e. when the specimen essentially has a very high effective Poissons ratio. For such a case, aspect ratios ranging from $R = 2$ to $R = 14$ were investigated and in all cases the shear stress distribution across the center of the specimen was found to be highly irregular. In practical applications such a situation occurs for $\pm$ 45° angle-ply specimens. This is illustrated in Figure 5 for Thornel 50 graphite-epoxy composites where the shear stress resultant N_{xy}, normalized by the average $\overline{N}_{xy}$, for a $\pm$ 45° laminate is compared to a [0° $\pm$ 45°] laminate with $R = 12$. Experimentally determined unidirectional ply properties are shown in Table I.

Table 1. Ply Properties.

	MII/2387*	Th 50/E-798**	Narmco 5505***
E_{11}	19.6 msi	21.2 msi	30.0 msi
E_{22}	1.2 msi	0.85 msi	3.0 msi
ν_{12}	0.30	0.31	0.35
G_{12}	0.59 msi	0.57 msi	1.2 msi

* Morganite II graphite-epoxy system
** Thornel 50 graphite-epoxy system
*** Boron-epoxy system

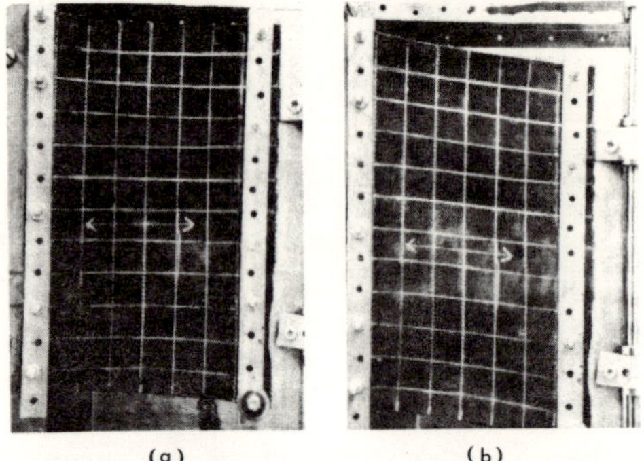

Figure 3. *Rail shear test on 90° unidirectional nylon reinforced tire-cord material, (a) undeformed specimen, (b) deformed specimen.*

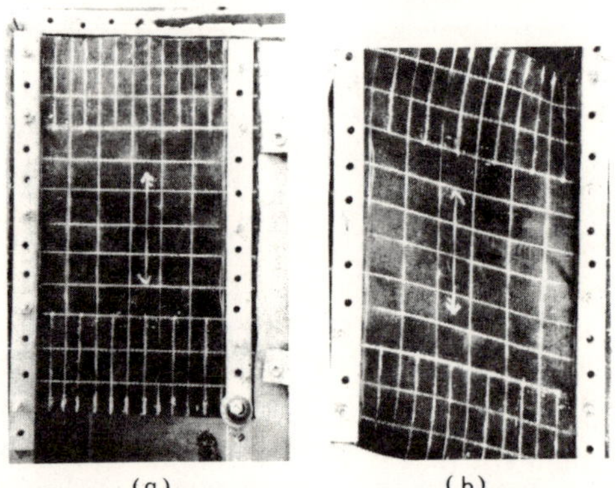

Figure 4. *Rail Shear test on 0° unidirectional tire-cord material, (a) undeformed specimen, (b) deformed specimen.*

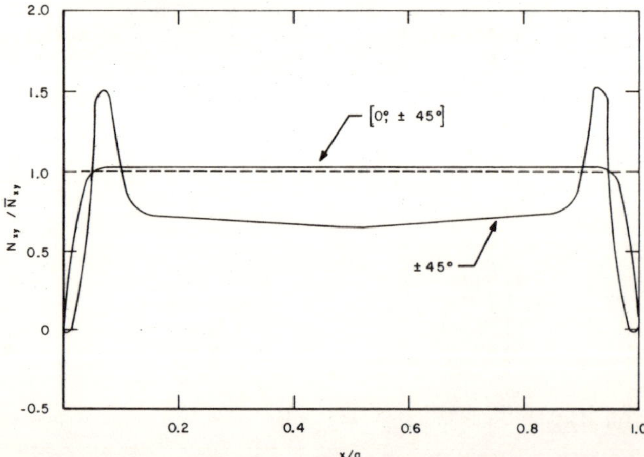

Figure 5. *Shear distribution across center (y = b/2) of Thornel 50 graphite-epoxy specimen, R = 12.*

Although Fourier series solutions are not necessarily valid on the boundaries, it is usually possible to obtain numerical results a short distance from the edges. Figure 6 shows the distribution of N_y at $y = b/100$ for the laminates of Figure 5. In general, N_y will be singular in the corners. This is not possible to show directly with the Fourier analysis; however, results obtained by Bogy [9] for the stresses at the interface of two-quarter planes of dissimilar materials indicate the presence of singularities at the corner. As pointed out previously, the existence of large normal stresses near the corners may initiate premature failures in a non-shear mode. Whether this occurs under actual test conditions can only be ascertained by considering rail shear data.

Experimental data is shown in Table II. Excellent agreement is found between shear modulus data and values predicted from lamination theory [10], even in the case of $\pm$ 45° laminates where the shear distribution is not uniform over the center section. Since the gage section is very narrow, however, the strain gages essentially measure an average value across the width which the analysis shows tends to nulify the effect of the non-uniform shear distribution.

The lack of reliable shear data on tubular specimens makes the rail shear strength data difficult to interpret. One possible approach, however, is to compare the data to existing failure criteria. In Table II the maximum strain criteria (MAX ϵ) of Reference [10], netting analysis (NA) in which filament failure within a ply is required, and Hill's criteria (HC) from Reference [11] are considered. As one might suspect from the stress analysis, very poor agreement between theory and experiment is observed in general, for the $\pm$ 45° laminates. Although better agreement is found between predicted failure and experiment for the $\pm$ 45° Thornel 50 graphite-epoxy composites,

Table 2. Rail Shear Data*.

	EXPERIMENT		*ANALYSIS*			
	N_{xy}/h	G_{xy}	*Max* ϵ	*HC*	*NA*	G_{xy}
MII/2387						
0° (8 ply)	14	0.6	14	14	—	0.6
±45° (8 ply)	30	4.9	65	58	105	5.0
[0°, ±45°]	35	2.5	36	41	74	2.8
Narmco 5505						
±45° (8 ply)	44	7.8	70	64	152	7.8
Th 50/E-798						
±45° (16 ply)	28	5.0	26	30	81	5.5
[0°, ±45°]	24	3.7	18	21	57	3.8

* Shear strength in ksi, modulus in msi. All properties based on average of three specimens.

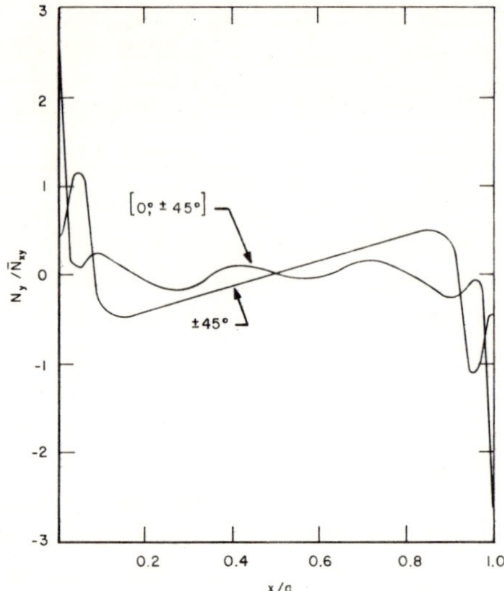

Figure 6. Normal stress distribution at y = b/100 of Thornel 50 graphite-epoxy specimen, R = 12.

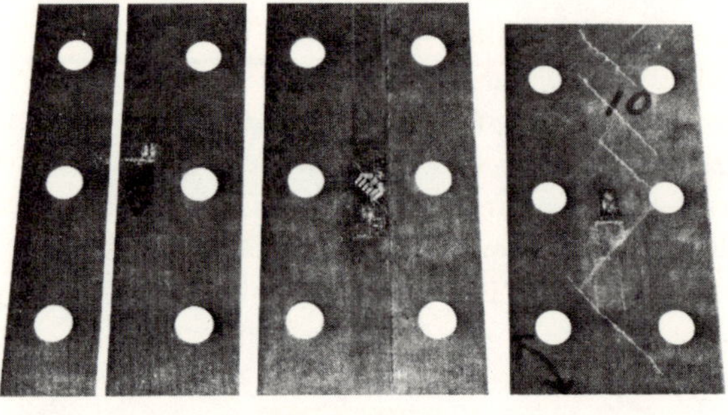

Figure 7. Failed specimens: (a) Morganite II, 0° unidirectional, (b) Thornel 50, [0°, ± 45°] laminate, (c) Narmco 5505, ± 45° laminate.

the results are misleading as this system is highly "transverse critical" (poor unidirectional shear and transverse tensile strengths) which leads to unacceptably low shear strength for laminate configurations. The high shear values predicted by netting analysis are probably not realistic as it neglects ply failures transverse to the filaments and unidirectional shear failures. This is somewhat substantiated by the [0°, ± 45°] failed specimen shown in Figure 7. Although the picture is not clear, failure was initiated in the 0° ply, which is essentially subjected to uniform shear. The shear strength of the 0° Morganite II specimen is in good agreement with short beam shear data [12].

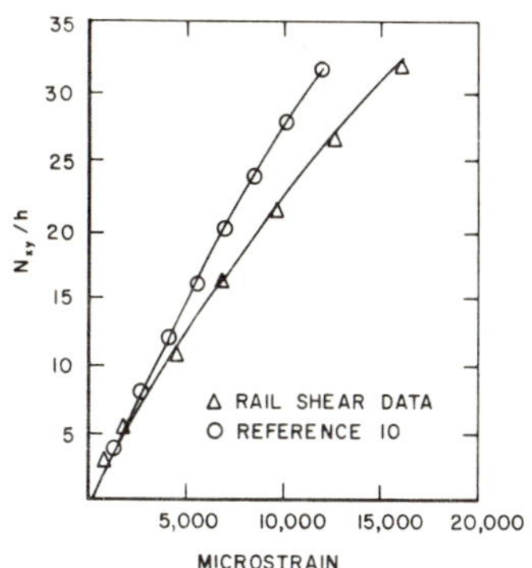

***Figure 8.** Stress-strain curve from rail shear test on Morganite II, [0° ± 45°] laminate.*

A typical stress-strain curve is shown in Figure 8 for a 0°, ± 45° laminate and compared to results predicted by the procedure in Reference [10].

CONCLUSIONS

The rail shear test method, as evaluated in this study, is valid for measuring the shear modulus of laminated composites. It would also appear that with certain limitations the method provides a useful means of determining in-plane shear strength. In particular, for laminates with $R \geqq 10$ a uniform shear stress is obtained a short distance away from the free edges provided the laminate does not have a high effective Poisson's ratio (i.e. $A_{12}/A_{11} \simeq A_{12}/A_{22} \simeq 1$). For laminates having an essentially uniform shear stress distribution, the free edge effects appear to be minimal as exemplified by the [0°, ± 45°] laminate strengths reported in Table II. In the case of laminates having a high effective Poissons ratio, such as ± 45° angle-ply composites, the shear stress distribution is very irregular, and leads to extremely low values of shear strength as determined by the rail shear test. Thus, if this test method is used to obtain design allowables in shear, it must be used in conjunction with the proper laminate orientations or the designer will pay an unnecessary penalty by using values which are far too conservative, resulting in over design of the structure.

APPENDIX

Since Equation (8) is not valid on the edges $x = 0$, $x = a$, differentiation with respect to x cannot be accomplished term-by-term [5]. Assuming $\overset{0}{u}_{,x}$ can be represented by a cosine-sine series, partial integration leads to the result

$$\overset{0}{u}_{,x} = \sum_{m=1}^{\infty} \sum_{n=1}^{\infty} \left[\frac{(2m-1)\pi}{a} A_{mn} + a_n \right] \cos \frac{(2m-1)\pi x}{a} \sin \frac{2n\pi y}{b} \qquad (10)$$

where a_n are the Fourier coefficients in the expansion of u^0 on the edges $x = 0$, a, *i, e.*

$$u^0(0,y) = u^0(a,y) = \frac{-a}{4} \sum_{m=1} a_n \sin \frac{2n\pi y}{b} \qquad (11)$$

A similar procedure applied to $\overset{0}{v}_{,y}$ yields constants b_n associated with the expansion of $\overset{0}{v}_{,x}$ on the edges $x = 0, a$; and constants c_m associated with the expansion $\overset{0}{v}_{,y}$ along the edges $y = 0, b$. All other necessary derivatives can be obtained through term-by-term differentiation.

The boundary conditions on the shear resultant N_{xy}, as given by Equation (6), yield the relationships

$$a_n = \frac{a}{2n\pi R} b_n \quad (n = 1, 2, \ldots)$$
$$b_0 = \frac{-8Rc}{a^2} \tag{12}$$

where R is the aspect ratio of the plate a/b.

Substituting the appropriate derivatives of Equations (8) and (9) into Equations (1), (2), taking Equation (12) into account, and solving the resulting algebraic equations yields

$$A_{mn} = \frac{-a^2(2m-1)}{2D_{mn}n\pi^2 R}\{[A_{11}A_{66}(2m-1)^2 + 4(A_{11}A_{22} - A_{66}A_{12} - A_{12}^2)n^2R^2]\,b_n + 4A_{22}(A_{11}A_{66})\,n^2R^2c_m\} \tag{13}$$

$$B_{mn} = \frac{-a^2}{D_{mn}\pi^2}\{[A_{11}A_{66}(2m-1)^2 - 4A_{12}A_{66}n^2R^2]\,b_n + [A_{22}A_{11}(2m-1)^2 + 4A_{22}A_{66}n^2R^2]\,c_m\} \tag{14}$$

where

$$D_{mn} = [A_{11}(2m-1)^2 + 4A_{66}n^2R^2]\,[A_{66}(2m-1)^2 + 4A_{22}n^2R^2] - 4(A_{12} + A_{66})^2\,(2m-1)^2n^2R^2$$

Equations (13), (14) in conjunction with the boundary conditions, Equation (5), yield the following two sets of algebraic equations for the constants b_n and c_m

$$A_{66}(A_{11}A_{22} - A_{12}^2)n^2R^2b_n \sum_{m=1}^{\infty} \frac{1}{D_{mn}} \frac{-A_{22}A_{66}}{4} \sum_{m=1}^{\infty}\left[\frac{A_{11}(2m-1)^2}{D_{mn}} - \frac{4A_{12}n^2R^2}{D_{mn}}\right] c_m = 0 \qquad \text{for} \quad n = 1, 2, \ldots \tag{15}$$

$$A_{66}\sum_{n=1}\left[\frac{A_{11}(2m-1)^2 - 4A_{12}n^2R^2}{D_{mn}}\right] b_n + A_{22}\left\{\frac{1}{2(2m-1)^2A_{66}} + \sum_{n=1}^{\infty}\left[\frac{A_{11}(2m-1)^2 + 4A_{66}n^2R^2}{D_{mn}}\right]\right\} c_m = \frac{4cR}{a^2(2m-1)^2} \qquad \text{for} \quad m = 1, 2 \ldots \tag{16}$$

Truncation of Equations (15) and (16) leads to $m + n$ equations in $m + n$ unknowns. Coefficients which are infinite series should be summed separately. Although these results cannot be expressed in closed form without considerable difficulty, sufficient terms can be summed to obtain a desired degree of convergence.

The truncated equations are solved on a digital computer for the coefficients b_n, c_m, and convergence established by increasing the order of the system. In general, the rate of convergence depends on the plate stiffness A_{ij} and the aspect ratio R. For the numerical examples considered in the present paper, it was not necessary to go beyond a 24 × 24 system of equations. Force resultants are calculated from the plate constitutive relations, Equation (3).

ACKNOWLEDGMENTS

The authors wish to acknowledge Mr. E. Guthrie of the Aeronautical Systems Division, Wright-Patterson Air Force Base, Ohio for the computer analysis and Mr. B. Maurer for his assistance in checking numerical calculations. We also wish to express our appreciation to Mr. W. Polley for his assistance in sample fabrication and experiments on the rubber tire-cord material.

REFERENCES

1. N. J. Pagano and J. M. Whitney, "Geometric Design of Composite Cylindrical Characterization Specimens", *J. Composite Materials,* Vol. 4 (1970), p. 360.
2. E. G. Coker, "An Optical Determination of the Variation of Stresses in a Thin Rectangular Plate Subjected to Shear", *Royal Society of London Proc.* (Series A), Vol. 86 (1912), p. 291.
3. S. B. Dong, R. B. Matthiesen, K. S. Pister, and R. L. Taylor, "Analysis of Structural Laminates", *Air Force Report ARL-76* (1961).
4. J. M. Whitney and A. W. Leissa, "Analysis of Heterogeneous Anisotropic Plates", *Journal of Applied Mechanics,* Vol. 36 (1969), p. 261.
5. A. E. Green, "Double Fourier Series and Boundary Value Problems", *Cambridge Phil. Society Proc.,* Vol. 40 (1944), p. 222.
6. J. M. Whitney and A. W. Leissa, "Analysis of a Simply-Supported Laminated Anisotropic Rectangular Plate", *AIAA Journal,* Vol. 8 (1970), p. 28.
7. J. M. Whitney, "The Effect of Boundary Conditions on the Response of Laminated Composites", *J. Composite Materials,* Vol. 4 (1970), p. 192.
8. L. H. Floeter and K. H. Boller, "Use of Experimental Rails to Evaluate Edgewise Shear Properties of Glass-Reinforced Plastic Laminates", U.S. Forest Products Laboratory, Madison, Wisconsin, *Task Force for Test Methods, Industry Advisory Group,* MIL-HDBK-17, 28 April 1967.
9. D. B. Bogy, "Edge Bonded Dissimilar Orthogonal Elastic Wedges Under Normal and Shear Loading", *J. Applied Mechanics,* Vol. 35 (1968), p. 460.
10. P. H. Petit and M. E. Waddoups, "A Method of Predicting the Nonlinear Behavior of Laminated Composites", *J. Composite Materials,* Vol. 3 (1969), p. 2.
11. S. W. Tsai and V. D. Azzi, "Strength of Laminated Composite Materials", *AIAA Journal,* Vol. 4 (1966), p. 296.
12. R. J. Dauksys and J. D. Ray, "Properties of Graphite Fiber Nonmetallic Matrix Composites", *J. Composite Materials,* Vol. 3, p. 684.

Moiré Analysis of the Interlaminar Shear Edge Effect in Laminated Composites*

R. BYRON PIPES, *General Dynamics Corporation, Fort Worth, Texas, 76101* AND
I. M. DANIEL, *IIT Research Institute, Chicago, Illinois 60616*

(Received January 16, 1971)

INTRODUCTION

The three-dimensional state of stress in a symmetric laminate ($B_{ij} = 0$) under uniaxial loading has been examined in recent studies [1, 2]. The results of these studies indicated that significant interlaminar stress components were necessary in regions near the laminate free-edge to accomplish shear transfer between laminae of the laminate. In addition, the interlaminar stresses were found to be restricted to a boundary-layer region near the edge approximately equal to the laminate thickness and the state of stress in interior regions of the laminate was found to be accurately predicted by the plane stress relations given in Laminated Plate Theory [3-5].

The boundary-value problem treated in Reference 1 was a rectangular laminate configuration typical of a uniaxial tensile coupon subjected to a uniform axial strain (Figure 1). The region of study was restricted to regions of the laminate removed from the areas of load introduction. The functional form of the axial displacement field was determined under this restriction by examining thte stress-displacement relations and enforcing the appropriate symmetry conditions

$$u = Cx + U\ (y,z)$$

where "C" is the applied axial strain, "ϵ_x", and is the same for each layer. In an effort to verify the functional form of the axial displacement field derived in Reference 1, the authors employed the Moiré technique to study the surface ($z = 2h_o$) displacement field of a four-layer, graphite-epoxy laminate under a uniaxial loading.

THE MOIRÉ TECHNIQUE

The Moiré effect is an optical phenomenon observed when two closely spaced arrays of lines are superimposed and viewed with either transmitted or reflected light [6-8]. When the two arrays differ in either spacing or orientation, optical interference between the arrays produces Moiré fringes. The concept offers a means for determining full-field surface displacement and strain fields.

* This study was jointly supported by the Independent Research and Development Program of the Fort Worth facility of the Convair Aerospace Division, General Dynamics Corporation and the IIT Research Institute Internal Research and Development Program.

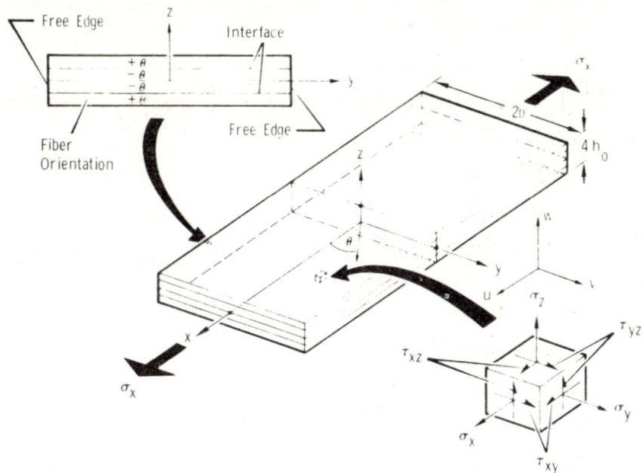

Figure 1. *Specimen geometry.*

Figure 2. *Failed ±25° graphite-Epoxy specimens.*

In actual practice, one array of lines is applied to the specimen surface by photographic etching or transfer techniques. A transparent reference array (master grid) is then placed near or in contact with the specimen array. Moiré patterns appear when the specimen is deformed under load and observed through the reference array. The fringes which make up the Moiré pattern may be interpreted as the loci of points having the same component of displacement normal to the direction of the lines in the reference array.

EXPERIMENTAL PROCEDURE

A 16-ply laminate fabricated from graphite-epoxy (Morganite II fiber and 4617 resin system) was employed in this investigation. Since the shear coupling term in the graphite-epoxy stiffness matrix, C_{16}, attains a maximum near twenty-five degrees, the laminate configuration of $[+25°_4/-25°_8/+25°_4]$ was chosen for this study. Three specimens 14.0 inch in length, 1.02 inch in width, and 0.155 inch in thickness were fabricated from the laminate with 4.0-inch, highly tapered glass end grips employed for load introduction. The failed specimens are shown in Figure 2, and a 100X photomicrograph of a transverse section through one half the specimen thickness is shown in Figure 3. The photomicrograph reveals that the specimens were fabricated from a high quality laminate.

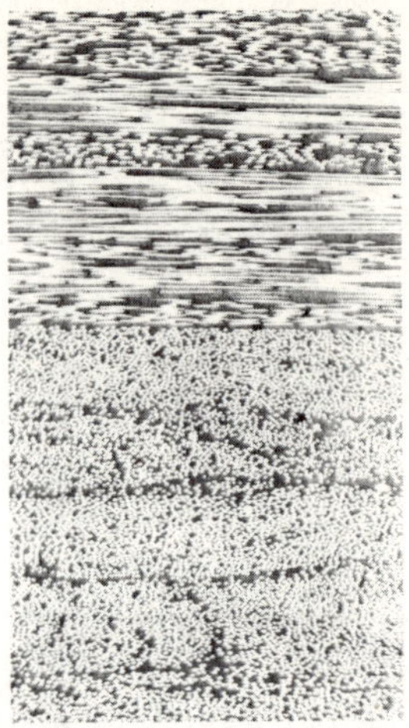

***Figure 3.** 100X Photomicrograph of the $[25_4°/-25_8°/25_4°]$ graphite-epoxy laminate.*

Two of the three specimens were employed in the Moiré analysis, and the third specimen served as a static control. In the preparation of the specimens for application of the Moiré arrays, a reflective aluminum coating was first vacuum deposited on one face of each specimen. Subsequently, an array of 1000 lines per inch perpendicular to the longitudinal axis of the specimen was photoprinted in this reflective surface by means of the Kodak Photoresist process (KPR). This was accomplished by the following procedure: The liquid photosensitive emulsion (KPR) was applied to the specimen as it rotated at a constant velocity on a rotating table and cured under heat lamps. The film master with an array of 1000 lines per inch was placed in contact with the coated surface in a vacuum frame and exposed to a carbon-arc light which contains ultraviolet radiation. Subsequently, the exposed surface of the specimen was developed and dyed.

RESULTS

Each of the three specimens was loaded to failure in an Instron testing machine employing Instron friction grips. Moiré fringe patterns were photographed at several load levels and analyzed to obtain the specimen axial strain field. These results and those of the control specimen are shown in Figure 4.

An enlarged photograph of the Moiré fringe pattern obtained at a stress level of 44,200 psi is shown in Figure 5. The fringes in the Moiré pattern have a characteristic S-shape with pronounced curvature near the specimen edges and their distribution indicates that the axial displacement field was accurately described in Reference 1 as $u = Cx + U(y,z)$.

Finally, Figure 6 shows a comparison of the axial displacement distribution at the specimen surface, $z = 2h_o$, determined by the solution to the governing elasticity field equations (Reference 1) and the Moiré grid analysis. These results indicate

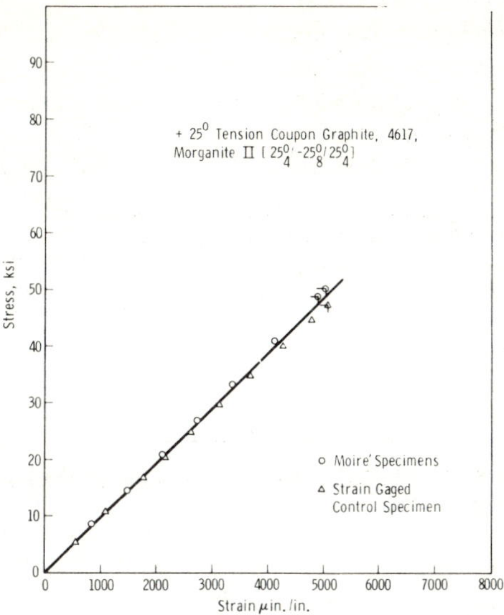

Figure 4. Specimen stress-strain response.

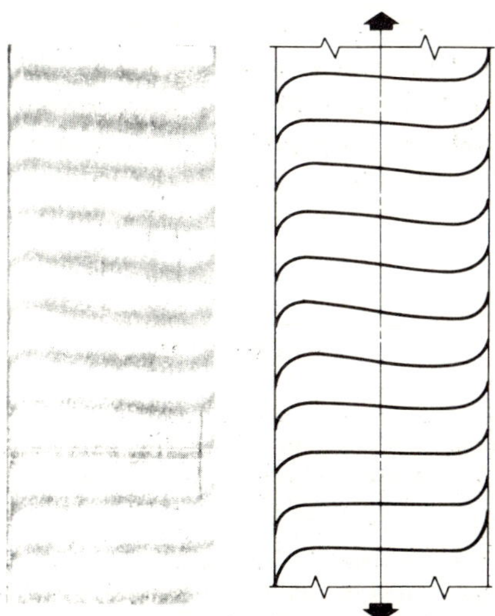

Figure 5. Moiré fringe pattern.

close agreement between analysis and experiment. It is also significant to note that the boundary layer region in which the edge effect is confined is shown to be approximately equal to the laminate thickness.

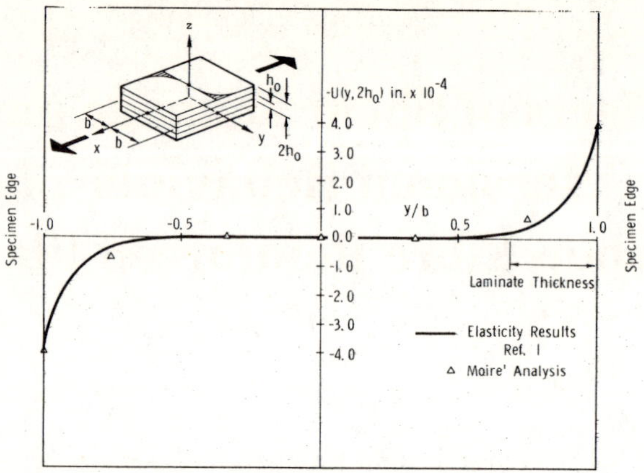

***Figure 6.** Axial displacement distribution at the laminate surface $z = 2h_o$.*

CONCLUSIONS

The form of the axial displacement field derived in Reference 1 has been verified through the application of the Moiré experimental technique. Moiré fringe patterns confirm the characteristic S-shape of the lines of constant axial displacement in the displacement field. In addition, the boundary layer region in which the interlaminar edge effect is confined is verified to be approximately equal to the laminate thickness.

REFERENCES

1. R. B. Pipes and N. J. Pagano, "Interlaminar Stresses in Composite Laminates Under Uniform Axial Extension," *J. Composite Materials,* Vol. 4 (1970), p. 538.
2. A. H. Puppo and H. A. Evensen, "Interlaminar Shear in Laminated Composites under Generalized Plane Stress," *J. Composite Materials,* Vol. 4 (1970), p. 204.
3. E. Reissner, and Y. Stavsky, "Bending and Stretching of Certain Types of Heterogeneous Aeolotropic Elastic Plates," *J. Applied Mechanics,* Vol. 28 (1961), p. 402.
4. S. B. Dong, K. S. Pister, and R. L. Taylor, "On the Theory of Laminated Anisotropic Shells and Plates," *J. Aerospace Sciences* (1962), p. 969.
5. J. E. Ashton, J. C. Halpin, and P. H. Petit, *Primer on Composite Materials: Analysis,* Technomic Publishing Co. (1969).
6. J. W. Dally, and W. F. Riley, *Experimental Stress Analysis,* McGraw-Hill (1965), p. 356.
7. A. J. Durelli and W. F. Riley, *Introduction to Photomechanics, Prentice-Hall,* 1965, p. 194.
8. A. J. Durelli and V. J. Parks, *Moiré Analysis of Strain,* Prentice-Hall, Englewood Cliffs, N.J., 1970.

Effects of Shear Damage on the Torsional Behaviour of Carbon Fibre Reinforced Plastics

ROBERT D. ADAMS AND J. E. FLITCROFT

Department of Mechanical Engineering
University of Bristol
Bristol, England

N. L. HANCOX AND W. N. REYNOLDS
Process Technology Div. *Nondestructive Testing Centre*

Atomic Energy Research Establishment
Harwell, England

(Received November 25, 1972)

ABSTRACT

The torsional modulus and damping capacity of uniaxial specimens of carbon fibre reinforced epoxy resin have been measured as a function of strain amplitude, proportion and type of fibre and of fibre surface treatment. The specimens were then twisted to shear failure and the measurements repeated. The values of shear strength observed were analysed in the light of the modulus and damping capacity values. The significance of such measurements in the mechanical testing of these materials has been demonstrated.

INTRODUCTION

IT IS REQUIRED to develop techniques for the inspection of carbon fibre reinforced plastics. Some of these tests must be nondestructive and should be applicable to the determination of such factors as the type and proportion of fibres present, their state of alignment, the state of cure of the resin matrix and the degree of adhesive bonding between fibres and matrix. It is also necessary to be able to detect porosity, cracking and delamination in the structure, either in the as manufactured state or after use in potentially damaging situations.

Some attempts have recently been reported to use mechanical vibrations and damping for this purpose. Schultz and Warwick [1] studied the fatigue of glass fibre/epoxy resin flexural specimens vibrated at approximately 200Hz and noted that fatigue cracking correlated with increased damping. Brown et al [2] made ultrasonic measurements at frequencies between 10 and 220KHz., of the compression and shear moduli and internal damping associated with the former, on solid rods of carbon fibre reinforced epoxy resin composite, before and after these had been tested to initial failure in a torsion rig. Although the cracks introduced in torsional testing had an obvious effect on the mechanical properties of the specimens when these were handled, internal damping could be either greater or less after failure, compared with the results on undamaged specimens, and showed no marked trend with volume loading of fibre, fibre surface treatment, or test frequency range.

In this paper we describe torsion pendulum measurements, made at a frequency of approximately 16Hz., of the shear modulus and internal damping of unidirectional carbon fibre epoxy resin composites, before and after testing to failure in the static torsion test apparatus. The vibrational apparatus used was the same as that employed by Adams et al [3] in an earlier study. This described the measurements of shear modulus and specific damping capacity ψ[1] between 15 and 25Hz and of flexural modulus and damping coefficient between 290 and 420Hz, and showed that the damping coefficients decreased as the glass or carbon fibre content and the elastic moduli increased. It was anticipated that measurements of damping at low frequencies and high strain amplitude would provide positive indication of fabrication defects or of damage in service by overload, fatigue or mechanical impact. The results obtained demonstrate the effects of shear cracks.

EXPERIMENTAL PROCEDURE

The dynamic torsion testing consisted of measuring the damping and shear modulus at a series of increasing and decreasing amplitudes at a test frequency of approximately 16Hz. The maintained torsional pendulum used has a low inherent damping capacity ($\psi = 0.05\%$) and has been in use for some years.

Test bars 330mm long by 6.3mm square were made from high modulus (type I) and high strength (type II,) carbon fibre, both surface treated and untreated, and an 828 type resin (liquid bisphenol A) by a wet lay up technique. The epoxy resin mix consisted of 100 parts by weight (p.b.w.) of resin to 80 p.b.w. of hardener (acid anhydride) to 1 p.b.w. of accelerator, cured for 3 hours at 120°C. The resin mix was degassed for a few minutes before and after the fibres were added. Fibre volume loadings of 30, 40, 50, 60 and 70 *v/o* were used. Two specimens approximately 150mm long were obtained from each bar.

---

[1] $\psi = 2\pi/Q = \Delta W/W$, where Q is the dynamic amplification factor, W the maximum stored energy and ΔW the energy lost in a cycle.

First, the 150mm long square cross section bars were subjected to the full dynamic torsional test programme. The middle 100mm of each bar was then machined to the maximum possible diameter (usually about 6.15mm) to produce circular section test bars. The machining consisted of grinding between centres using light cuts. These round bars were also subjected to the full dynamic torsional test programme.

The bars were then twisted in a static torsion test apparatus until they just cracked, and the damping and dynamic shear modulus again measured.

RESULTS

Typical results obtained for modulus and damping capacity at 16Hz are illustrated in Figures 1(a) and (b) respectively. For uncracked specimens, the modulus was almost independent of stress level except for type I untreated specimens which showed a decrease of 2–10% at the highest stress amplitudes. The elastic moduli were also measured in the torsional strength testing apparatus under conditions of slowly applied stress. The overall behaviour was similar but the numerical values obtained were some 5–10% lower than by the dynamic method. The differences were generally larger for specimens using untreated fibres than for those with surface treatment, and the specimens showing the highest damping were amongst those with the largest static-dynamic modulus difference.

The variation of modulus and damping capacity with composition is discussed by Adams and Bacon [4]. Results of this kind for both carbon and glass fibres have already been given by Adams et al [3] who compared the moduli with theoretical values. The steadily decreasing tendency of damping capacity with fibre content is also in line with their results, but in the present work anomalously high values have been obtained with type I fibres at the higher volume loadings, particularly with untreated fibres. The variation of modulus with fibre type and surface treatment is also similar to that previously reported (Reynolds and Hancox [5]).

After cracking, the dynamic moduli were lower and showed a marked decrease with increasing stress generally falling by 10–20% of the new value at the highest stress for treated specimens, and by 20–40% for untreated. The actual decrease like the much increased value of the damping capacity depended on the amount of damage induced in each case and was therefore not in itself very significant. However the damping capacity generally increased with stress amplitude, maximum values above 80% for type I and 60% for type II being recorded. In some cases, especially with high volume fractions of type II fibre, there was a decrease of ψ with stress, usually after an initial increase. On re-testing at lower stress amplitudes, there was usually found a further decrease in modulus and increase in damping, suggesting that the test itself had caused damage propagation at the higher stresses employed.

The values of shear strength obtained are illustrated in Figure 3 together with the mean results of Reynolds and Hancox [5] on a larger number of specimens. The main point to observe is that whereas the specimens made from untreated

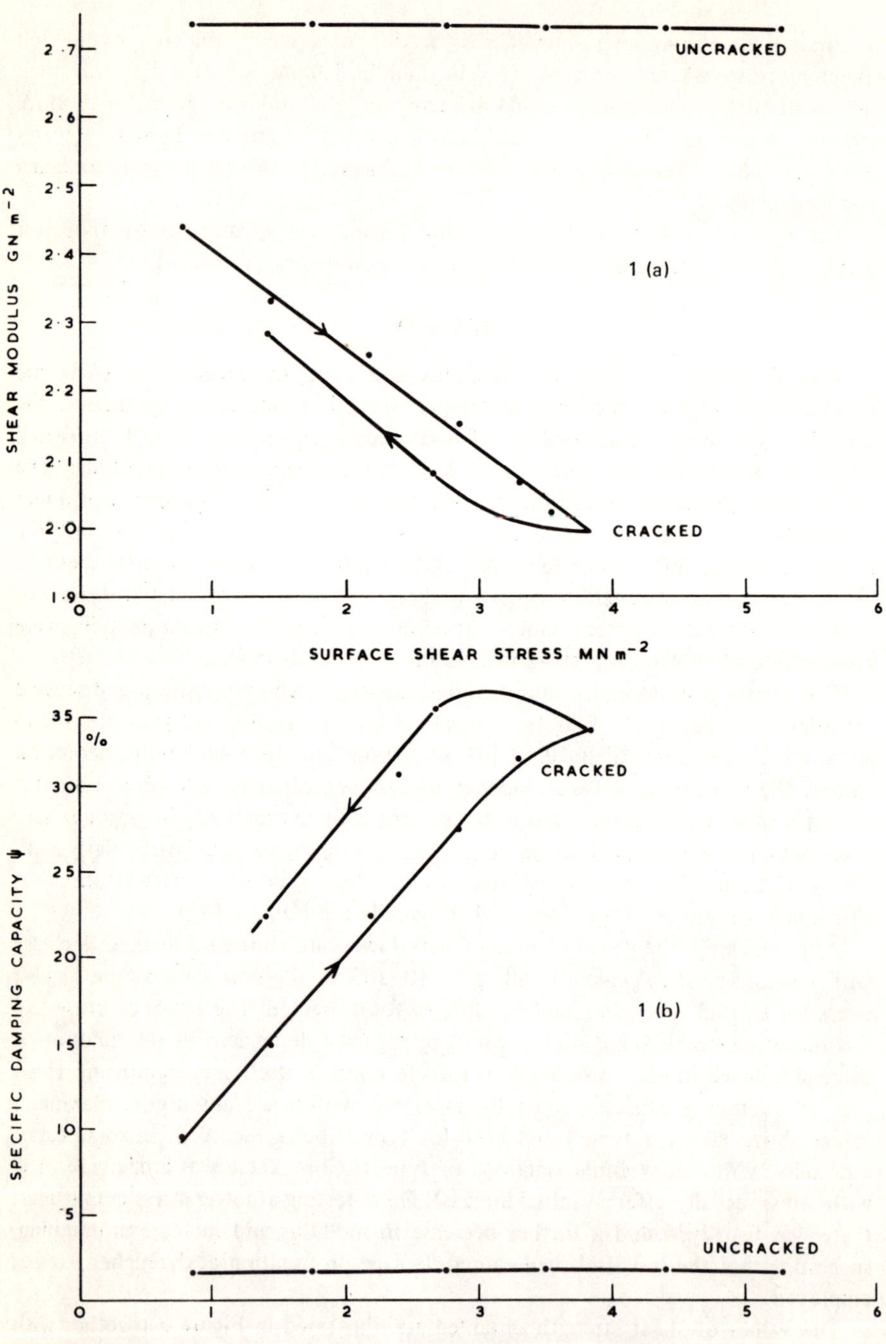

Figure 1. (a) Shear modulus; (b) Damping capacity of a 40% type I treated composite as measured before and after torsional failure.

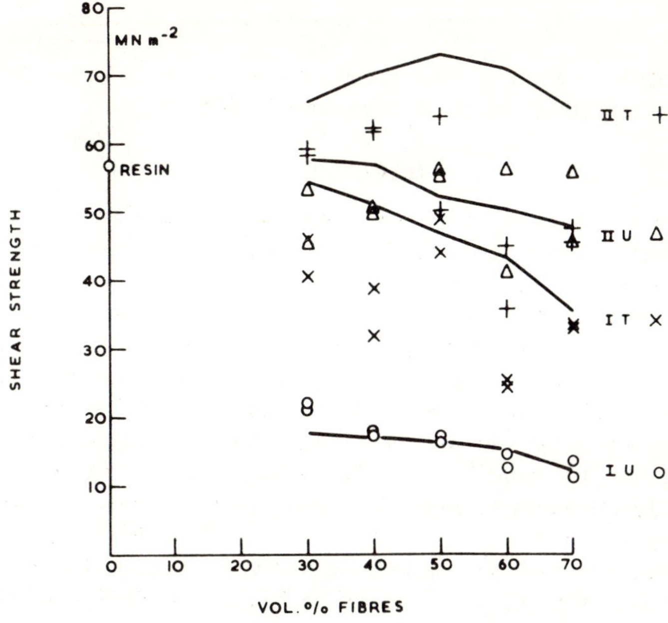

Figure 2. Measured shear strength of specimens. Solid lines from Reynolds and Hancox [5].

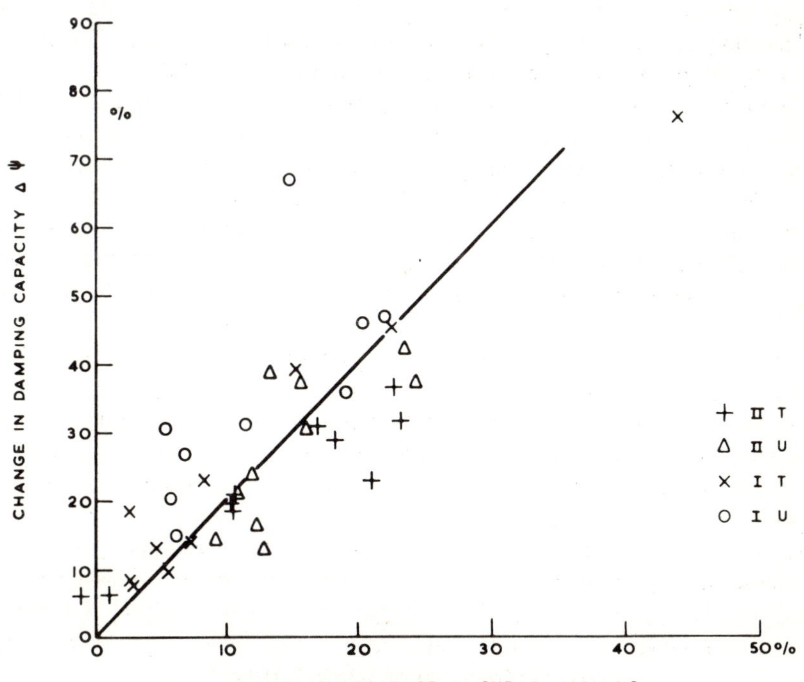

Figure 3. Change in damping capacity, Δψ, as a function of relative decrease in modulus after cracking.

fibres are up to strength on the average, the surface treated fibres are all of relatively low strength, in some cases being down by 40%.

DISCUSSION

From the results presented in the previous section, a number of important inferences may be drawn. In the first place, the damping capacity of the composite normally lies between the values for the separate components, in a similar way to the overall behaviour of the shear modulus. As the proportion of fibre is increased, the stored strain energy is increasingly removed from the lossy resin and the total damping falls. Moreover, for a given constitution the damping increases in the order Type II treated < Type II untreated < Type I treated < Type I untreated, and the total range at a particular fibre volume loading is about 1%. In general the values lie between that for pure resin (3½%) and that for pure fibre (<0.1%). However, a few specimens (Type I untreated 60 and 70% and Type I treated 60%) showed much higher values. It may be that some of those specimens with a damping capacity greater than the resin value already possessed undetected delaminations since, as has already been noted the effect of torsional cracking was always to produce greatly enhanced damping. It is also possible that untreated type I fibres cause frictional loss by slipping in the matrix. However, although the specimens with initially high damping tended to show low strength, many other specimens which did not show high damping before cracking also showed comparatively low strength. This applied to virtually all this batch of specimens made with surface treated fibres.

When a rod is twisted the shear stress is a maximum in the surface and falls to zero along the axis so that shear cracks are formed in the surface and propagate inwards. Dye penetrant and radiographic tests on our specimens have indicated that surface delamination cracks are formed which run the full length of the specimen without failure in individual fibres. It is therefore presumed that the decrease in modulus G is caused by an effective decrease in specimen radius r ($\Delta G/G = 4\ \Delta r/r$) and the increase in internal damping $\Delta\psi$ is due to dynamic friction between the opposite sides of the crack. In Figure 3, the change in damping capacity $\Delta\psi$ due to cracking is plotted for all the specimens as a function of relative decrease in shear modulus $\Delta G/G$ at a fixed value of surface shear stress (250 p.s.i.) There is a good deal of scatter partly because the cracks are not perfectly radial, but are partly circumferential in orientation. The relative orientation varies between specimens and as the circumferential part will contribute to the frictional losses without affecting the torsional modulus, an exact correlation is not expected. The line drawn through the origin shows the clear difference between type I untreated specimens and type II treated, and also, to a lesser extent that between type I treated and type II untreated. This is consistent with the analysis of torsional failure already given by Reynolds and Hancox [4] in that type I untreated specimens fail at the fibre-resin interface and type II treated fail in the resin. Thus in the first case the friction causing the damping is mainly between fibre and resin, whereas in the second case it is between resin and resin. The other cases fall in between, but the inference is that resin-fibre friction is higher than resin-resin. At low stresses, parts of the crack are held together by static friction but as the stress level is increased this is gradually overcome, causing additional friction losses and decrease of modulus. This additional increase in damage indication shows similar characteristics to those illustrated in Figure 3 except that at high stress levels some specimens show a region of decreased damping. This may be due to opening of the cracks at a high degree of twist.

Measurements of permanent twist after fracture gave values of up to 56°, with a fairly consistent decrease with increasing fibre fraction and a steady increase with bonding strength from type I untreated through to type II treated. Both these effects correlate with the shear strain to failure estimated from modulus and strength, and can therefore be explained on the basis of the plastic flow in the matrix.

CONCLUSIONS

The following definite conclusions emerged from this study

(a) In comparing the original square section bars with circularly ground ones before cracking, no significant change in modulus was found, but some specimens showed a measurable increase in damping. This suggests that even the surface damage incurred during the light machining cuts used may have a significant effect.

(b) It should be emphasised that a single measurement of damping is not a good guide to strength in itself, as can be seen from the fact that for our specimens the damping on uncracked specimens tended to decrease with increasing fibre content, as did the measured strength.

(c) The measurement of shear modulus by static or vibrational methods (or better still by both) was a clearer and more discriminating test of quality in the sense of fibre/resin ratio, type of fibre and fibre surface treatment than measurement of damping.

(d) After cracking, the damping was a more sensitive indication of the presence of damage than the modulus. The significant features were values of damping capacity considerably in excess of the pure resin value, and a large rate of increase with strain amplitude. Sufficiently high amplitudes cause propagation of the delamination cracks, and for nondestructive testing purposes measurements would best be made in the range 3–10% of failure stress.

REFERENCES

1. A. B. Schultz and D. N. Warwick, "Vibrative response: a nondestructive test for fatigue crack damage in filament-reinforced composites," *J. Composite Materials*, Vol. 5 (1971), p. 394.
2. C. Brown, N. L. Hancox and W. N. Reynolds, "Nondestructive testing of carbon fibre

reinforced plastics: evaluation of an ultrasonic resonance method," *J. Phys. D. (Applied Physics)*, Vol. 5 (1972), p. 782.

3. R. D. Adams, M. A. O. Fox, R. J. L. Flood, R. J. Friend and R. L. Hewitt, "The dynamic properties of unidirectional carbon and glass fibre reinforced plastics in torsion and flexure," *J. Composite Materials*, Vol. 3 (1970), p. 594.
4. R. D. Adams and D. G. C. Bacon, "The Dynamic Properties of Unidirectional Fibre Reinforced Composites in Flexure and Torsion," *J. Composite Materials*, Vol. 7 (1973), p. 53.
5. W. N. Reynolds and N. L. Hancox, "Shear strength of the carbon-resin bond in carbon fibre reinforced epoxies," *J. Phys. D. (Applied Physics)*, Vol. 4 (1971), p. 1747.

In-Plane Shear Stress-Strain Response of Unidirectional Composite Materials

DAVID F. SIMS, *Air Force Materials Laboratory*
Wright-Patterson Air Force Base, Ohio 45433

(Received December 13, 1972)

The experimental determination of the in-plane or longitudinal shear stress-strain response of a unidirectional composite material usually presents a difficult task. Several test techniques have been proposed for determining this quantity:

(1) Pure torque applied to a tubular specimen (Adams and Thomas [1]),
(2) Cross-sandwich beam test (Shockey and Waddoups [2]),
(3) Rail shear test (Whitney, Stansbarger, and Howell [3]), and
(4) ±45° balanced angle-ply laminate subjected to uniaxial tension (Petit [4] and Rosen [5]).

The purpose of this note is to present experimental results of a comparison of three of these test methods: the tube test, the rail shear test, and the ±45° balanced angle-ply laminate under uniaxial tension. From a practical standpoint, the tube test has the disadvantages of expensive specimen fabrication and requires special test equipment. These disadvantages can be eliminated in the rail shear test and ±45° laminate test since flat specimens are used.

EXPERIMENTAL PROCEDURE

The materials used in this investigation were glass-epoxy (commercially known as Scotchply) and HTS/ERLA 2256 graphite-epoxy. An 8-ply 0/90° balanced laminate was made from unidirectional glass-epoxy material. Specimens for the rail shear test and ±45° laminate test were cut from the 0/90° panel. Uniaxial tension specimens were also made from an 8-ply ±45° graphite-epoxy panel.

A. Rail Shear Test

The test fixture currently in use at the Air Force Materials Laboratory for the rail shear test is illustrated in Figure 1. This test fixture has advantages over the fixture used in a previous investigation [3]. It is believed that a better approximation to a state of pure shear can be obtained with the fixture shown in Figure 1 since the compressive load is applied parallel to the clamped edges of the specimen. Another advantage is that two independent calculations of the shearing stress-strain response can be made from one specimen. Two 350 ohm strain gages were placed on the panel at an angle of 45° to the load axis as shown schematically in Figure 1.

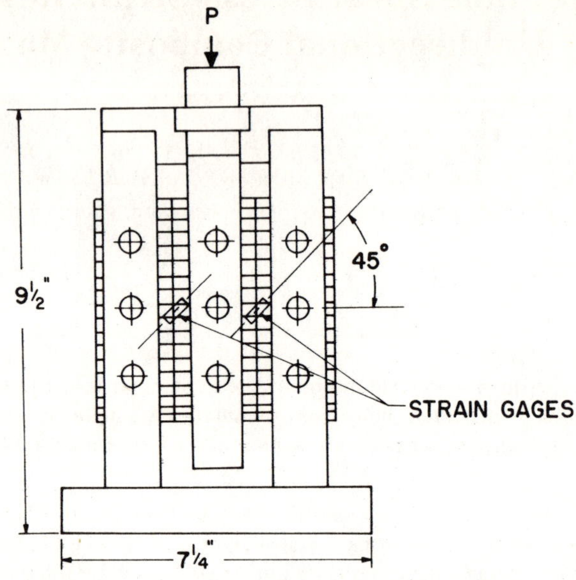

Figure 1. Rail shear test on 0/90° laminate.

The 0/90° rail shear specimen was placed in the test fixture and a torque of 40 ft-lbs was applied to each of the nine bolts. A compressive load of 20 lbs/sec was applied to the center part of the fixture and the load-strain response was monitored until permanent deformation was visible.

B. ±45° Laminate Test

Two 350 ohm strain gages were placed on the ±45° angle-ply specimens. The longitudinal and transverse strains were monitored for a tensile load of 20 lbs/sec until the specimen failed.

DATA REDUCTION

Rail shear data was reduced by assuming that a pure shear state of stress existed in the specimen. The in-plane unidirectional shearing stress was calculated from the equation:

$$\tau_{12} = P/2A \tag{1}$$

where τ_{12} is the unidirectional in-plane shearing stress, P is the total compressive load, and A is the specimen cross-sectional area. The shearing strain γ_{12} was computed from:

$$\gamma_{12} = 2\epsilon_{45^\circ} \tag{2}$$

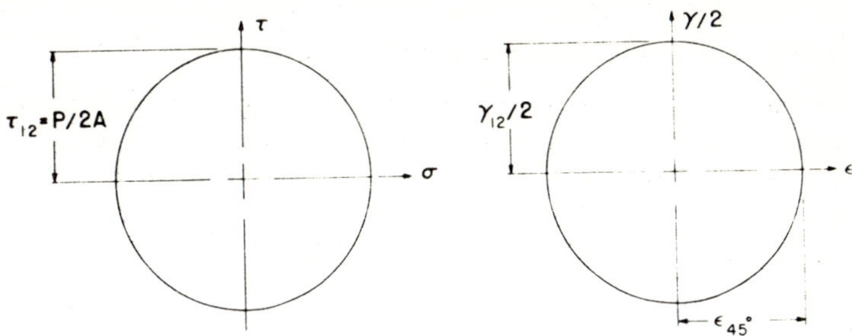

***Figure 2.** Mohr's circle for stress and strain, respectively. Rail shear test.*

Equations (1) and (2) can be derived from Mohr's circle for stress and strain as shown in Figure 2.

The ±45° tensile test data was reduced by the method suggested by Rosen [5]:

$$\tau_{12} = \bar{\sigma}/2 = P/2A \tag{3}$$

where $\bar{\sigma}$ is the applied uniaxial stress on the laminate, and

$$\gamma_{12} = (\epsilon_x - \epsilon_y) \tag{4}$$

where ϵ_x is the normal strain in the longitudinal direction, and ϵ_y is the transverse strain. Equations (3) and (4) can be derived from Mohr's circle for stress and strain as shown in Figure 3.

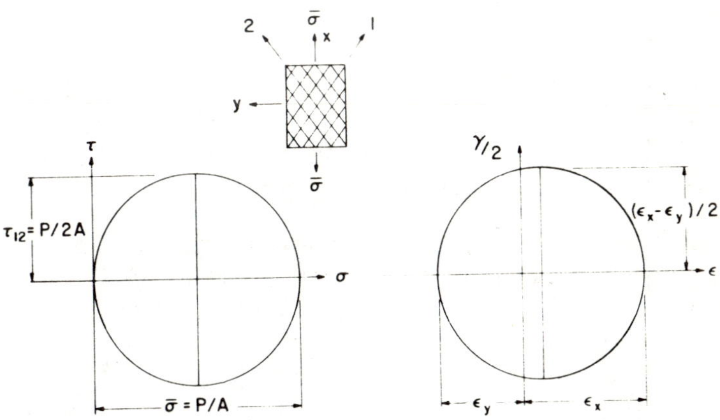

***Figure 3.** Mohr's circle for stress and strain, respectively. ±45° laminate under uniaxial tension.*

RESULTS AND CONCLUSIONS

The in-plane shear stress-strain response for glass-epoxy is illustrated in Figure 4. The rail shear results were obtained from two 0/90° panels and the ±45° laminate results were obtained from three tensile tests. It was observed that the shear stress-strain response from both methods agreed very well.

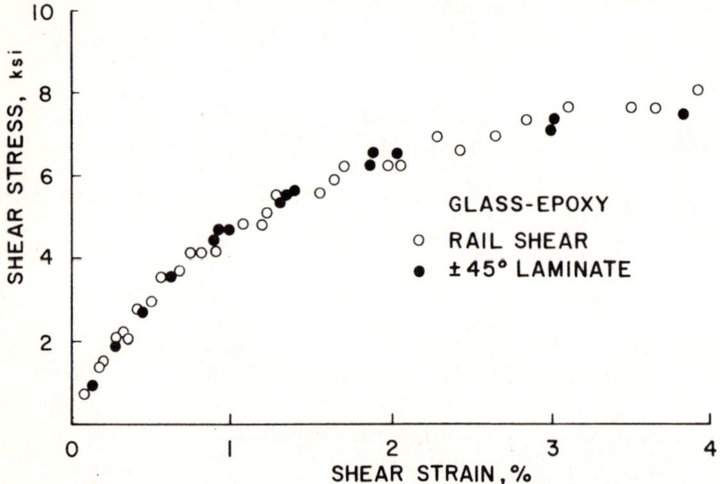

Figure 4. Comparison of unidirectional glass-epoxy shear stress-strain response from rail shear test and ±45° laminate test.

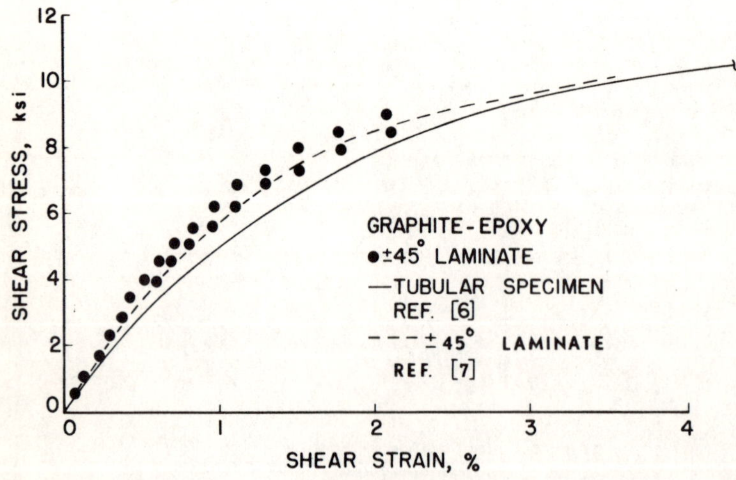

Figure 5. Comparison of unidirectional graphite-epoxy shear stress-strain response from ±45° laminate test and tube under torsion.

The in-plane shear stress-strain response for graphite-epoxy is illustrated in Figure 5. The experimental results obtained in the present study for two ±45° graphite-epoxy specimens are compared to results reported by Grimes, et al ([6] and [7]) for a torsion test on a 4-ply unidirectional tubular specimen and 4-ply ±45° laminate tests. In Reference [6], the fibers were parallel to the longitudinal axis of the tube. Along the initial portions of the curves, agreement was good. The deviation between the response for the ±45° laminate and the tube can be attributed to the difference in fiber volume content. The ±45° laminate specimens had a fiber volume content of 59% and the tubular specimen had a fiber volume content of 48%.

It appears that the rail shear test and the ±45° laminate test give good results for the in-plane shear stress-strain response of a unidirectional material for useful strain levels.

ACKNOWLEDGMENTS

This study was made at the Air Force Materials Laboratory while under a National Research Council Associateship. I would like to thank J. C. Halpin for his helpful suggestions in this study, and R. Ditmer and J. Camping for their assistance in specimen fabrication and testing.

REFERENCES

1. D. F. Adams and R. L. Thomas, "Test Methods for the Determination of Unidirectional Composite Shear Properties," *Advances in Structural Composites*, Society of Aerospace Material and Process Engineers, Vol. 12, Oct. 1967.
2. P. D. Shockey and M. E. Waddoups, "Strength and Modulus Determination of Composite Materials with Sandwich Beam Tests," General Dynamics, Ft. Worth FZM 4691, American Ceramic Society Meeting, Sept. 1966.
3. J. M. Whitney, D. L. Stansbarger, and H. B. Howell, "Analysis of the Rail Shear Test – Applications and Limitations," *J. Composite Materials*, Vol. 5 (1971), p. 24.
4. P. H. Petit, "A Simplified Method of Determining the In-Plane Shear Stress-Strain Response of Unidirectional Composites," *Composite Materials: Testing and Design. ASTM STP 460*, American Society for Testing and Materials, 1969.
5. B. W. Rosen, "A Simple Procedure for Experimental Determination of the Longitudinal Shear Modulus of Unidirectional Composites," *J. Composite Materials*, Vol. 6 (1972), p. 555.
6. G. C. Grimes, P. H. Francis, G. E. Commerford, and G. K. Wolfe, "An Experimental Investigation of the Stress Levels at which Significant Damage Occurs in Graphite Fiber Plastic Composites," AFML-TR-72-40, May 1972.
7. G. C. Grimes, P. H. Francis, G. E. Commerford, and G. K. Wolfe, "An Experimental Investigation of the Stress-Strain Behavior of Graphite Fiber Composites to Assess the Stress Levels at which Significant Damage Occurs," R & D Interim Technical Report No. III, Southwest Research Institute, April 1972.

Interlaminar Shear Strength of a Carbon Fibre Reinforced Composite Material Under Impact Conditions

K. H. Sayers, *Societe Nationale Industrielle Aerospatiale, Toulouse, France [formerly Senior Research Engineer, Rolls-Royce (1971) Ltd., Derby, England]*,
B. Harris, *School of Applied Sciences, University of Sussex, Brighton, England*

(Received November 29, 1972)

The object of this note is to present the results of an investigation conducted to determine the interlaminar shear strength of a carbon fibre composite material under high rates of loading. Because of the fact that significant viscoelastic and creep effects can occur in the shear loading of composite materials, it was anticipated that rate effects might be important in interlaminar shear tests.

The interlaminar shear test has been used on many types of composite. References 1 and 2 treat the effect of specimen geometry and non-uniformity of shear stress, whilst References 3 and 4 show the effects of fibre volume fraction, void content, span-to-depth ratio and roller diameter on apparent interlaminar shear strength. It is shown that all these variables can be significant in testing a given material.

The standard interlaminar shear (ILS) test takes about 30–60 seconds to produce interlaminar shear cracks. During the impact of birds onto composite fan blades in a turbofan engine, interlaminar shear cracks may occur in a fraction of a millisecond. The occurrence of such cracks does not necessarily imply incipient failure, especially in suitably reinforced blades. Nevertheless, it is necessary to know what bird size will lead to initiation of damage and the interlaminar shear strength must thus be determined under more realistic impact conditions.

MATERIAL

The material used was Hyfil[1] carbon fibre (28×10^6 Lb/in^2 modulus) in HR4C epoxy resin [6]. Three plates, two unidirectional and one 0-90° crossplied, were produced from unidirectional pre-impregnated fibre with dimensions 8.5 × 4.0 × 0.3 inches. Each plate contained thirty plies and was moulded for 60 minutes at 165°C followed by 16 hours postcuring at 180°C. Estimated fibre content was 60%. Specimens with dimensions 2.0 × 0.3 × 0.3 inches were cut from each plate.

[1] Registered trademark for carbon reinforced resin.

STATIC TESTS

These were conducted on a standard interlaminar shear fixture at room temperature using an Instron compression testing machine in the laboratories of Rolls-Royce (1971) Ltd. Test details were:

- Span $L = 1.50$ inches ($L/d = 5.0$)
- Rollers 0.25 inch diameter
- Machine rate 0.1 cm/min.

Average test duration was about 50 seconds. *X-Y* recorder plots of applied load versus test machine travel were obtained. Specimens failed in the mode indicated in Figure 1a. The initial shear crack corresponded to the maximum load obtained. The secondary shear cracks occurred at progressively lower loads, giving a saw-tooth

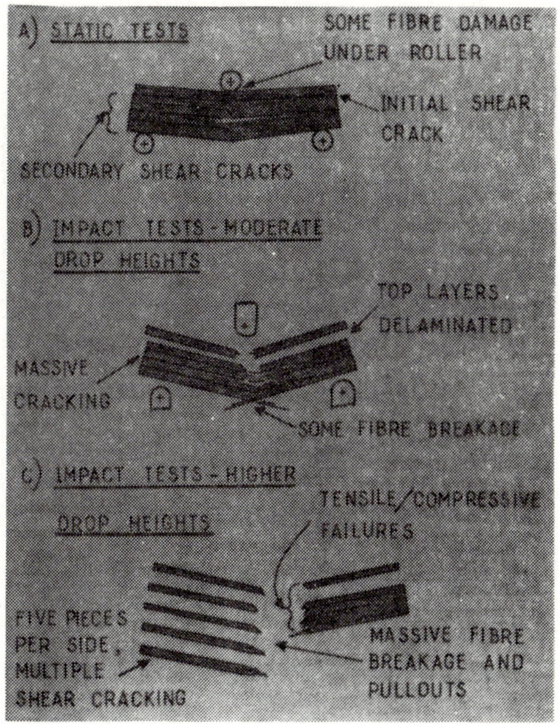

Figure 1. Specimen failure modes.

type of load-deflection curve. Apparent interlaminar shear strengths were computed using:

$$\tau = \frac{3P}{4bd} \qquad \text{[reference 2]}$$

where τ = apparent interlaminar shear strength (ILS), or the mean "engineer's theory of bending" shear stress at the beam mid-depth.

P = maximum load obtained during test (first crack)

b = specimen width

d = specimen depth.

The average strength of unidirectional specimens was 7,150 Lb/in^2 and that of the 0-90° crossplied specimens was 4,970 Lb/in^2. The individual points are plotted on the left-hand side of Figure 2.

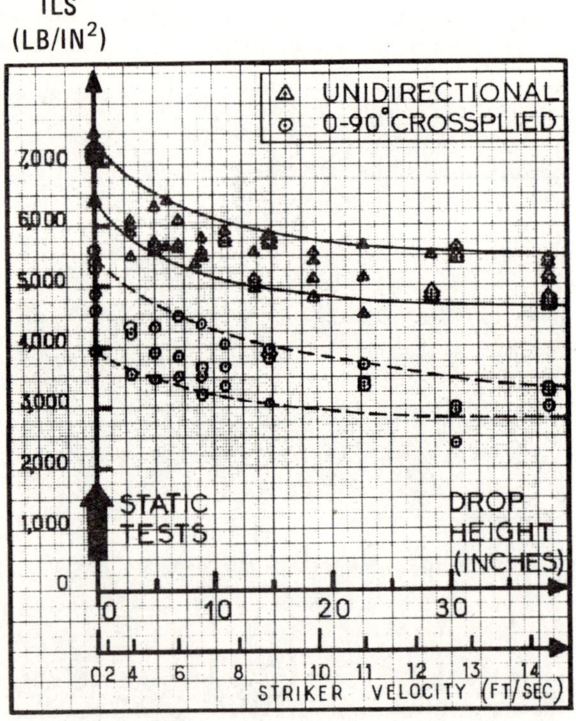

Figure 2. Interlaminar shear strength versus drop height and striker velocity.

IMPACT TESTS

These were conducted in the Applied Sciences Laboratory at the University of Sussex in a drop-weight machine. The specimen is supported near its ends on radiused supports as in the standard ILS test. A falling platform carrying a radiused striker falls down guide rods and hits the specimen at its centre.

The striker carries a conventional strain-gauge bridge and the out-of-balance signal from this is amplified and fed to a storage oscilloscope. The oscilloscope records a trace of transient load versus time which may then be traced or photographed. From the peak of the trace, the breaking load, estimated interlaminar shear strength and time to failure may be obtained. Test details are as follows:

- Span L = 1.57 inches (L/d = 5.25)
- Support and striker radii — 0.10 inch
- Weight of falling platform 14.31 lb.

The bridge and time-base are recalibrated at each testing session, the former by static dead loading. Probable error in the calibrated load measuring system is about ± 5%.

Specimens were tested with the platform falling from different heights. Those specimens which were tested at small drop heights failed in a manner similar to those tested statically. Times to maximum load for these were about 0.6–0.8 millisecond. Specimens tested with greater drop heights were broken in a more complete way (several ILS cracks – Figure 1b) although the failure mode was essentially an intensified form of the damage seen previously. Similar results may be observed statically if loading is continued beyond initial (first crack) failure. Times to maximum load for these specimens were about 0.2–0.5 millisecond.

At greater drop heights, multiple cracking and bending failures occurred, breaking the specimens into more pieces as height increased (Figure 1c).

The estimated interlaminar shear strengths (ILS) are plotted as a function of drop height and theoretical striker velocity in Figure 2. The static results are included at height = 0 inch for comparison. Both the unidirectional and the cross-plied material show the same trend – a reduction of ILS strength with increasing drop height and striker velocity.

The ranges of ILS values corresponding to each range of approximate times to maximum load are given in Table 1. There is a steady decrease in strength as rate is increased (i.e. the time to maximum load decreases from about 1.0 millisecond to about 0.3 millisecond).

Table 1. Variation of ILS with Time to Failure

UNIDIRECTIONAL			0-90° CROSSPLIED		
DROP HEIGHT (INCHES)	APPROX. TIME TO MAX. LOAD (M.SEC)	ILS RANGE (LB/IN²)	DROP HEIGHT (INCHES)	APPROX. TIME TO MAX. LOAD (M.SEC)	ILS RANGE (LB/IN²)
0 (STATIC)	~55 × 10³	6,400-7,500	0 (STATIC)	~40 × 10³	3,950-5,620
3-6	1.4-0.6	5,500-6,400	2-4	0.9-0.6	3,550-4,300
6-27	0.6-0.4	4,530-6,100	4-18	0.6-0.3	3,070-4,510
27-39	0.4-0.3	4,650-5,650	18-39	0.3-0.2	2,400-3,700

For a time-to-ILS failure of 0.3 millisecond, examination of the results showed that the "impact" ILS is about 5,100 Lb/in^2 for unidirectional material compared with 7,150 Lb/in^2 "static" and about 3,500 Lb/in^2 for 0-90° crossplied material compared with 4,970 Lb/in^2 "static." "Impact" ILS strengths are thus about 70% of the "static" values in both cases for the material used in this study.

CONCLUSIONS

1. "Impact" ILS is about 70% of "static" ILS for both unidirectional and crossplied material. A possible explanation for this behaviour is as follows:

   Conventional rate dependence of strength in a viscoelastic material would lead to higher strengths at greater rates. In the case of the shear loading on the finite-width ILS specimen, the non-uniformity of the shear stresses on the plane where initial cracking occurs results in substantially concentrated edge stresses which may be partially relieved by creep in a "static" test but not in a test under impact conditions. Failures may initiate in these areas at a lower applied loading, thus giving lower ILS values.

2. "Impact" ILS strengths and the associated damage modes approach the "static" strengths and damage modes at the lowest rates.

3. It seems logical that any further investigation should concentrate on the comparative testing of other material systems for which the merit rating under "impact" ILS loading may not be the same as for "static" loading. Testing of specimens typical of composite materials for construction of fan blade leading edges should be undertaken. Further basic work to determine the cause of the reduction of ILS strength with loading rate is required.

ACKNOWLEDGMENTS

The authors wish to thank C. D. Ellis who carried out the drop-weight tests.

REFERENCES

1. K. T. Kedward, "The short beam test method of estimating shear strengths." Paper presented at the Institute of Physics Conference on "Testing fibrous composites for mechanical properties," N.P.L. Teddington, England, July 1970. Also: *Fibre Science and Technology*, Vol. 5 (1972), p. 85.
2. S. A. Sattar and D. H. Kellogg, "The effect of geometry on the mode of failure of composites in the short-beam shear test," *Composite Materials: Testing and Design*, ASTM STP 460, American Society for Testing and Materials, 1969, p. 62.
3. G. A. Booty, M. Hall, F. P. Mallinder, and A. Maybury, "Flexural testing of fibre reinforced materials." Paper presented at the Institute of Physics Conference on "Testing fibrous composites for mechanical properties," N.P.L., Teddington, England, July 1970.
4. W. N. Reynolds and N. L. Hancox, "Mechanical and nondestructive testing of carbon fibre reinforced plastics "Paper 12, Seventh International Reinforced Plastics Conference," The British Plastics Federation, 20-22 October 1970.
5. C. C. Chamis, M. P. Hanson, and T. T. Serafini, "Impact resistance of unidirectional composites," *Composite Materials: Testing and Design* (Second Conference), ASTM STP 497, American Society for Testing and Materials, 1972, p. 324.
6. Anon., "Hyfil Carbon Fibre," property data sheet for Hyfil 2750, Rolls-Royce "Composite Materials" Ltd. 1972.

Structure-Property Relations and Reliability Concepts

JOHN C. HALPIN

Air Force Materials Laboratory
Dayton, Ohio 45433

(Received February 28, 1972)

ABSTRACT

Typical procedures currently employed in engineering design are reviewed and applied to the evaluation of composite technology as it exists at this time. Special attention is directed to the topics of stiffness, static strength, specific weight, fatigue, structural reliability and adhesive joining. Verification of the achievement of a safe composite design by a proof-test concept is introduced. Evaluation of composites vis-a-vis metallic technology yields a positive attitude toward the future of composite technology. Significant support is found to justify the long-term potential of advanced filamentary composites.

INTRODUCTION

THE DECADE OF the Fifties opened with an uninhibited enthusiasm for the extraordinarily high strength and stiffness properties primarily available in single-crystal whiskers but also in continuous filament forms. While the realization of the potential of defect-free "whiskers" in engineering composites has seeped away in the murk of technological reality, striking progress has been achieved in advanced filament reinforcement. The Sixties witnessed the development of boron and graphitized polymeric fibers as well as several other potential reinforcements. In fact, the technology has developed to the point that "proof-of-concept" is now being demonstrated in operational engineering structural items [1, 2]. At this point, it is interesting to note that our understanding of composite solids has altered in a subtle but nonetheless radical manner. The initial emphasis upon netting analysis, high fiber properties, has been tempered by the realization that the engineer cannot, nor will not, execute design based solely upon longitudinal filament properties. Technical composites are multiphased, anisotropic bodies. Efficient and reliable use of such materials necessitates optimization of a laminated composite with respect to all of its directional properties (particularly, transverse tensile shear, and longitudinal compression, as well as the earlier emphasis upon longitudinal properties).

Employing current design engineering characterization and design procedures, we are in a position today to evaluate the performance capability of composites for a variety of filaments ranging from E-glass through the very high-modulus, high-strength graphite systems.

LAMINATE PROPERTIES – STRENGTH AND STIFFNESS

Current attitudes regarding composite materials [3, 4] emphasize the relationship of structural performance to the properties of a ply. A "ply" is a thin sheet of material consisting of an oriented array of fibers embedded in a continuous matrix material. These plies are stacked one upon another, in a defined sequence and orientation, and bonded together yielding a laminate with tailored properties. The properties of the laminate are related to the properties of the ply by the specification of the ply thickness, stacking sequence, and the orientation of each ply. The properties of the ply are, in turn, specified by the properties of the fiber and matrix, their volumetric concentration, and geometric packing in the ply. Generally, the ply material is preformed and can be purchased in a continuous compliant tape or sheet form which is in a chemically semi-cured condition. Fabrication of structural items involves using this "prepreg" material and either winding it onto a mandrel or cutting and stacking it onto a mold after which either heat and pressure or tension is applied to complete the chemical hardening process.

The basis for engineering design for such a material is then – the properties of a cured ply or lamina *as it exists in a laminate.* For example, consider the ply illustrated in Figure 1. This ply is treated as a thin, two-dimensional item and is mechanically characterized by its stress-strain response to loadings in: (1) the direction of the filaments which exhibit a nearly linear response up to a rather large fracture stress; (2) in the direction transverse to the filament orientation which exhibits a significantly decreased moduli and strength; and (3) the response of the material to an in-plane shear load not illustrated in Figure 1.

By contrast with isotropic metallic materials, an oriented ply, in the form of a thin sheet, is anisotropic and requires four elastic (plane stress) constants [3, 4] to specify its stiffness properties in its natural orientation:

$$\begin{aligned} \sigma_1 &= Q_{11}\epsilon_1 + Q_{12}\epsilon_2 \\ \sigma_2 &= Q_{12}\epsilon_1 + Q_{22}\epsilon_2 \\ \sigma_6 &= Q_{66}\epsilon_6 \end{aligned} \qquad (1)$$

The elastic stress-strain properties of a laminate are related to the ply properties through the assumptions:

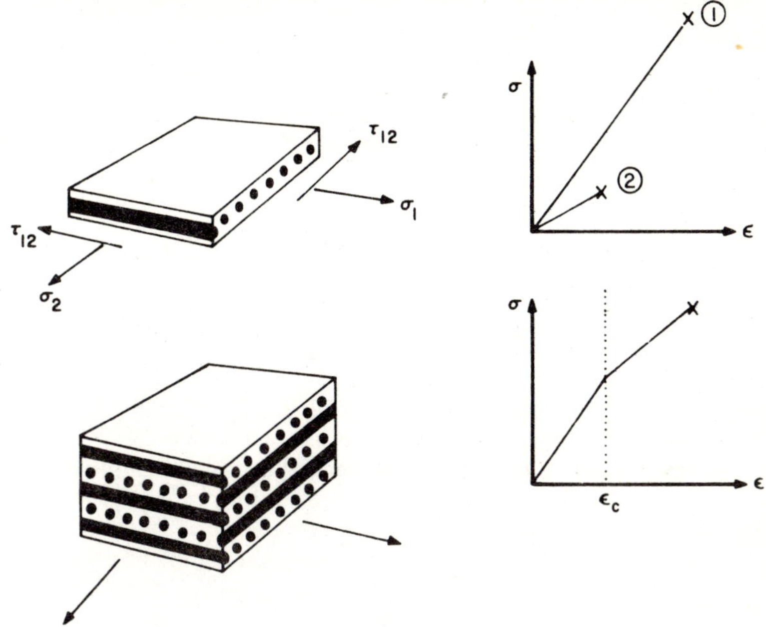

Figure 1. Illustration of lamina characterization and strength properties at a laminate.

1) All layers are in a state of plane stress; the stress transverse to lamination plane are neglected;

$$\sigma_3 = \sigma_4 = \sigma_5 = 0 \tag{2}$$

2) All layers are bonded together and the strain components, in the 1-2 plane are linear functions of z, the distance through the depth of the laminate;

$$\epsilon_i = \epsilon_i + z\,k_i \tag{3}$$

where ϵ°= in-plane or mid-plane stretching and k=plate curvature, $i = 1, 2$ refers to the normal components, and $i = 6$ refers to the shear component, the engineering strain component.

3) All layers obey generalized Hooke's law; (see Equation 1). The stress-strain relation for the assumed plane-stress condition, including the thermal effect for each layer is:

$$\sigma_i = Q_{ij}\,(\epsilon_j - \alpha_j\,T), \quad i,j = 1,2,6 \tag{4}$$

where α_i = anisotropic thermal expansion matrix

T = temperature increase from a reference (stress-free) temperature. Stress resultants (N_i) and stress couples (M_i) can be defined as

$$[N_i, M_i] = \int_{-h/2}^{+h/2} \sigma_i \, [1,z] \, dz \tag{5}$$

Substituting Equations (3) and (5) into (6) renders:

$$\bar{N}_i = N_i + N_i^T = A_{ij}\epsilon_j^o + B_{ij}k_j \tag{6}$$

$$\bar{M}_i = M_i + M_i^T = B_{ij}\epsilon_j^o + D_{ij}k_j \tag{7}$$

where

$$[A_{ij}, B_{ij}, D_{ij}] = \int_{-h/2}^{+h/2} Q_{ij} \, [1,z,z^2] \, dz \tag{8}$$

$$[N_i^T, M_i^T] = \int_{-h/2}^{+h/2} Q_{ij} \, \alpha_j \, T \; [1, z] \, dz \tag{9}$$

The brackets above are symbolic rather than operational; the equality applies to the corresponding terms in the bracket. A_{ij} is the in-plane stretching stiffness; D_{ij} is the bending stiffness and B_{ij} is a coupling stiffness matrix due to nonsymmetric lamination.

If the ply illustrated in Figure 1 is used in the construction of a balanced and symmetrical 0/90 laminate and is mechanically tested, a bi-linear stress-strain curve is observed up to rupture. This initial slope of the 0/90 curve or moduli is the sum, through the thickness, of the plane-stress stiffness of each layer. As the laminate is deformed, each ply possesses the same in-plane strain, ϵ°, and when the strain on the 90° layers reaches the strain level at which ply failure was experienced, the 90° layers crack (craze), shown in Figure 2. In a glass/resin system the occurrence of crazing is associated with audible noise similar to the creaking of a wooden mast, and a whitening of the material. The failure of the 90° layers in the laminate prevents the 90° layers from carrying their share of the load. This load is transferred to the unbroken layers, the 0° layers for our illustration, and results in a loss of laminate stiffness or modulus. Continual loading will ultimately produce a catastrophic failure of the laminate when the strain capability of the unbroken 0° layers is exceeded. For a 0/90 construction, employing the glass/epoxy material specified in Table 1, the ratio of the ultimate failure stress to the crazing stress, the knee in Figure 1 is 6.1. Experimental data and a theoretical stress-strain curve based upon the data of Table 1 are shown in Figure 3 for a $\pi/4$ (0°/± 45°/90°) glass/epoxy laminate. Note the

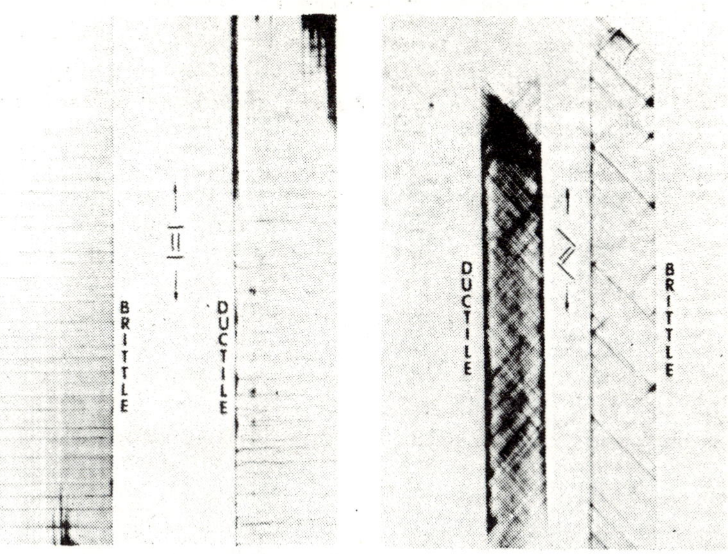

Figure 2. Crazing in glass-epoxy specimens for both brittle and ductile matrix systems, after Ishai and Lavengood [5].

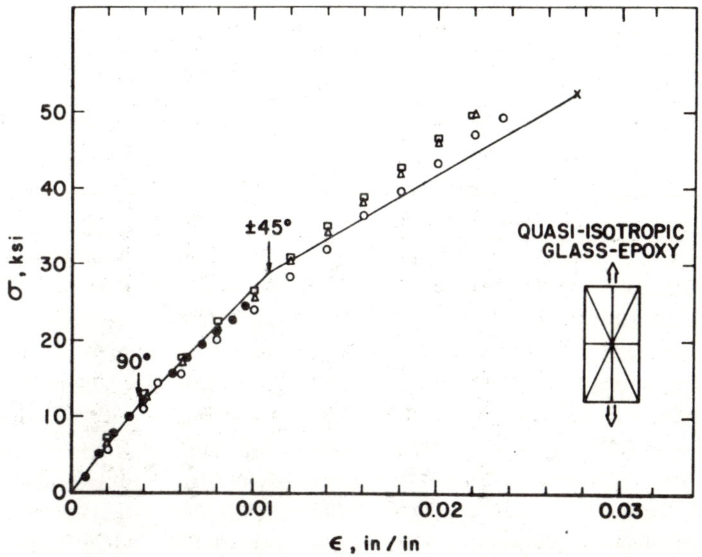

Figure 3. Comparison of theory and experiment for a glass-epoxy laminate.

Table 1. Unidirectional Lamina Design Properties as a Function of Material

Lamina Material (Density)	Engineering Moduli, 10^6 psi			Poisson's Ratio	Maximum Strain, percent			Expansion, 10^{-6} in/in/°F	
	E_{11}	E_{22}	G_{12}	ν_{12}	$\pm\epsilon_1$	$\pm\epsilon_2$	ϵ_6	α_1	α_2
Glass Epoxy (0.064 #/in³)	5.6	1.2	.6	0.26	2.75 1.58	.38 1.42	2.80	3.5	11.4
PRD 49-1 Epoxy (0.050 #/in³)	15	.83	.6	0.21	1.01 0.25	.25 .90	1.00	−0.6	–
High-Strength Graphite-Epoxy (0.054 #/in³)	21	1.7	.65	0.28	.72 .98	.45 .90	1.50	−0.4	16-18
High-Modulus Epoxy (0.058 #/in³)	27.6	1.1	.8	0.21	.45 .28	.30 2.60	.70	−0.55	–
Boron Epoxy (0.073 #/in³)	30	2.7	.65	0.21	.60 1.13	.35 1.13	1.95	2.5	13.0
Boron Aluminum (0.0975 #/in³)	33	21	7.0	0.23	.50 .55	.10 .10	.15	3.2	10.6
HM-HS Graphite Epoxy (0.052 #/in³)	46.0	1.0	.6	0.25	.16	–	–	−(0.5-.65)	(17-33)

change in stiffness as the 90° and then the 45° layers craze and the correspondence of the theoretical ultimate strength of 52,000 psi with the experimental results of 50,500 psi. The ratio of ultimate stress to 90° ply failure is 4.5. While the strain for transverse ply failure is constant from laminate to laminate, the stress required to craze the system, as well as final failure load, is a function of laminate geometry (Figure 4) because the construction of the laminate specifies the stiffness properties (crazing stress = stiffness × transverse ply failure strain). Therefore, lamination permits the engineer to tailor a fixed prepreg system to meet the conflicting stress/strain demands at different points in a structure. A further point, the crazing stress, or threshold, is generally at or below the creep fracture or fatigue limit for all classes of composites (for glass/epoxy the fatigue limit lies between 0.25-.30 of static ultimate strength). Boron and graphite are fatigue insensitive filaments, thus no fatigue damage is realized below first ply failure.

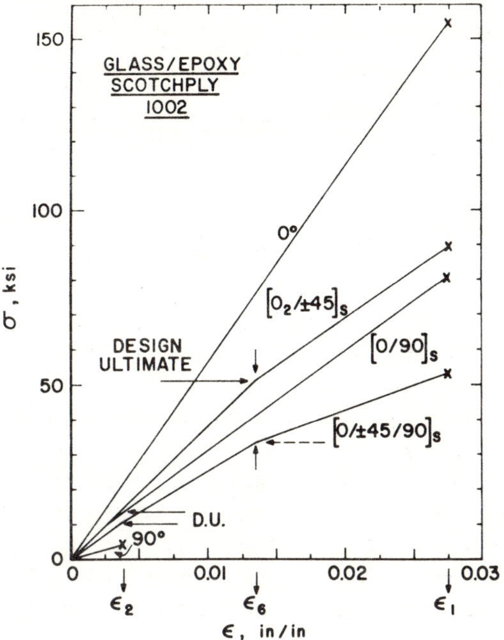

Figure 4. *Influence of lamination upon the uniaxial tensile strength of a composite material.*

Thus, the material properties of a laminate can be specified in terms of the ply engineering moduli; E_{11}, E_{22}, ν_{12} and G_{12}; the engineering strains to failure; ϵ_1, ϵ_2 and ϵ_6; and the thermal expansion coefficients; α_1 and α_2. Engineering properties of currently commercially available prepreg materials are tabulated in Table 1 for reference purposes.

DESIGN PHILOSOPHY

In order to evaluate high-modulus, fiber-composite technology, we must commit ourselves to a design concept. Do we design to the finial catastrophic failure of a laminate or to the first occurrence of mechanical damage [6] in the laminate? Because current research efforts indicate that the stress levels required for safe-life fatigue; for restricted environmental degradation on the composite filaments through a crazed matrix; and for controlled stiffness properties are in the vicinity of the stress levels predicted for first-ply failure, we have adopted this condition for illustration purposes.

Design ultimate loads shall result in a stress that does not exceed the *design ultimate stress* for the laminate used, where *design ultimate stress* is the maximum laminate stress attainable without rupture of any lamina.

Material stiffness and density also enter into an evaluation scheme. This density-stiffness sensitivity is a result of the eternal requirement of enhanced engineering property levels at reduced weight. Material stiffness allows control of the product shape and structural performance under buckling, flexural and vibrational loading conditions. Thus, we are concerned with the interaction of these variables in a material evaluation scheme.

MATERIALS EVALUATION

Emphasis upon high-strength, high-modulus fiber properties has characterized the composite development philosophy of the Fifties and early Sixties. This emphasis upon enhanced filament properties was strongly influenced by the belief that enhanced fiber properties were sufficient to guarantee engineering acceptance. In Figure 5, we have presented a plot of the modulus and strength in the filament direction for prepregged advanced composite materials. The strengths are the product of $E_{11} \cdot \epsilon_1$ of Table 1. Note that while the actual strengths may not be as high as some people report, these values represent engineering conservatism in that they are the *lower bound statistical limits on the ply properties.* The important point is that the strength is roughly constant, with the exception of the graphite fibers, with increasing filament stiffness. A typical relationship between graphite fiber stiffness and strength is illustrated in Figure 6.

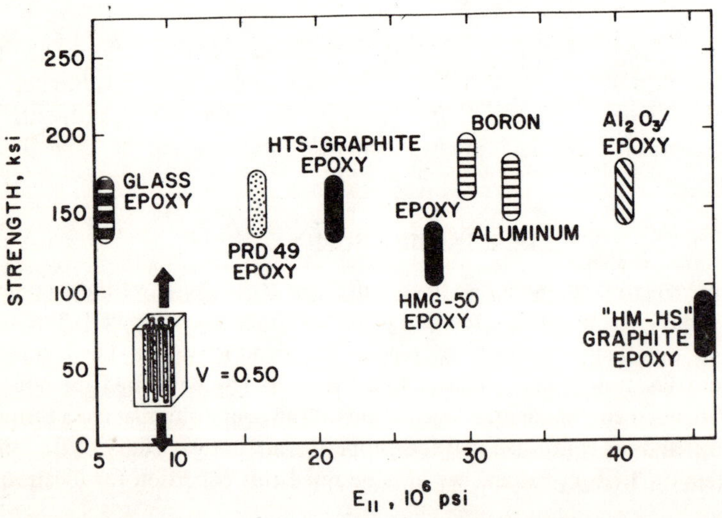

Figure 5. Comparison of longitudinal engineering strength against longitudinal modulus for typical prepreg materials.

Employing the data of Table 1 and the maximum strain criterion, uniaxial tensile properties were computed for two engineering laminates: a 0/90 system which

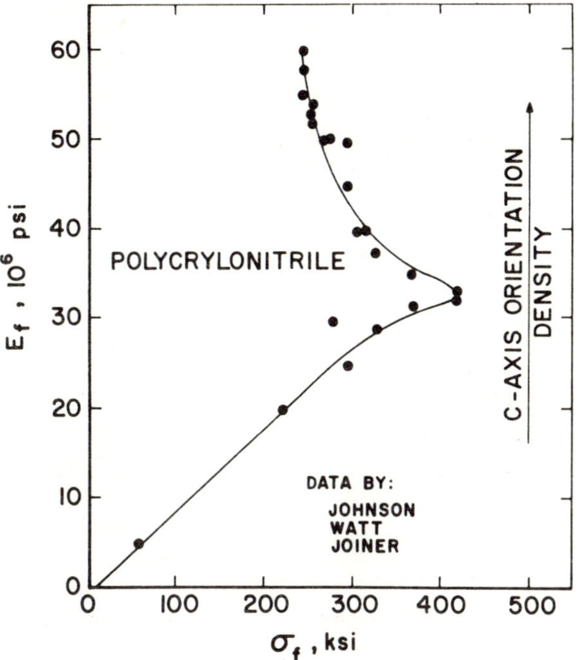

Figure 6. Illustration of the interrelationship between fiber modulus and strength for a typical graphite system.

possesses equal biaxial properties, and a $0_2/\pm 45$ system which has increased uniaxial strength properties and shear stiffness with reference to the 0/90 system. Both laminates possess similar initial stiffness and ultimate strengths, see Tables 2 and 3 and Figure 7. The *design ultimate load,* first ply failure, is the useful engineering capability of these systems. Increased design level of $0_2/\pm 45$ family over 0/90 family may be attributed to the increased strain capability of $\pm 45°$ layers, as compared to 90° layers. The relative ordering of design ultimate strength properties around the indicated dash lines is a direct reflection of transverse strain and modulus capability, ϵ_2 and E_{22} of Table 1. Both the DuPont PRD-49 fiber and the high modulus graphite fiber, "prepreg," have transverse properties which are partially attenuated by high filament anisotropy. The poor showing by the boron/aluminum system is the result of a low transverse strain at yield which offsets the large transverse stiffness advantage of the system. Note that the uniform ultimate strength capability exhibited in Figure 5 limits the engineering utilization of advanced filaments possessing stiffness properties in excess of boron. The boron/aluminum results also illustrate the limitation imposed by matrix properties. For example, a high modulus resin matrix for boron would yield a loss in composite properties if transverse strain capability is decreased. The dash line in Figure 7 roughly correlates with a transverse strain of 0.3%.

For illustration purposes, let us compare the composite properties illustrated in

Table 2. Tensile Properties of $0_2/\pm45$ Laminates

	1st Ply Failure			Laminate Failure		
Material System	Stress, ksi	Strain, percent	Modulus, 10^6 psi	Stress, ksi	Strain, percent	Modulus, 10^6 psi
Glass/Epoxy	50.1	1.34	3.7	89.6	2.75	2.8
PRD 49-1/Epoxy	54.7	.60	9.2	87.7	1.01	8.0
HTS Graphite/Epoxy	85.7	.72	11.9	87.0	.83	12.0
H.M. Graphite/Epoxy	62.3	.40	15.4	68.6	.45	13.8
Boron/Epoxy	99.9	.60	16.6	105.3	1.05	1.2
Boron/Aluminum	30.2	.11	26.5	94.4	.50	16.5

Table 3. Tensile Properties of 0/90 Laminates

	1st Ply Failure			Laminate Failure		
Material System	Stress, ksi	Strain, percent	Modulus, 10^6 psi	Stress, ksi	Strain, percent	Modulus, 10^6 psi
Glass/Epoxy	13.0	.38	3.4	79.4	2.75	2.8
PRD 49-1/Epoxy	20.8	.60	8.4	81.9	1.01	8.0
HTS Graphite/Epoxy	51.2	.45	11.4	79.6	.72	10.5
H.M. Graphite/Epoxy	43.1	.30	14.4	63.8	.45	13.8
Boron/Epoxy	56.6	.35	16.4	94.8	.60	15.0
Boron/Aluminum	27.2	.10	27.0	93.5	.50	16.5

Figure 5 on a specific basis: property level divided by density. It is assumed in material comparison that laminate geometry and item geometry are fixed, and we wish to establish the "thickness" of the item required to satisfy the design load at a constant weight. For laminate strength, we would normalize composite properties as σ_i/ρ. First sizing of an aircraft item is performed on a strength basis. The second step in the sizing of a laminate would be on a stiffness basis for control of buckling and occasionally vibrational resonance stability. Vibration control at first resonance [7] would scale as $E_{11}/\rho^{3/2}$. Stiffness critical applications involves the bending rigidity of a laminate which varies as a cube of the thickness [8], or E_{11}/ρ^3. A plot of

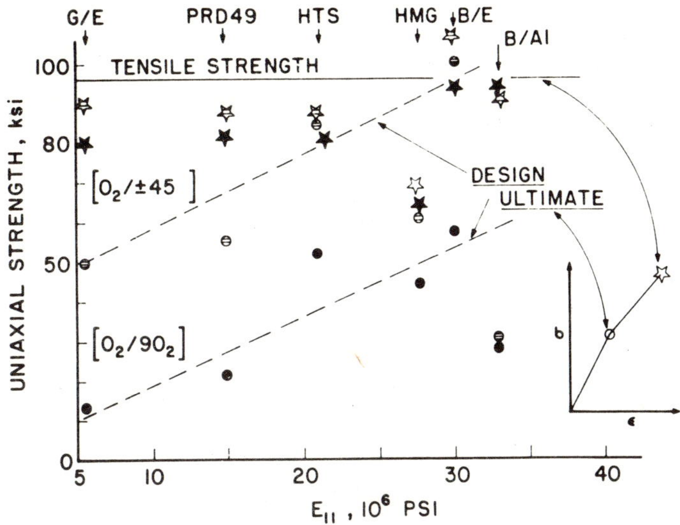

Figure 7. Interrelationship between design ultimate laminate capability and longitudinal prepreg stiffness; circles represent first ply failure; stars, laminate failure; solid symbols, 0/90 laminates; open symbols, $O_2/\pm 45$ laminates.

specific tensile strength against specific bending stiffness for the laminates considered in Figure 5 and Table 2 is shown in Figure 8. It is rather curious to observe that the graphite fiber which possesses a lower modulus is, in fact, the optimum material. The density difference between H.T.S. and HMG fiber is just sufficient to offset the modulus advantage of high modulus graphite fibers in stiffness critical applications. Also, observe that the density of the boron filament inflicts a penalty upon this composite. This density penalty may be sufficient to mitigate the compressive strength ($-\sigma_1 = E_x \cdot [-\epsilon_1]$ of Table 1) advantage of boron over graphite. A similar comment is pertinent to a sapphire-resin composite ($\rho \sim 0.110$ lbs/in^3). In this regard, the respectable properties of PRD-49 are lost in compression due to an inordinately low compressive failure strain $-\epsilon_1 \sim 0.0025$ or stress (−35,000 psi). It is this worker's opinion that highly anisotropic fibers possess poor compression properties for the analogous reason that increased anisotropy decreases the buckling load in an unidirectional lamina [7].

It is also interesting to note that the analysis presented here emphasizes operating strain requirements comparable to the engineering experiences in the metallic tradition. Traditional metallic design is based upon a 0.4 to 0.5% elastic strain capability, combined with a capacity for plastic deformation of 4-20% at high stress concentrations, (see Figure 9). A "safe-life" fatigue design results in limit strains of 0.2-0.3%. Accordingly, we would expect a lamina to possess a minimum strain

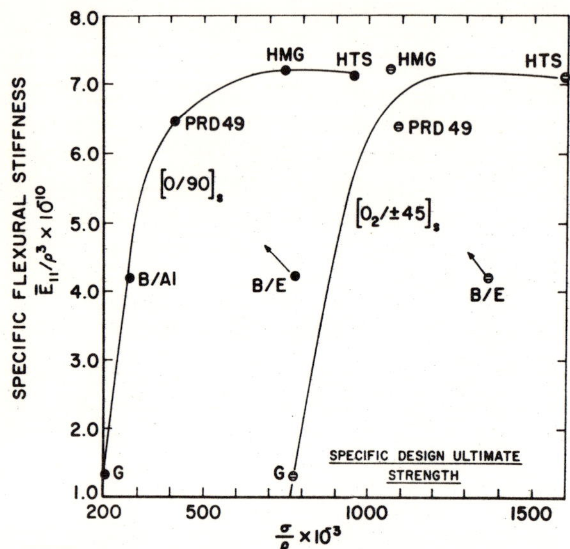

Figure 8. Specific flexural versus design ultimate properties for composite laminates in Figure 7.

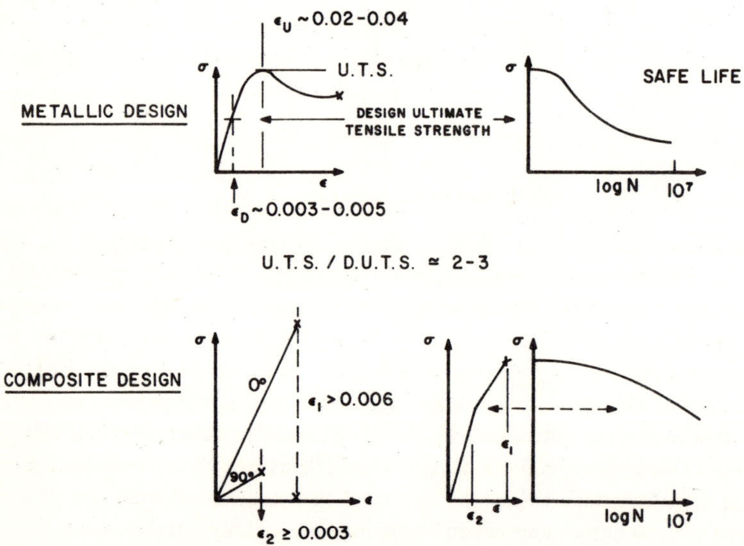

Figure 9. Comparison of metallic design requirements against composite design requirements.

allowable of this magnitude or:

$$\epsilon_2 \geq 0.003$$

The transverse lamina response is dominated by the matrix response, filament concentration, etc.; consequently, this allowable is a criteria to judge a specific matrix system performance in composite applications. Volumetric concentration of filaments is not a true design variable as balanced engineering and fabrication requirements fix this at roughly 50-60%. Longitudinal properties must exceed the transverse response as the longitudinal property provides the insurance or overload capability for the laminate. In addition, a balanced response in compression and tension is required for general engineering utilization or:

$$\epsilon_1 \geq \pm 0.006$$

Note that longitudinal tensile strain is essentially that of the reinforcement fibers. Experience also indicates that the shear strain should be comparable to/or greater than the longitudinal allowable:

$$\epsilon_6 \geq \epsilon_1$$

These criteria clearly demand a balanced materials evaluation program with emphasis placed upon quality control from fiber generation through prepregging to fabrication. The emphasis upon reproducible quality design allowables is the force favoring dry prepreg technology over the traditional wet filament winding technology.

Metallic matrices have not demonstrated the same performance capability available at ambient conditions currently available with resin matrix composites. Fabrication technology, limited by the chemical sensitivity of the reinforcement to the matrix at elevated temperatures, constitutes a formidable obstacle to general engineering acceptance. Extensive effort is being expended to develop metal matrix systems to the same acceptable performance levels.

DIMENSIONAL STABILITY

Composite materials find application where high specific properties, chemical inertness, low maintenance, dielectric capability and dimensional stability are required. Problems related to dimensional stability are particularly well suited for composite application, as currently available filaments possess a wide range of technically useful properties. An inspection of Table 1 will indicate that the PRD 49-1 and the graphite filament family yield negative thermal expansion coefficients combined with high axial stiffness along the fiber axes. Combining fibers of this type within a matrix, we may generate a stiff, strong laminated solid with exceedingly small, if not zero in-plane, expansion. This result is achieved when the fibers are of sufficient stiffness to dominate the axial response of the laminate.

Referring to Equation (4) we may express the thermal expansion properties of lamina in terms of fiber and matrix by Schapery [9] as:

$$\alpha_1 = \frac{E_m \alpha_m V_m + E_f \alpha_f V_f}{E_m V_m + E_f V_f} \tag{10}$$

and

$$\begin{aligned} \alpha_2 = (1 + V_m)\,\alpha_m V_m + (1 + V_f)\,\alpha_f V_f \\ - \alpha_1 (V_f V_f + V_m V_m) \end{aligned} \tag{11}$$

For anisotropic filaments, axial and transverse fiber properties are employed in the respective calculation for α_1 and α_2. Once the expansion coefficients of a single layer of fiber reinforced material have been computed from constitutive material properties, the response of a laminate can be determined through the use of lamination theory [3].

$$\begin{aligned} \bar{\alpha}_1 &= \frac{A_{22}R_1 - A_{12}R_2}{A_{11}A_{22} - A_{12}^2} \\ \bar{\alpha}_2 &= \frac{A_{11}R_2 - A_{12}R_1}{A_{11}A_{22} - A_{12}^2} \end{aligned} \tag{12}$$

where

$$R_1 = J_1 h + J_2 \sum_{n=1}^{n} h_n \cos 2\theta_n$$

$$R_2 = J_1 h - J_2 \sum_{h=1}^{n} h_n \cos 2\theta_n$$

$$J_1 = (U_1 + U_4) W_1 + 2U_2 W_2$$

$$J_2 = U_2 W_1 + 2W_2 (U_1 + 2U_3 - U_4)$$

$$W_1 = (\alpha_1 + \alpha_2)/2$$

$$W_2 = (\alpha_1 - \alpha_2)/4$$

The laminate calculation is somewhat surprising in that it suggests that at certain angles, the laminate will contract in the longitudinal direction when heated. This prediction has been confirmed by Halpin and Pagano [3, 10] and Fahmy and Ragai [11] and these results are illustrated in Figure 10 for both positive and negative α_1. In Figure 11 is shown photographs of a model system undergoing swelling expansions due to absorbed liquids. The angle ply $\pm\theta$ in (b) is shown before an expansion. Note

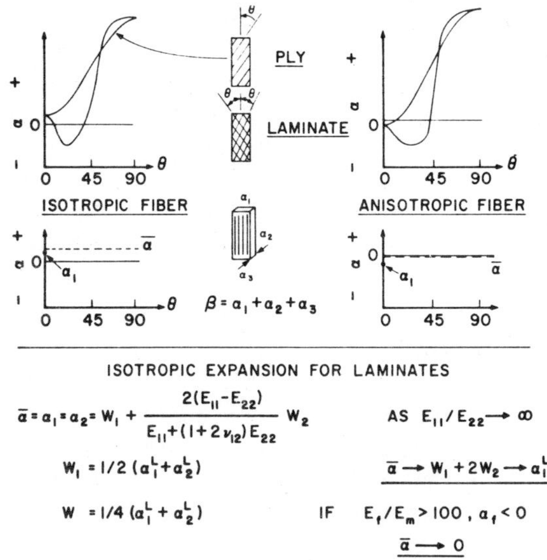

Figure 10. *Influence of lamination upon thermal expansion properties.*

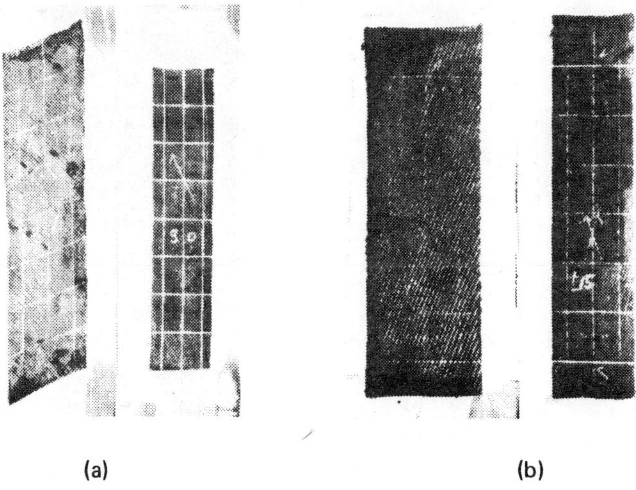

Figure 11. ***Off-axis and angle-ply laminate expansion experiment.***

the unusual negative longitudinal contraction and the correspondingly large transverse expansion. In (a) of the figure is shown the expansion response of a unidirectional off-axis ply. Compare the large longitudinal and transverse expansion as compared with the angle ply. Also, note the shear response due to the angular rotation. Results of this type have obvious technological significance as they specify the material parameters and geometric construction necessary to yield structural elements with zero or minimal expansion coefficients.

For example, if it is desired to fabricate a plate possessing a zero expansion coefficient in any direction within the plane of the plate one solves the equivalent problems of the "quasi-isotropic laminate." Halpin and Pagano [3, 12] have obtained a solution for the "quasi-isotropic expansion strain" in terms of the ply properties. This solution asserts the following general principle: for any symmetric laminate ($B_{ij} = 0$) with equal stiffnesses in the two in-plane directions ($A_{11} = A_{22}$), the expansional strain field is isotropic in this plane and the normal strain in all directions is simply given by:

$$\bar{\alpha} = \alpha_1 = \alpha_2 = W_1 + \frac{2W_2(E_{11} - E_{22})}{E_{11} + (1 + 2\nu_{12})E_{22}} \tag{13}$$

The experimental proof is shown in Figure 12 for the same material shown in the earlier figure. Note the very small dimensional changes of the expanded item (b). This very small expansion strain is in accord with theory which states that as $E_{11}/E_{22} \rightarrow \infty$, the isotropic expansion strain becomes the same in the longitudinal

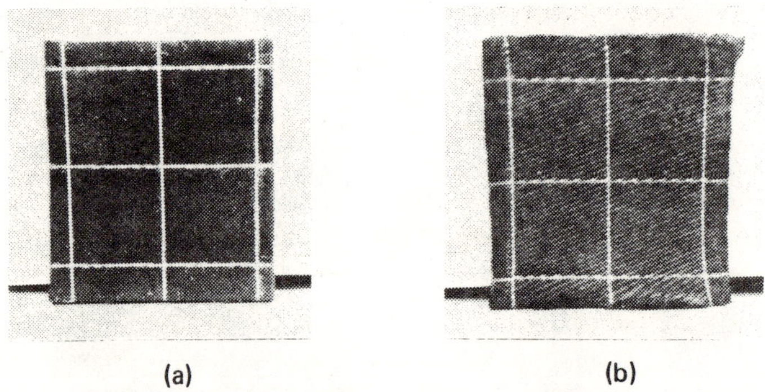

(a) (b)

Figure 12. Experimental verification of minimization of expansion through lamination.

unidirectional ply $e \rightarrow e_1^L$. Thus, for very highly anisotropic composites (high ratio of longitudinal to transverse properties), the isotropic strain approaches the expansion strain of a unidirectional ply parallel to the filaments. To achieve a zero coefficient of expansion one should have:

$$E_f / E_m > 100 \tag{14}$$

$$\alpha_f < 0 \tag{15}$$

Zero, isotropic expansion coefficients have been technically achieved in this manner. Specialized, commercial applications demanding dimensional stability will generate a market potential for an anisotropic filament reinforcement, in addition to the market for primary and secondary structural applications.

RELIABILITY AND SAFETY OF COMPOSITE STRUCTURES

The structural reliability of an item is dependent on the likelihood of a static or a fatigue failure. The specific goal of a reliability analysis is to determine the expectation of first damage due to static or fatigue failure processes in structures or specific structural detail. This goal is confronted with the confounding elements of material variability, fabrication, quality control, design errors, assembly variables, statistical load environment and physical environments as illustrated in Figure 13. Structural safety in aircraft design is implemented with two sets of safety factors and a design philosophy.

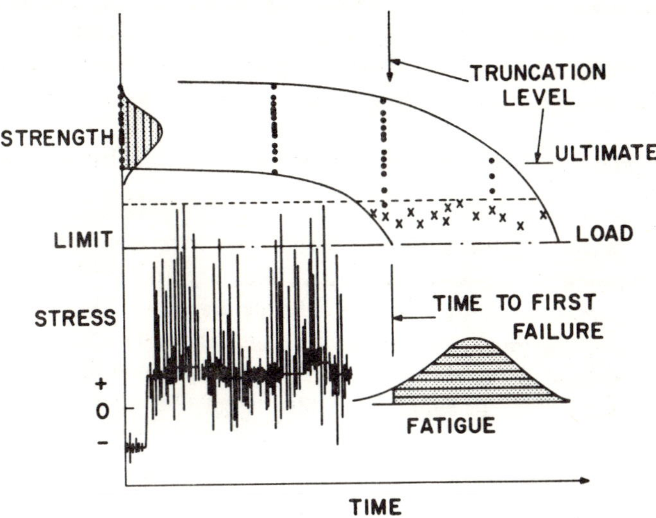

Figure 13. Illustration of relationship between static fracture, engineering load history and fatigue.

Ultimate Factor of Safety: A margin of strength capacity is desired in excess of the major portion of the load spectrum (generally the most severe gust loads [13]) to insure against static fracture of the structures due to the above indeterminate variables. This operating load level is known as limit load, see Figure 13, and represents a safe operating load condition. For historical reasons the ratio of the limit load to the ultimate design load is 1.5. It is generally recognized that infrequent peak asymmetric maneuver loads exceed the limit load by a factor of 1.2. Due to the low static strength variability inherent in metallic structures, these overload conditions are acceptable. The loads at fatigue critical locations are often reduced by an additional safety factor of 1.15–1.30 on the limit load condition (for example, (1.5) times (1.3) gives 1.95 on the Design Ultimate Load).

Indefinitely Long Life: Based upon the assumption that no perceptible fatigue failure occurs if peak stresses are less than some value called a fatigue limit (the plateau in the *S-N* diagram illustrated in Figure 9).

This approach assumes the existence of a fatigue limit exhibited by most steels, but not always exhibited by aluminum alloys, and a structural weight inherent in a design limited by extremely low stresses under infrequent high loads. These problems led to the concept of:

1. *Safe Life Design.* Admission of a *finite* but usefully long trouble free life time after which the structural item is to be retired or repaired. Implementation involves the use of the scatter factor, probabilistic load history characterization, relatively low fatigue allowables and rigorous quality control to minimize materials variation.

2. *Scatter Factor.* Due to the fact that fatigue lifetimes exhibit statistical scatter and the mean lifetime is only approximately known, the load level in a fatigue critical element is reduced until the expected mean lifetime is in excess of the desired lifetime by a factor of 2 to 4. If the dispersion in the fatigue lifetime data is small, as it is in classical aluminum alloy structures, this procedure proves to be operationally useful.

Technical problems and structural weight considerations have led to a third approach called:

3. *Fail-Safe Design.* Admission to the possibility of detectable fatigue crack development, combined with the commitment that such occurrences shall be prevented from becoming catastrophic by nondestructive damage detection, inspection, and preventive maintenance prior to catastrophic failure. These steps are combined with design such that small, detectable cracks permit safe operation between inspection periods. The successful implementation of this design philosophy presupposes the definition of a safe-life precrack time (a statistical variable) and a safe inspection period before the option of preventive maintenance can be exercised.

The adoption of a *Fail-Safe Design* philosophy is in effect a tacit admission that advanced structural design load levels (combined with increased yield stress metals) are at such a point that optimized lightweight metallic structural systems are not dominated by static fracture, but by the probability of fatigue cracking. These trends which have produced an increase in the occurrence of premature service fatigue failures [14], substantially increased maintenance expenditures, and a vocal discontent where driven by the dual need for weight and cost savings. See Table 4 for typical fatigue data. The demonstrable technical need for materials compatible with advanced structural requirements is the driving force behind the developing composite technology as summarized in Figure 9.

The fatigue characteristics of laminated composites are such that technically acceptable specific load levels are accessible on the basis that no fatigue failures in composite laminates are acceptable during the vehicle lifetime. This situation results from the fatigue insensitivity of the graphite and boron filaments combined with the

Table 4. Comparative Fatigue[1]

Material	Maximum Stress per Cycle at 10^7 cycles (ksi)
Boron or Graphite	
0	110
Boron or Graphite	
Epoxy	
$0_3/45$	85
$0_2/90_2$	58
D6AC Steel	
220-240HT	38
6-4 Titanium	35
2024T 851 Aluminum	13

[1] Constant amplitude fatigue
$k_t = 5$, $R=0.1$ room temperature

fact that laminated composite solids are internally redundant structures. Consequently, simple monolithic crack growth characteristics, such as those exhibited in metallic solids, are not easily observed in laminated composites.

This internal redundancy is balanced by the occurrence of a "static-notch effect" in laminated materials but this notch effect, in contrast with metallic response, disappears under fatigue conditions [15]. Thus, a safe-fatigue life offers protection against static failures. These behavioral characteristics are summarized by the term "quasi-brittle" and the following statements:

1. Static strength distributions for laminates controlled by filament failures are dictated by static strength distribution of the 0° prepreg direction

$$P(\sigma) = \exp-[\sigma/\hat{\sigma}]^{\alpha} \tag{16}$$

where $P(\sigma)$ is probability of N specimens out of m initial specimens surviving a static stress, σ, α is a shape parameter, and, $\hat{\sigma}$ is called a location parameter where the form of the equation is assumed to be Weibull.

2. Effect of notches is to reduce the static strength without a change in the shape of the distribution; a dependence of $\hat{\sigma}$ upon notch effects but not α; see Table 5 and [15, 16].

3. Although fatigue insensitive, laminated composites will accumulate damage and ultimately fail due to this damage accumulation. In lieu of a refined nondestructive testing capability, we [17] may study the effects of this damage accumulation by assuming that the rate of damage C is

Table 5. Comparison of Weibull Parameters for HTS-Epoxy Laminates[1]

Angular Content θ	Shape Parameter α	Location Parameter $\hat{\sigma}$ (ksi)
0	10.11	151.8
0/90	10.91	76.5
0/±45	10.8	66.9
0/±45/90	11.46	49.4
0/±45/90		
with round hole	11.5	39.2
with a slit	10.8	39.8
90	7.54	5.3

[1] Properties based upon early commercial prepreg. Laboratory and other commercial prepreg yields α values of 16-20 in magnitude.

$$\frac{dc}{dt} \simeq M C^r; \quad r \geq 1.0 \tag{17}$$

where M is a history dependent quantity and r is a material constant. In addition, assume that laminate failure occurs when

$$\sigma \sqrt{c} = K \tag{18}$$

where K is an apparent toughness or work parameter.

Employing Equations (16), (17) and (18) we can show that the laminate will possess a fatigue lifetime curve of the form:

$$\hat{t}_b \, \sigma_b^{2r} = \text{material constant} = B; \tag{19}$$

and a continuous change in static strength of the ith specimen expressed as:

$$t = \frac{1}{(r-1)\,M} \cdot \frac{1}{K^{2(r-1)}} \left[\sigma_i(0)^{2(r-1)} - \sigma_i(t)^{2(r-1)} \right] \tag{20}$$

when the strength $\sigma_i(t)$ reduces to the peak stress in the load history illustrated in Figure 13, the specimen breaks producing a fatigue failure, t. If the initial static strength distribution, Equation (16), is assumed to represent a distribution in the damage parameter C, Equation (18), and each damage site develops according to Equation (17) that the fatigue lifetime distribution will be given as:

$$P(t) = \exp - [t/\hat{t}]^{\frac{\alpha}{2(r-1)}} \tag{21}$$

The material term r is obtained from Equation (19) or from the derivative of (20):

$$\frac{d\sigma_i(t)}{dt} = \left[\frac{1}{2} MK^{2(r-1)}\right] \sigma_i(t)^{-(2r-3)} \tag{22}$$

which is an indirect measure of Equation (17). Preliminary material constants α and r are presented in Table (6).

Table 6. Comparison of Damage Rate Parameters

	r	α *(Static)*	$\alpha 2/(r-1)$ *(Fatigue)*
E-Glass/Epoxy	5.64	12.3	∿ 1.32
Boron/Epoxy	9-11	18-23	0.9-1.1
Graphite/Epoxy	8.5-10	17-19*	0.9-1.1
Aluminum	4-6	30-45	4-5

*High-quality prepreg

The results of these operations are illustrated in Figure 13 where we observe that the static and fatigue lifetime shape parameters are causally related to each other. In addition, note that the distribution of residual static strengths broadens in time and is truncated in the vicinity of the maximum load peaks in the variable history. Thus, the weakest numbers of the static strength population produces the first fatigue failures. The specification or location of the weakest member in the population, for a given load history, specifies the magnitude of the expected safe-life of the item. Truncation of the static strength distribution results in a truncation in the lifetime distribution function.

For example, consider the proof test experiment demonstrated in [16] for the 0/±45/90° glass epoxy laminate treated in Figure 3. Following the procedure and results of Figure 2 in [16], the stated truncated distribution is defined as:

$$P_p(\sigma) = \frac{P(\sigma)}{P(\sigma_p)} \tag{23}$$

where σ_p is the stress level below which no specimen will fail after truncation. For this illustration $P(\sigma_p)$ was equal to 0.812 (24 out of 30 specimens remained after the truncation). These specimens were then fatigued with a Gaussian load history until fatigue failure occurred. The truncated lifetime [17] distribution is expressed as:

$$P_p(t) = \frac{P(t)}{P(\sigma_p)} = \frac{\exp-[t/\hat{t}]^{\alpha/2(r-1)}}{\exp-[\sigma_p/\hat{\sigma}]^{\alpha}}$$

and the results are illustrated in Figure 14. Note that the requirement for a useful proof test is that the surviving specimens are not noticeably altered in their fracture properties. In this illustration we have an extreme case in that we have truncated the static specimens at 47 ksi, while the design ultimate for this laminate construction is only 10–11 ksi. Obviously, serious effort is required in this research area paralleling the work of Tiffany [18].

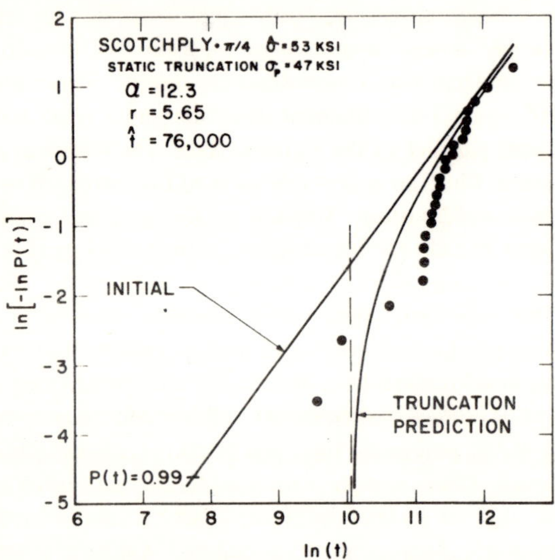

Figure 14. Truncation of fatigue lifetimes by a static-proof test [14].

The ranking of structural items based upon static response characteristics also contains the implicit assumption that the delayed time dependent fracture or fatigue characteristics are an extension of static fracture modes. In fact, this supposition is not applicable to metallic structures joined with mechanical fasteners [19]. It is often observed, in metallic structures [14], that structural fatigue is initiated by fretting around the fasteners and accelerated by galvanic corrosion induced in the structural detail by the fasteners. Primary organic adhesive bonding is currently being introduced into airframe construction in an effort to minimize these effects which contribute to the indeterminate [19] nature of structural safety in metallic structures. Composite structural technology is also seeking to exploit the potential advantages of adhesive joining technology. Quality control in adhesive bonding is being developed and the techniques outlined above will play a serious role in this evaluation as conventional, nondestructive techniques are not sufficient for the job. Well made adhesive joints are reliable, exhibit predictable scatter properties and yield fatigue failure modes which are related to static failure modes. Thus, the results presented here [17], by Halpin, Kopf and Goldberg [16], by Fruedenthal [19], and

Tiffany [18] will provide the analytical bases for a composite structural technology which exhibits a significantly higher degree of reliability than its equivalent metallic counterpart.

PROJECTIONS AND CONCLUSIONS

It is this worker's personal opinion that the (civilian as well as aircraft structures) general use of advanced composites will center around a filamentary prepreg composite which exhibits a longitudinal composite strain capability at failure of at least ±0.007, a transverse ply strain capability of at least 0.003, a shear strain capability equal to or greater than the longitudinal capability, fatigue insensitivity (75-80% of ultimate at 10^6 cycles) in a filament direction and a small but negative thermal expansion coefficient parallel to the filament axis. The technical preference for a high-strength, high-strain fiber, an acceptable ratio of high strength to modulus, is augmented by economic considerations. Implicit in this personal opinion is the argument that over the next decade the unsubsidized commercial market will only sustain one high-modulus, *general purpose* prepreg in addition to the existing E-glass/epoxy system. The economic scale of a materials system is not only constrained by development-production costs and market volume, but by the time and costs required in the development of a behavioral and engineering data base necessary for utilization of a materials system. At this specific time carbonization graphitization technology (both polymeric fiber and pitch precursors) yielding intermediate modulus high-strength fibers is attracting considerable serious interest. Such a system possesses a wide range of product application which cannot be duplicated by existing metallic technology. Augmenting this material will be a series of other composite systems which fulfill specific needs in modern technology. Future developments may produce a high modulus ribbon or flake composite which would reduce the fabrication requirements for complex laminate geometry and a matrix system with increased thermal properties and decreased moisture absorption. The ultimate nonmetallic construction material may not be a heterogeneous material, but a homogeneous polymeric material highly oriented (on the molecular scale) in at least two in-plane directions and yielding mechanical properties comparable to that of high-strength, graphite-epoxy systems discussed here.

The insertion of composite solids into engineering design and structures will alter structural design philosophy. Obviously structural design geometry must reflect some changes in material characteristics, but more importantly, composite technology complemented by advanced structural reliability concepts will emphasize the attainment of safe, reliable, predictable structures. Preliminary evidence indicates that some necessary aspects of structural reliability begins at the materials development level. This effort will be supported by increased emphasis upon education and the further development of a reproducible, high-production-volume material.

REFERENCES

1. A. M. Lovelace and S. W. Tsai, "Composites Enter the Mainstream of Aerospace Vehicle

Design," *Aero and Astro*, Vol. 8 (1970), p. 54.

2. G. P. Peterson and S. W. Tsai, "Status of Development of Advanced Composites," Proceedings 4th Symposium on Comp. Mat.; Tokyo, Japan, August (1971).
3. J. E. Ashton, J. C. Halpin, and P. H. Petit, *Primer on Composite Materials: Analysis*, Technomic Pub. Co., (1969).
4. S. W. Tsai, J. C. Halpin, and N. J. Pagano, *Composite Materials Workshop*, Technomic Pub. Co., (1968).
5. O. Ishai and R. E. Lavengood, "The Mechanical Performance of Bidirectional Fiber-Glass Polymeric Composites," *Israel Journal of Technology*, Vol. 8 (1970), p. 101.
6. C. W. Rogers, "Structural Design with Composites," *Fundamental Aspects of Fiber Reinforced Plastic Composites*, edited by R. T. Schwartz and H. S. Schwartz, Interscience, N. Y. (1968).
7. J. E. Ashton and J. M. Whitney, *Theory of Laminated Plates*, Technomic Pub. Co., (1971).
8. J. M. Whitney, "Effects of Constituent Material Properties on the Stability of Fiber Reinforced Composite Plates," Ref. 6 above.
9. R. A. Schapery, "Thermal Expansion Coefficients of Composite Materials Based on Energy Principles," *J. Comp. Mat.*, Vol. 2 (1968), p. 380.
10. J. C. Halpin and N. J. Pagano, "Consequences of Environmentally Induced Dilation in Solids," *Proc. 6th Ann. Meeting, Society of Engineering Science*, Springer (1969).
11. A. A. Fahmy and A. N. Ragi, "Thermal Expansion of Graphite-Epoxy Composites," *J. Appl. Phys.*, Vol. 41 (1970), p. 5112. .
12. J. C. Halpin and N. J. Pagano, "The Laminate Approximations for Randomly Oriented Fibrous Composites," *J. Comp. Mat.*, Vol. 3 (1969), p. 720.
13. F. R. Shanley, "Historical Note on the 1.5 Factor of Safety for Aircraft Structures," *J. Aero. Sci*, (1962), p. 243.
14. H. J. Grover, *Fatigue of Aircraft Structures*, Navair 01-1A-13 (1966).
15. M. E. Waddoups, J. R. Eisenmann and B. E. Kaminski, "Macroscopic Fracture Mechanics of Advanced Composite Materials," *J. Comp. Mat.*, Vol. 5 (1971), p. 446.
16. J. C. Halpin, J. R. Kopf and W. Goldberg, "Time Dependent Static Strength and Reliability for Composites," *J. Comp. Mat.*, Vol. 4 (1970), p. 462.
17. J. C. Halpin, T. Johnson, and K. Jerina, unpublished results.
18. C. F. Tiffany, "On the Prevention of Delayed Time Failures of Aerospace Pressure Vessels," *J. Franklin Inst.*, Vol. 290 (1970), p. 567.
19. A. M. Freudenthal, "The Material Aspects of Reliability," *Structural Fatigue in Aircraft*, ASTM STP 404, 1966.

Dynamic Mechanical Properties of Graphite-Epoxy and Carbon-Epoxy Composites

T. HIRAI* AND D. E. KLINE

Department of Material Sciences
The Pennsylvania State University
University Park, Pennsylvania 16802

(Received February 28, 1973)

ABSTRACT

Dynamic elastic moduli and damping of carbon- and graphite-filled epoxy composites have been measured over the temperature range of 85 to 300°K. The epoxy resin matrix was cured with diethylenetriamine. Dynamic elastic moduli increase with filler content in both cases; however, graphite powder is more effective than carbon powder in increasing the modulus. For the graphite composite, experimental data agree with Wu's equation for disk-shaped filler particles with a modulus of 56×10^{10} dyne/cm^2. For the carbon composite, experimental data cannot be effectively compared to any theoretical model, probably because the carbon powder is composed of agglomerated particles. The effect of voids on the dynamic modulus of a particulate composite is also discussed. The β damping peak (~ 250°K) does not change in its intensity with filler content, while the intensity of the γ peak (~ 150°K) is decreased as the amount of filler is increased. The relation between the filler content and the intensity of the β peak suggests that damping in the composites includes a component due to friction at the interface between the filler and matrix. The decrease in the γ peak with filler content may be explained by interaction between the filler surface and unreacted epoxides.

I. INTRODUCTION

THE ADDITION OF fillers to polymers can produce composite materials with a number of desirable properties [1]. In general, fillers usually increase the modulus of composites. A large number of theories and equations have been

*Present address: Oarai Branch, Research Institute for Iron, Steel and Other Metals, Tohoku University, Oarai-mach, Ibaraki-ken, Japan.

developed to explain the effect of fillers on mechanical properties. These are reviewed by Hashin [2], Nielsen [3] and others; however, almost all the equations apply to materials filled with spherical particles. The effect of filler-particle shape on the elastic modulus of a composite is not well understood; however, a recent theoretical treatment by Wu [4] predicts that randomly-oriented disk-shaped particles should give rise to a higher modulus than needle- or spherical-shaped particles. As yet, it does not appear that Wu's equations have been compared with experimental data.

Epoxy resins form some of the most useful polymers for composite matrices, and thus much research has been carried out on dynamic mechanical properties of epoxy resins filled with aluminum powder [5, 6], copper powder [6], glass beads [7], silica powder [8], titanium oxide powder [9, 10], and glass filaments [11]. However, there seems to be very little work on the low-temperature dynamic mechanical behavior of epoxy resins filled with graphite and carbon powders.

In the present paper we report results of an investigation of the dynamic elastic modulus and internal friction of an epoxy resin cured with diethylenetriamine and filled with natural graphite and amorphous carbon powders. The temperature range of tests was 85 to 300°K. Dynamic moduli were analyzed using Wu's equations [4]. The dynamic mechanical properties of the matrix resin system have been reported in previous papers [12, 13].

II. EXPERIMENTAL

A. Sample Preparation

Graphite "*G*" and carbon "*C*" powders were dispersed into a mixture of diglycidyl ether of bisphenol *A* (Epon 828) and diethylenetriamine (DETA). Scanning electron microscopic observations revealed the "*G*" powder to be composed of flake-like particles having a mean diameter of ~ 30μ. It is estimated that their average thickness is ~ 0.5μ. Observations suggest that the "*C*" powder spherical particles were ~ 250 Å in diameter, with agglomerates ~ 250μ in diameter. The conditions of sample preparation are shown in Table 1. The maximum volume percentages of "*G*" and "*C*" powders used in composite samples were limited to 29.8 and 8.89 vol %, respectively, because of the high viscosity of the mixtures. In the present experiments, as shown in Table 1, 59% of stoichiometric composition (SC) was used as an epoxy resin matrix to prepare composites. The reason for the choice of this matrix resin and the 59% SC is that this system has a β peak at ~ 250°K and a γ peak at ~ 150°K with approximately equal intensities [12, 13]. It is convenient to compare changes in these intensities with the addition of fillers.

The mixtures of Epon 828, DETA and "*G*" or "*C*" powders were stirred for 5 min. Bubbles were removed by vacuum pumping for 15 min. The loss of DETA during the pumping process is negligible. The mixtures were then drawn into a Pyrex tube (8 mm in diameter, coated on the inside with petroleum jelly), and allowed to cure for 2~4 months at room temperature. The cured samples were cut in lengths of ~ 11.1 cm. Some samples were then heat-treated at 120°C for 25 hr [13], under vacuum of $\sim 10^{-2}$ torr to examine the heat-treatment effect on the dynamic mechanical properties of the composites.

Table 1. Conditions of Sample Preparation

Graphite powder[1]	0.01~50 g (Volume %: 0.0085~29.8)
Carbon powder[2]	0.01~10 g (Volume %: 0.0098~ 8.89)
Epon 828[3]	50 cm^3
DETA (diethylenetriamine)[4]	4 cm^3 (59% of stoichiometric composition)

[1] Spectroscopic graphite powder, grade SP-1, from National Carbon Co., Density; 2.26 g/cm^3, C_{004} = 6.71Å.
[2] Spheron 6, D-4, from Cabot Co., Density; 1.97 g/cm^3, C_{004} = 6.88Å.
[3] Shell Chemical Co. product, epoxide equivalent ~ 185.

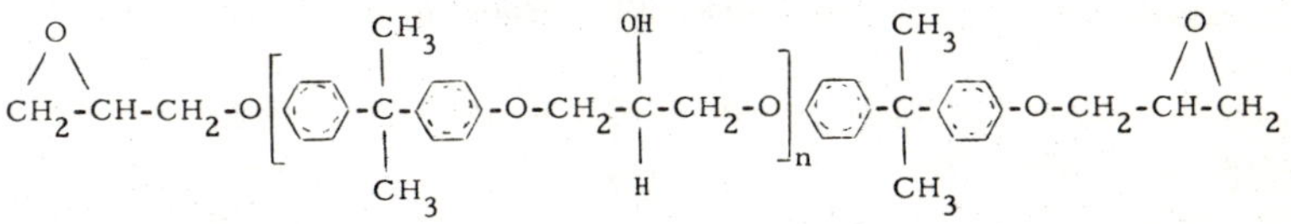

```
  O              CH3              OH                  CH3              O
 / \              |                |                   |              / \
CH2-CH-CH2-O[(○)-C-(○)-O-CH2-C-CH2-O]n(○)-C-(○)-O-CH2-CH-CH2
                  |                |                   |
                 CH3               H                  CH3
```

[4] A room temperature curing agent; $H_2N—CH_2—CH_2—NH—CH_2—CH_2—NH_2$.

B. Measurements of Dynamic Mechanical Properties

The apparatus used is similar to that first described by Förster [14] and later refined by Kline [15]. Each specimen, in the form of a rod, was suspended horizontally by two threads. One thread is attached to a magnetostrictive transducer which drives the specimen. The second thread is attached to a piezoelectric crystal pickup cartridge, which measures the specimen response. The specimen was vibrated in its fundamental flexural mode. The measurements were made in the temperature range from 85 to 300°K at 5°C increments, and in the frequency range from 700 to 2000 Hz. The heating rate was maintained at about 1 C°/min, which is sufficient to achieve relatively uniform temperature conditions for the duration of a measurement.

It can be shown that for a bar-shaped specimen of length L, diameter a, mass W, and resonant frequency f_r, the dynamic modulus is

$$E' \,(\text{dyne/cm}^2) = 1.606\,(L/a)^4\,(W/L)\,f_r^{\,2} \tag{1}$$

where units of L, W, a, and f_r are cm, g, cm, and Hz, respectively. The internal friction, Q^{-1}, was determined from the relation

$$Q^{-1} = \frac{\Delta f}{f_r} \tag{2}$$

where Δf is the separation of the frequencies above and below the resonant frequency f_r of the fundamental vibration mode at which the sample's response is 3 db down from the maximum.

III. RESULTS

A. Density

The densities of "G" and "C" samples are shown in Figure 1, which includes the results of the samples cured at room temperature above 2~4 months, d(RC), and followed by the heat treatment of 25 hrs at 120°C, d(HT). Figure 1 also indicates the theoretical densities with dotted lines. The composite experimental density, d_{comp}, was obtained by the mass-volume method, and the theoretical composite density, D_{comp}, for a void-free sample, was calculated by:

$$D_{comp} = d_m + (d_f - d_m) C_f \tag{3}$$

Here d_m and d_f are the densities of the matrix and filler, respectively, and C_f is the volume fraction of the filler.

In the case of the epoxy matrix used in the present experiments (59% of SC), the density is slightly decreased by heat treatment as described in a reference [13]. This small decrease in the density with heat treatment is evident in Figure 1.

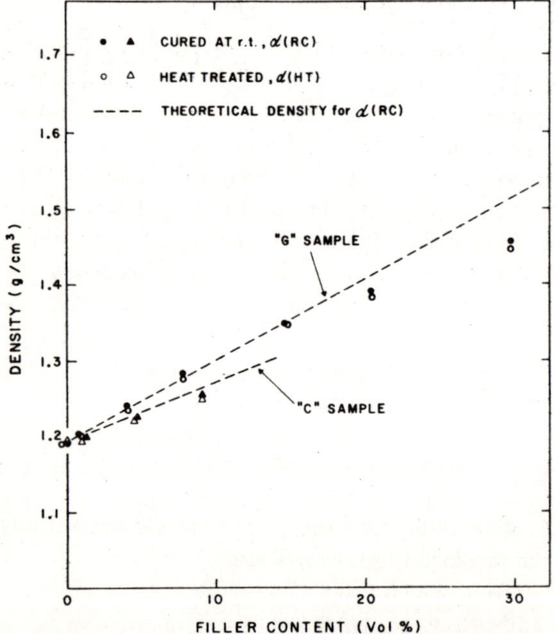

Figure 1. Experimental and theoretical densities of graphite – and carbon-filled epoxy composite.

As shown in Figure 1, the densities of "*G*" and "*C*" samples increase with filler content. For the high filler contents of 20.3 and 29.8 vol % for "*G*" samples and 8.9 vol % for "*C*" samples, the experimental densities show lower values than the theoretical densities. This difference is ascribed to voids. The highest-density sample, with 29.8 vol % of "*G*" filler, has 3.6 vol % of voids.

B. Dynamic Elastic Moduli at 85 and 297°K

Results for the dynamic modulus as a function of filler content are given in Figure 2. Values of the modulus measured at room temperature, E′297, are shown

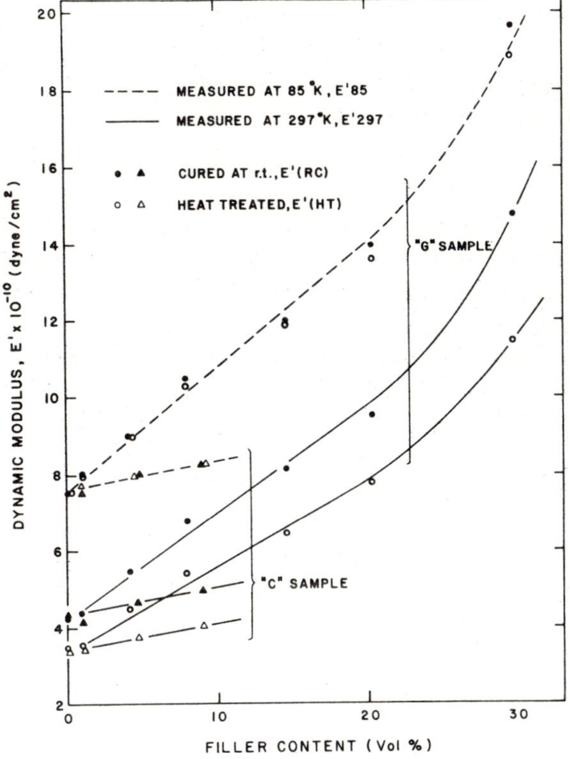

Figure 2. Dynamic elastic moduli as a function of filler content (carbon and graphite fillers in epoxy matrix).

by solid lines. For each vol %, the results for samples cured at room temperature, E′(RC)297, and the results for heat-treated samples, E′(HT)297, are presented. For "*G*" samples, E′(RC)297 and E′(HT)297, indicated by ● and *o* respectively, are increased as the amount of "*G*" powder is increased, especially at high filler contents. For "*C*" samples, E′(RC)297 and E′(HT)297, ▲ and Δ, are slightly increased by the addition of "C" powder. The addition of "G" powder is more effective in increasing E′297 than "C" powder. The heat treatment appreciably lowers the dynamic modulus in both "G" and "C" samples. This phenomenon has been discussed in a study of the effects of heat treatment on the properties of the epoxy-DETA matrix [13].

In Figure 2, values of the dynamic modulus measured at 85°K, E′(RC)85 and E′(HT)85, are also shown by dotted lines. At 85°K, the dynamic moduli show higher values than those at 297°K in both cases. The decrease in E′85 with heat treatment is relatively slight for the high filler contents for "G" samples, and is not observed for "C" samples. The relative change in E′85 with filler content is similar to that for E′297.

C. Temperature-Dependent Dynamic Elastic Modulus and the Relative Dynamic Elastic Modulus

Figures 3 and 4 show the temperature dependence of the dynamic moduli for "G" and "C" samples, respectively, over the temperature range of 85 to 300°K. In these figures, solid and dotted lines represent results for the room-temperature-cured samples, E′(RC), and the heat-treated samples E′(HT), respectively. For the room-temperature-cured samples, E′(RC) decreases rather steadily with increasing temperature. For the heat-treated samples, however, E′(HT) decreases somewhat more sharply with increasing temperature above 150~200°K. As a result of heat treatment, the dynamic modulus decreases at 297°K as shown earlier (Figure 2), but it should be noted that E′ increases in the range 120~200°K.

D. The Heat-Treatment Effects

By heat treatment, absolute values of the dynamic modulus are lowered at 85 and 297°K (Figure 2) and the temperature dependence of the dynamic modulus and the relative dynamic modulus are markedly changed, particularly for the high filler contents (Figures 3 and 4). The heat-treatment effect on the dynamic modulus can be expressed by the ratio, E′(HT)/E′(RC), which is shown in Figure 5. The ratio at each temperature is almost independent of amount of filler content. This probably indicates that heat treatment contributes only to structural changes of the matrix as discussed in an earlier paper [13] and may not appreciably affect the nature of the interface between the filler and the matrix.

E. Damping

Values of the internal friction at 85°K for the room-temperature-cured samples, Q^{-1}(RC)85, and the heat-treated samples, Q^{-1}(HT)85, are shown in Figure 6. For both the "G" and "C" samples, Q^{-1}(RC)85 and Q^{-1}(HT)85 seem to be independent of filler content, although the data have a rather wide spread. Values of Q^{-1} at 85°K are definitely lowered by heat treatment.

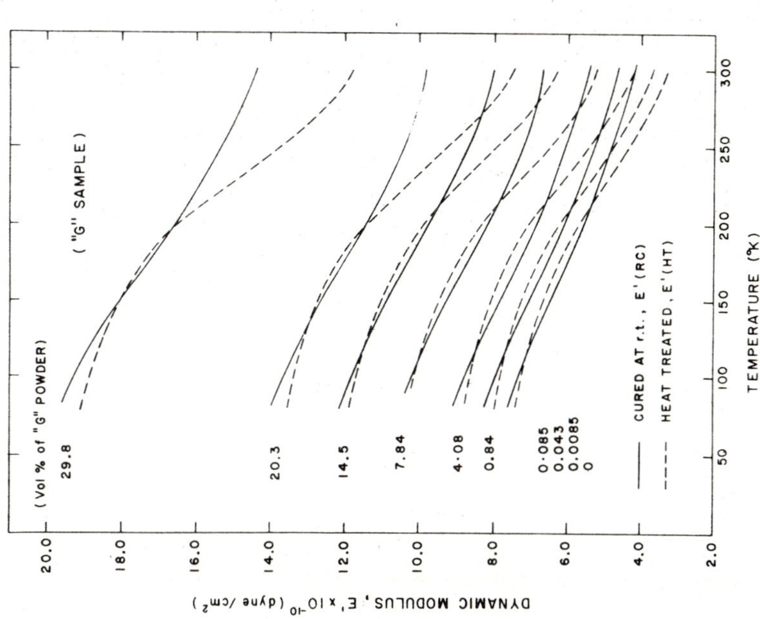

Figure 3. *Temperature dependence (85~300°K) of the dynamic elastic modulus for graphite-filled epoxy composite.*

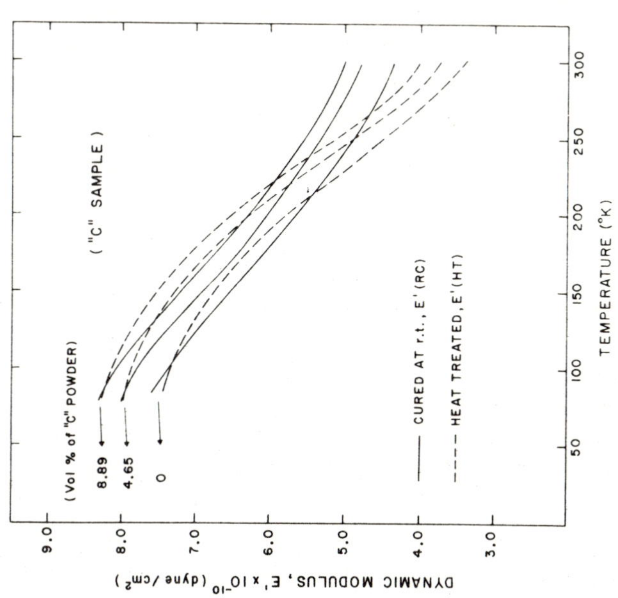

Figure 4. *Temperature dependence (85~300°K) of the dynamic elastic modulus for carbon-filled epoxy composite.*

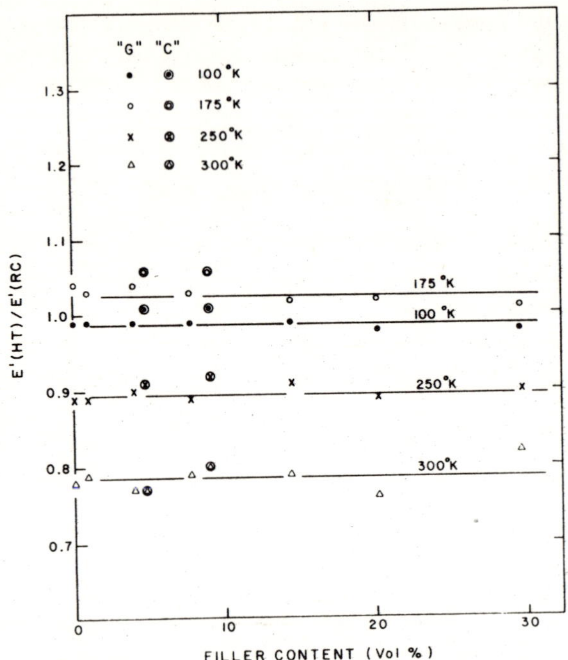

Figure 5. E'(HT)/E'(RC) plotted as a function of filler content (carbon and graphite fillers in epoxy matrix).

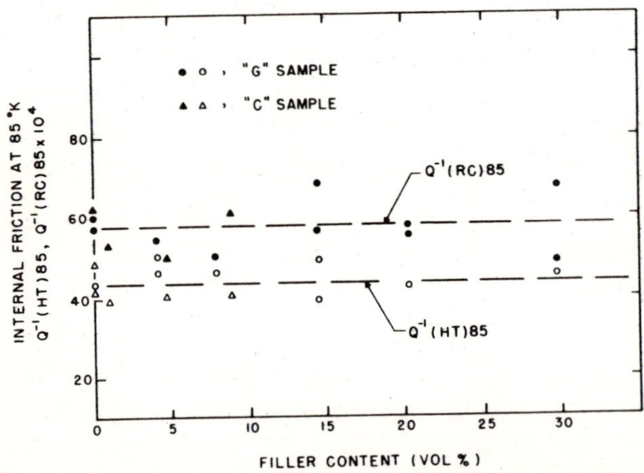

Figure 6. Internal friction measured at 85°K as a function of filler content, for the room-temperature-cured (solid symbols – RC) and heat-treated (open sumbols – HT) samples.

As described in the previous papers [12, 13], resonance curves are sometimes broad at 297°K, especially for the heat-treated samples. Therefore, precise values of Q^{-1}(RC)297 and Q^{-1}(HT)297 were not obtained. Table 2 gives approximate values of Q^{-1}297 for the matrix resin (no filler) and for composites with "*G*" and "*C*" fillers. From this it appears that Q^{-1}(RC)297 and Q^{-1}(HT)297 do not change by amounts greater than the data spread with changing amounts and types of filler.

Table 2. Values of Q^{-1} at 297°K for the Matrix Resin and Composites

	matrix resin	*"G" sample*	*"C" sample*
Q^{-1}(HT)297	$90 \sim 120 \times 10^{-4}$	$80 \sim 110 \times 10^{-4}$	$80 \sim 150 \times 10^{-4}$
Q^{-1}(RC)297	$200 \sim 400 \times 10^{-4}$	$300 \sim 500 \times 10^{-4}$	$300 \sim 400 \times 10^{-4}$

Filler contents are 0.0085 ~ 29.8 vol % for "*G*" sample and 0.0098 ~ 8.89 vol % for "*C*" sample. RC and HT indicate the room-temperature-cured and the heat-treated samples, respectively.

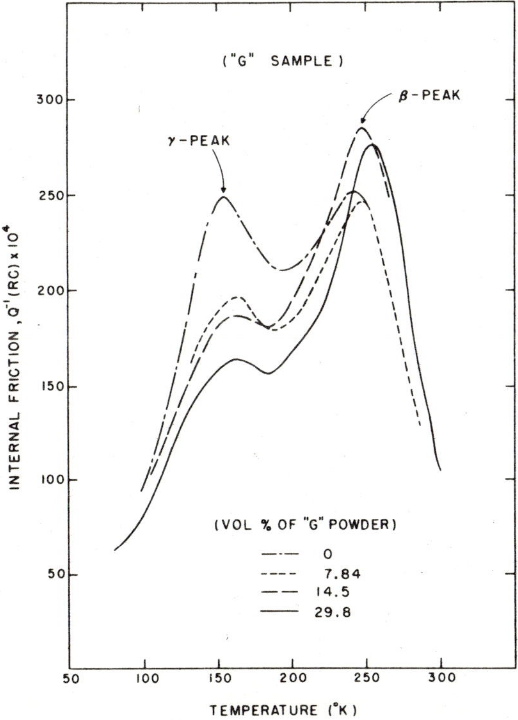

***Figure** 7. Internal friction of the room-temperature-cured samples as a function of temperature for "G" material.*

Figure 7 shows typical results of the temperature dependence of Q^{-1}(RC) for "*G*" samples cured at room temperature. Two secondary peaks (β and γ peaks) are observed near 250 and 150°K, respectively, on all samples. Similar results are obtained on "*C*" samples. These two peaks result from the matrix resin, as shown earlier [12].

Values of Q^{-1} for the β and γ peaks, $Q^{-1}(RC)\beta$ and $Q^{-1}(RC)\gamma$, as a function of filler content are shown in Figure 8. The amount of change in $Q^{-1}(RC)\beta$ with filler content is not clear because of the broad and weak resonance curves at ~ 250°K. However, the results suggest that $Q^{-1}(RC)\beta$ is independent of filler content for "*G*" and "*C*" samples. On the other hand, $Q^{-1}(RC)\gamma$ definitely decreases as the amount of filler is increased.

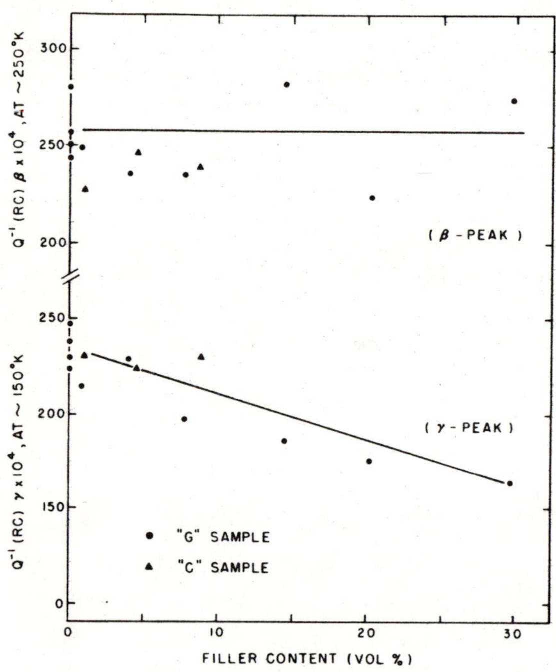

***Figure 8.** Intensities of the β (~ 250° K) and γ (~ 150° K) peaks as a function of filler content.*

IV. DISCUSSION

A. Damping

It would appear that the intensity of the mechanical loss depends on the nature and concentration of fillers and on friction at the interface between filler and matrix. If both the filler and matrix cause damping, then the simplified relationship may be given:

$$Q^{-1}_{comp} = C_m Q^{-1}_m + C_f Q^{-1}_f \tag{4}$$

where C is the volume fraction and the subscripts *comp*, *m* and *f* refer to the properties of the composite, matrix, and filler, respectively. The "*G*" powder used in the present experiments has damping peaks in the region of 130 ~ 160°K and ~ 250°K [16], near where the γ and β peaks of the matrix resin appear, as shown in Figure 7 and the previous paper [12]. However, the results of the internal friction for graphite show that contributions to Q_{comp} from $C_f Q_f^{-1}$ are small and lie within the experimental error. Using $C_m = 1 - C_f$, this leads to the reduced relationship [17],

$$Q^{-1}_{comp}/Q^{-1}_m = 1 - C_f \tag{5}$$

Lewis and Nielsen [7] examined the damping ratio of the system of glass beads and Epon 828 cured with triethylene tetramine (TETA), and observed that for all of the experimental data $Q^{-1}_{comp}/Q^{-1}_m > 1 - C_f$. They suggested from this that the damping increase resulting from the addition of filler was due to friction at the interface between matrix and filler.

Figure 9 presents the experimental data for "*G*" samples and shows Equation (5) with a dotted line. Although the data points spread widely above the below the $Q^{-1}_{comp}/Q^{-1}_m = 1$ line, the damping ratio of the β peak appears to be independent of filler content. The ratio of the γ peak decreases with increasing filler content.

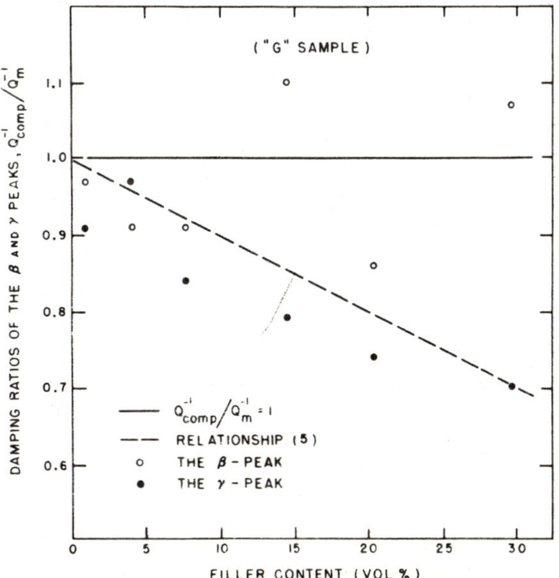

Figure 9. Relation between the damping ratios (β and γ peaks) and filler content.

The values of Q^{-1}(RC) at 85°K (Figure 6) and 297°K (Table 2) are almost independent of filler content for "*G*" samples. These values should instead decrease as the amount of filler is increased if the filler has a low damping value. The present results thus indicate that damping may result from friction at the interface between matrix and filler, as suggested [7]. This explanation also appears reasonable for the change in Q^{-1}(RC)β with filler content. Thus for 85 and 297°K, and for the β peak, the would-be decrease in the damping ratio with increasing volume fraction of low-damping filler may be compensated for by the increase in friction at the interface accompanying the increasing surface area of filler. The wide spread of the data indicates that this interface-slippage contribution to the internal friction is unpredictable, and perhaps depends on some parameter which was not controlled in the sample-preparation process.

For the γ peak, the damping ratio decreases with increasing filler content, suggesting that at this temperature there is little contribution to damping by friction at the interface. The concentration of unreacted epoxide ends may be related to the intensity of the γ peak [12]. It would appear that addition of filler does not alter the γ peak relaxation mechanism or induce interface friction at the corresponding temperature.

As shown in Figure 6 and Table 2, Q^{-1} is lowered by heat treatmentt (120°C), however, Q^{-1}(HT)85 and Q^{-1}(HT)297 still appear to be independent of filler content. Figure 10 shows typical results of the temperature dependence of Q^{-1}(HT) for the heat-treated "*G*" sample, and also includes a curve from a room-temperature-cured sample of the unfilled matrix. The temperature dependence of Q^{-1}(HT) is remarkably different from that of Q^{-1}(RC). For the heat-treated samples, the separate β and γ peaks are not observed. This was also shown earlier for unfilled resins [13], and it is assumed to be related to heat-treatment curing which affects the previously unopened epoxide ends. The Q^{-1}(HT) curves in Figure 10 are not continued through the temperature range of 200°K to room temperature because of excessively high values. A high-damping region (termed β') is present. The temperature dependence of Q^{-1}(HT) for the heat-treated "*C*" samples is almost the same as that of the "*G*" samples.

As discussed in a previous paper [13], heat-treated samples are considered to have the β' peak at about 270°K. Lewis and Nielsen [7] observed a broad secondary peak at about 230°K (using a torsion pendulum) on a system of glass beads and Epon 828-TETA heat treated at high temperatures. Since the difference between the β' peak temperatures of 230°K (by Lewis and Nielsen) and 270°K (by Kreahling and Kline [18] and the present experiments) may be attributed to the different frequencies and methods used in both the experiments, as well as slightly different matrices, these two peaks are considered to be essentially the same. In the Lewis and Nielsen experiments, the β' peak decreased with increasing amount of filler. From the present results, it also seems that the β' peak decreases with increasing filler content. However, since the maximum intensity of the β' peak, Q^{-1}(HT)β', was unobtainable in the present experiments, the magnitude of the change in the β' peak with filler content cannot be compared with earlier results.

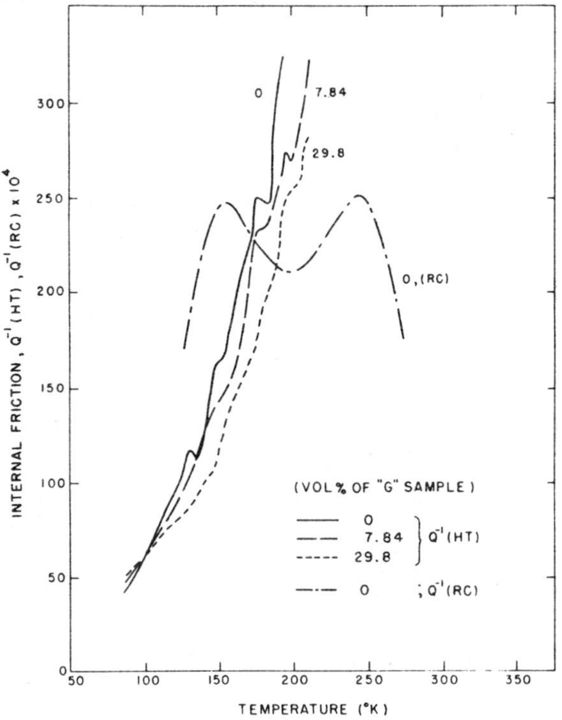

Figure 10. Internal friction of the heat-treated samples as a function of temperature for "G" material.

B. Analysis of the Dynamic Modulus of Composites

As shown in Figure 2, the addition of "*G*" powder, as compared to "*C*" powder, is more effective in increasing the dynamic modulus. This behavior is general. Plate-like particles such as those of the natural graphite powder are more effective than spherical particles such as those of the carbon powder [4]. In the present experiments, moreover, the difference in the moduli of "*G*" and "*C*" particles apparently affects the increases in the dynamic moduli of the composites.

Pedawer and Beecher [19] suggest that quite high moduli should be achieved in parallel-flake composites. Thus graphite flakes oriented parallel to the sample axis should result in a modulus of the order of the graphite crystal's *a*-axis modulus, for which Kelly [20] gives a value of $1.06(10^{13})$ dyne/cm^2. The relatively low moduli observed in the present samples suggests that there is very little coherent axial orientation of the flakes. There may not be complete randomness, either, but it is of interest to compare the data with the results predicted by Wu's theory [4].

For a particulate composite with randomly-oriented disk-shaped filler particles, Wu [4] presents the relationship:

$$\frac{1}{G_{comp}} = \frac{1}{G_m}\left\{1 + \frac{C_f}{5}\left[\frac{2}{1+A} + 1 + \frac{2+(2-4R/3)A+2(3-4R)B}{1+A+(3-4R)B}\right]\frac{G_m - G_f}{G_{comp}}\right\} \quad (6)$$

where

$$A = \frac{G_f}{G_{comp}} - 1$$

$$B = \frac{1}{3}\left(\frac{K_f}{K_{comp}} - \frac{G_f}{G_{comp}}\right)$$

$$R = \frac{3G_{comp}}{3K_{comp} + 4G_{comp}}$$

C_f = volume fraction of the filler

G = the shear modulus

and K = the bulk modulus.

In these equations, the composite, the filler, and the matrix resin are specified by the subscripts of *comp*, *f*, and *m*, respectively. The elastic constants E, G, and K are related by the equation:

$$E = 2G(1 + \nu) = 3K(1 - 2\nu) \quad (7)$$

where

E = Young's modulus

and ν = Poisson's ratio

In order to simplify these relations, ν is assigned the value of 0.2, as Wu did for an illustrative example in his paper [4]. (The actual value of ν for the matrix is probably closer to 0.3, but Wu's formula is relatively insensitive to changes in ν.) Taking $\nu = 0.2$ for the matrix, filler, and composite:

$$\frac{G_{comp}}{G_m} = \frac{E_{comp}}{E_m}, \frac{G_f}{G_m} = \frac{E_f}{E_m},$$

and Equation (6) can be rearranged to give:

$$\frac{E_{comp}}{E_m} = \frac{2 - C_f + C_f(E_f/E_m)}{2 - C_f + C_f(E_m/E_f)} \tag{8}$$

Up to the present time, little experimental data appear to be available to support Wu's equation. In the present experiments, the data at 297°K for the epoxy resin filled with natural graphite powder fit Equation (8) if E'_f/E'_m is assigned the value of 13, which gives the upper curve in Figure 11. Since $E'_m = 4.3 \times 10^{10}$ dyne/cm^2, this implies $E'_f \cong 56 \times 10^{10}$ dyne/cm^2 as a value of the dynamic modulus of the natural graphite powder.

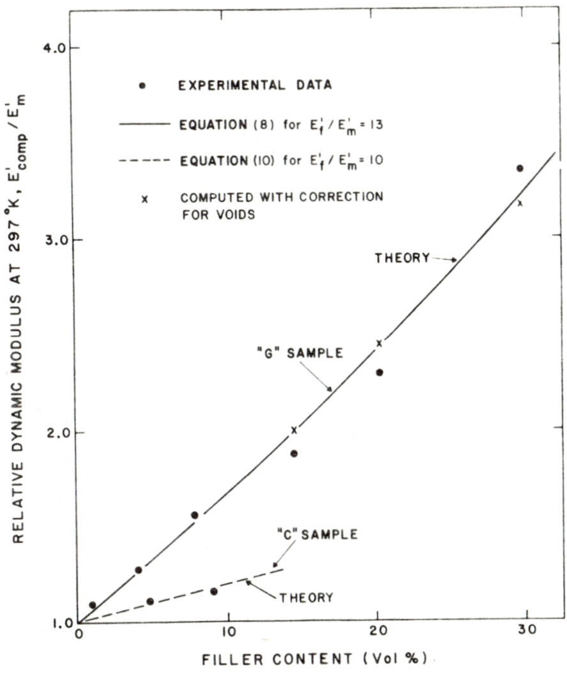

Figure 11. Theoretical and experimental values of the relative dynamic elastic modulus (carbon and graphite fillers in epoxy matrix).

The graphite flakes are anisotropic crystallites whose modulus varies with the direction relative to the principal crystal axes. Several models have been proposed for average aggregate properties of randomly-oriented crystalline particles; generally they assume equal stress or equal strain along all the crystal axes. The equal-stress and equal-strain estimates bracket the effective average modulus. However, because of the extreme anisotropy of the graphite crystal, its equal-stress and equal-strain averages are quite far apart. Moreover, the elastic constants of the graphite crystal are not known to high precision and their values are continually being revised as new information is obtained. Thus the significance of an average value based on available data is questionable. For thin flat graphite flakes the flat faces, which account for most of the area of contact between the matrix and the filler particles, are orthogonal to the c axis of the crystal. This may cause the graphite flakes to be stressed principally in tension/compression along the c axis and/or basal-plane shear. Kelly [20] has given values of 3.9×10^{10} dyne/cm^2 for the basal-plane shear modulus and 36.5×10^{10} dyne/cm^2 for the c-axis modulus. These are lower than most of the other elastic moduli of the graphite crystal, so the effective average value for randomly-oriented graphite flakes may be slightly higher. The value estimated above appears to be of the right order of magnitude.

For spherical fillers (corresponding to "C" samples), the shear modulus of a composite is given [4] by:

$$\frac{1}{G_{comp}} = \frac{1}{G_m}\left[1 + \frac{C_f}{1 + (2 + 4R/3)(A/5)} \cdot \frac{G_m - G_f}{G_{comp}}\right]. \tag{9}$$

Again taking $\nu = 0.2$ for the matrix, filler, and composite, we obtain the following equation:

$$\frac{E_{comp}}{E_m} = \frac{1}{2}\left[(1 - 2C_f)(1 - \frac{E_f}{E_m}) + \left((1 - 2C_f)^2 (1 - \frac{E_f}{E_m})^2 + 4\frac{E_f}{E_m}\right)^{1/2}\right]. \tag{10}$$

As shown in Figure 11 by a dotted line, Equation (10) with $E'_f/E'_m = 10$ fits the data at 297°K for the carbon-powder-filled epoxy resins. However, "C" powders are in agglomerates. Assuming that the value of $E_f/E_m = 10$ applies to the system of epoxy resin and "C" powder agglomerates, we obtain $E'_f = 43 \times 10^{10}$ dyne/cm^2 for the modulus of the agglomerates. This is doubtful because the 250μ agglomerates are composed of loosely-bonded 250Å carbon particles. Therefore, it is assumed that the equations cannot be directly applied to analyze the modulus of a composite with aggregated fillers, such as the "C" samples.

C. Effects of Voids on the Dynamic Modulus

As shown by comparing experimental and theoretical densities in Figure 1, voids are included in the samples with the higher filler contents. The elastic modulus usually decreases with increasing void content [1, 21].

Following the approach of Cohen and Ishai [21], assuming that the presence of the filler does not affect the elastic properties of the porous matrix, and that the voids are of spherical shape, Equation (10) can be modified to describe a matrix including voids. Considering the filler particles to have the properties of voids, we find that $A = -1$, $B = 0$, and $C_f = C_v$. Then with $\nu_{comp} = 0.2$, Equation (10) becomes:

$$E_{m(v)} = E_m(1 - 2C_v) \tag{11}$$

where $E_{m(v)}$ is the modulus of the matrix-void system and C_v is the volume fraction of voids. The void-corrected relative dynamic modulus of "*G*" samples was calculated, and is shown by "*X*" in Figure 11. These results show that the effect of voids on the dynamic modulus is negligible in the present experiments (maximum void content 3.6 vol. %). This agrees with the report by McCrum [22] who found that the dynamic shear modulus is very nearly independent of void space if the voids occupy less than 4 vol. %.

V. CONCLUSIONS

The behavior of "*G*" and "*C*" fillers in an epoxy resin matrix cured with diethylenetriamine was studied for filler concentrations up to ~30% and ~9%, respectively. In particular it was observed that:

1. Two secondary peaks (β and γ peaks) were observed in the matrix resin at the temperatures of ~250 and ~150°K, respectively. This is consistent with earlier results.
2. The internal friction values of 85 and 297°K for the room-temperature-cured and heat-treated samples were almost independent of filler content.
3. The intensity of the γ peak decreased with increasing filler content, while that of the β peak appeared to be independent of the amount of filler.
4. From the damping behavior with increasing volume fraction of filler, it appears that internal friction losses occur at the interface between filler and matrix, as reasoned by Lewis and Nielsen [7] in similar work for fillers of glass beads.
5. In heat-treated samples, the β and γ peaks were not observed, but another peak (β') was observed near 270°K. The magnitude of Q^{-1} tends to decrease with added filler content as noted by Lewis and Nielsen [7] for experiments with glass beads.
6. The dynamic modulus of samples with "*C*" filler does not appear to conform to any theory because the filler particles are agglomerates of smaller particles.
7. The dynamic modulus of samples with "*G*" filler appears to conform to Wu's theory [4] if the filler particles are assigned a modulus value of $56(10^{10})$ dyne/cm^2. The appropriate value is not known to high precision because the "*G*" particles are extremely anisotropic and the values of the graphite-crystal elastic constants are continually being revised. The assumed value appears to be of the right order of magnitude.

8. A void content of a few volume % has negligible effect on the dynamic elastic modulus of the composites.

ACKNOWLEDGMENT

This research was supported in part by the U.S. Atomic Energy Commission, Contract No. AT(30-1)-1710. The authors are very indebted to Dr. J. R. Jenness, Jr. for comments and discussion of the manuscript.

REFERENCES

1. L. Holliday, *Composite Materials*, Elsevier Publishing Co., New York (1966).
2. Z. Hashin, "Theory of Mechanical Behavior of Heterogeneous Media," *Appl. Mech. Rev.*, Vol. 17 (1964), p. 1.
3. L. E. Nielsen, "Mechanical Properties of Particulate-Filled Systems," *J. Comp. Mat.*, Vol. 1 (1967), p. 100.
4. T. T. Wu, "The Effect of Inclusion Shape on the Elastic Moduli of a Two-Phase Material," *Int. J. Solids Structures*, Vol. 2 (1966), p. 1.
5. D. E. Kline, "Dynamic Mechanical Properties of Polymerized Epoxy Resins," *J. Poly. Sci.*, Vol. 47 (1960), p.237.
6. G. E. Nevill, Jr., J. M. Loyd, Jr. and H. R. Fuehrer, "Dynamic Properties of Metal Powder-Epoxy Resin Composites," *J. Comp. Mat.*, Vol. 3 (1969), p. 174.
7. T. B. Lewis and L. E. Nielsen, "Dynamic Mechanical Properties of Particulate-Filled Composites," *J. Appl. Poly. Sci.*, Vol. 14 (1970), p. 1449.
8. J. M. Lifshitz and A. Rotem, "Determination of Reinforcement Unbonding of Composites by a Vibration Technique," *J. Comp. Mat.*, Vol. 3 (1969), p. 412.
9. I. Galperin, "Dynamic Mechanical Properties of a TiO_2-Filled Crosslinked Epoxy Resin from 20–90°C," *J. Appl. Poly. Sci.*, Vol. 11 (1967), p. 1475.
10. T. K. Kwei and C. A. Kumins, "Polymer-Filled Interaction: Vapor Sorption Studies," *J. Appl. Poly. Sci.*, Vol. 8 (1964), p. 1483.
11. A. B. Schultz and S. W. Tsai, "Dynamic Moduli and Damping Ratios in Fiber-Reinforced Composites," *J. Comp. Mat.*, Vol. 2 (1968), p.368.
12. T. Hirai and D. E. Kline, "Relation between Dynamic Mechanical Properties and Structural Features of Non-Stoichiometric, Amine-Cured Epoxy Resin," *J. Appl. Poly. Sci.*, (December 1972?).
13. T. Hirai and D. E. Kline, "Effects of Heat Treatment on Dynamic Mechanical Properties of Non-Stoichiometric, Amine-Cured Epoxy Resin," *J. Appl. Poly. Sci.*, (January 1973?).
14. R. Förster, "Ein neues Messverfahren zur Bestimmung des Elastizitatsmoduls und der Dampfung," *Z. Metallk.*, Vol. 29 (1937), p. 109.
15. D. E. Kline, "A Recording Apparatus for Measuring the Dynamic Mechanical Properties of Polymers," *J. Poly. Sci.*, Vol. 22 (1956), p. 449.
16. R. E. Taylor, D. E. Kline and P. L. Walker, Jr., "The Dynamic Mechanical Behavior of Graphites," *Carbon*, Vol. 6 (1968), p.333.
17. L. E. Nielsen, "Creep and Dynamic Mechanical Properties of Filled Polyethylenes," *Trans. Soc. Rheology*, Vol. 13 (1969), p. 141.
18. R. P. Kreahling and D. E. Kline, "Thermal Conductivity, Specific Heat, and Dynamic Mechanical Behavior of Diglycidyl Ether of Bisphenol A Cured with m-phenylenediamine," *J. Appl. Poly. Sci.*, Vol. 13 (1969), p. 2411.
19. G. E. Padawer and N. Beecher, "On the Strength and Stiffness of Planar Reinforced Plastic Resins," *Polymer Engineering and Science*, Vol. 10 (1970), p. 185.
20. B. T. Kelly, "Mechanical Properties of Graphite," The Pennsylvania State University Conference on the Scientific Base of Carbon Technology (March 1972).
21. L. J. Cohen and O. Ishai, "The Elastic Properties of Three-Phase Composites," *J. Comp. Mat.*, Vol. 1 (1967), p. 390.
22. N. G. McCrum, "Torsion Pendulum Method for Determining Crystallinity and Void Content of Tetrafluorethylene Resins," ASTM Bulletin No. 242 (December 1959), p. 80.

Effect of Fibre Orientation and Laminate Geometry on the Dynamic Properties of CFRP

R. D. ADAMS AND D. G. C. BACON

Department of Mechanical Engineering
University of Bristol
Bristol. BS8 1TR, U.K.

(Received April 2, 1973)

ABSTRACT

Theoretical predictions have been made of the effect of fibre orientation and laminate geometry on the flexural and torsional damping and modulus of fibre reinforced composites. Materials with fibres at $+\theta$ (off-axis), $\pm\theta$ (angle-ply), 0/90° (cross-ply) and a general plate were investigated.

In almost all cases, very good agreement was obtained between the theoretical prediction and the experimental results. Some limitations of plate theory for the torsion of angle-ply laminates were revealed and discussed.

INTRODUCTION

IN GENERAL THE damping properties of materials do not fall into the well defined categories which are amenable to analysis, such as viscoelastic or hysteretic behaviour. Most metals behave in a hysteretic manner at stress levels up to the fatigue limit [1], while plastics are generally characterised as being viscoelastic [2]. However, the high performance resin systems used for CFRP and GFRP have low damping, which is not particularly frequency or temperature dependent except near the glass transition temperature [3]. This difficulty of classifying the type of damping precisely for all conditions has led to an approach, reported here, whose only criteria is that the damping should be low (< 50% S.D.C.[1]).

[1] Specific damping capacity, S.D.C., is defined as $\Delta U/U$ where ΔU is the energy dissipated during a stress cycle and U is the maximum strain energy. The symbol used for S.D.C. is ψ.

The cyclic stress-strain relationship for real materials is not single-valued but forms a closed loop. In general, the shape of the hysteresis loop will change with stress so that ψ will be a function of the maximum stress in the cycle,

$$\psi = \psi(\sigma).$$

Using this definition for the damping, an analysis has been developed for the prediction of the damping of laminated plates, based on the strain energy of each layer. This method has the advantage over viscoelastic or complex modulus techniques that materials with stress dependent damping can be accommodated and the energy dissipation throughout the laminate can be examined in detail.

Work by other investigators on the damping of laminated composites has been inconclusive. Schultz and Tsai [4] tested beams of different lamination geometry in flexure but there were large differences between their theoretical and experimental values. Clary [5] investigated the effect of fibre orientation on the flexural vibration of plates and beams and measured the damping capacity. He could find no apparent relationship between damping and fibre orientation but did comment that the damping values were small. It is possible that the results quoted in [4] and [5] do not truly represent the damping properties of the materials because the test techniques were not sufficiently sensitive for the low values of damping which can be obtained with these materials.

The purpose of this work was to examine the effect of fibre orientation and lamination geometry on the flexural and torsional damping and dynamic moduli of carbon fibre reinforced plastics. Theoretical and experimental values are compared.

APPARATUS

Flexure

The flexural apparatus as reported in [6] was subsequently improved and has been described in [7]. Beams were tested in a free-free flexural mode with a central mass. Only symmetric modes of vibration were induced, these having even numbers of nodes, and the beam was supported at two such nodes equidistant about the middle. The central mass took the form of a coil which was clamped to the beam via cylindrical surfaces and moved between the poles of two electromagnets. The apparatus was shown to be capable of exciting beams at high cyclic amplitudes over the temperature range – 50°C to 200°C with a low background damping of 0.06%. It was found for low-damping high-modulus specimens, such as unidirectional 0° CFRP and Duralumin, that even at small amplitudes aerodynamic damping was significant and it was necessary for all these specimens to be tested *in vacuo* (0.5 Torr was usually sufficient). In air, the central amplitude was measured using an image shearing microscope and this was used to calibrate the coil/magnet pair employed as a pick-up in the vacuum system. Damping was measured at constant cyclic stress/amplitude by monitoring the energy input to the system.

Torsion

The torsional apparatus has been described in [6] and consisted, essentially, of a rigid frame, into the top of which the test specimen was clamped: the lower end was clamped to an inertia bar. A dummy specimen made from Duralumin was attached between the inertia bar and the base of the frame, to restrain flexural motion. Oscillations were maintained and detected using coil/magnet pairs and the apparatus was shown to have extraneous damping losses of less than 0.05%.

MATERIALS AND SPECIMENS

The specimens were made using the "pre-preg" method. Commercial material, 0.010 in. thick, was used and consisted of high modulus treated fibre (HM-S), pre-impregnated with the Shell Epikote epoxy resin system DX209 (828/DDM), cured with BF_3. Later in the project, other carbon fibres pre-impregnated with Shell Epikote DX210 epoxy resin cured with BF_3 were employed. The sheets were cut to size and laminated in closed moulds in a heated press. The manufacturers recommended cure cycles were followed. Specimens were cut from the laminates and the width ground to size. Typical dimensions for a flexural specimen were 0.500 in. wide by 0.100 in. thick and approximately 9 in. long. Some 0° unidirectional specimens were made 11 in. long. Torsion specimens (0° unidirectional) were moulded 8 in. long by 1/2 in. square. At each end, 1 in. was left for the grips and the middle section was turned to 0.400 in. diameter. Rectangular torsion specimens were fabricated from beam specimens by bonding steel end pieces, 1 in. long, 1/2 in. wide and 0.2 in. thick to each side of the laminate to increase the thickness to 1/2 in. A gauge length of 4 in. was used with these specimens.

THEORY

Theoretical Prediction of Damping for Laminated Composites

In this analysis, where the composite is in the form of thin unidirectional layers, the material is assumed to exhibit the elastic symmetry of two-dimensional orthotropy. Each lamina has four independent moduli C_{11}, C_{22}, C_{66}, C_{12} where the stress-strain relationship for the kth lamina is

$$\sigma_i^k = C_{ij}^k \epsilon_j^k ,$$

$$\text{and} \quad \epsilon_i^k = S_{ij}^k \sigma_j^k ,$$

C_{ij} and S_{ij} being respectively the lamina stiffness and compliance matrices. The components of C_{ij} are related to the engineering constants E_L, E_T, G_{LT} and ν_{LT} by the following:

$$C_{11} = \frac{E_L}{1 - \nu_{LT}^2 \dfrac{E_T}{E_L}} \qquad C_{22} = \frac{E_T}{1 - \nu_{LT}^2 \dfrac{E_T}{E_L}}$$

$$C_{12} = \nu_{LT} C_{22} \qquad\qquad C_{66} = G_{LT}$$

It has been shown experimentally that the general laminate, constructed from layers of homogeneous anisotropic material, satisfies the constitutive equations of anisotropic plate theory [8]. These equations can be written, using the usual notation, as

$$\begin{bmatrix} N \\ M \end{bmatrix} = \begin{bmatrix} A & B \\ B & D \end{bmatrix} \begin{bmatrix} \epsilon^{\circ} \\ \kappa \end{bmatrix} \tag{1}$$

$$\begin{bmatrix} \epsilon^{\circ} \\ \kappa \end{bmatrix} = \begin{bmatrix} A' & B' \\ B' & D' \end{bmatrix} \begin{bmatrix} N \\ M \end{bmatrix} \tag{2}$$

where $[N]$ and $[M]$ are the force and moment resultants acting on a section.

$$N_i = \int_{-h/2}^{h/2} \sigma_i \, dz \quad (i = 1, 2, 6)$$

$$M_i = \int_{-h/2}^{h/2} \sigma_i \, z \, dz$$

ϵ_i^o and κ_i are the in-plane strain and curvatures where the total strain in a layer is given by

$$\epsilon_i = \epsilon_i^o + z \, \kappa_i$$

The four submatrices of Equation (1) can be written as

$$[A], [B], [D] = A_{ij}, B_{ij}, D_{ij} = \int_{-h/2}^{h/2} (1, z, z^2) \, C_{ij} \, dz$$

Equation (2) is the complete inversion of Equation (1).

For the vibration of a general laminate, in a single mode, a cross-section is subjected to a system of cyclic forces and moments at circular frequency p: these can be represented as $\begin{bmatrix} N \\ M \end{bmatrix} \sin pt$, where $\begin{bmatrix} N \\ M \end{bmatrix}$ are the peak values.

Under this system, the stresses in the kth layer are

$$\sigma_i^k = \bar{C}_{ij}^k\,[(A'_{j\ell} + zB'_{j\ell})N_\ell + (B'_{j\ell} + z\,D'_{j\ell})M_\ell\,]\ \sin pt \tag{3}$$

(neglecting thermally induced stresses)

$\bar{C}_{ij}^k$ is the transformed stiffness matrix of the kth layer whose fibre axis is at angle θ to the plate axis.

The resultant stress field in the fibre axis can be evaluated using the usual transformation

$$\begin{bmatrix} \sigma_x \\ \sigma_y \\ \sigma_{xy} \end{bmatrix} = \begin{bmatrix} m^2 & n^2 & 2mn \\ n^2 & m^2 & -2mn \\ -mn & mn & m^2-n^2 \end{bmatrix} \begin{bmatrix} \sigma_1 \\ \sigma_2 \\ \sigma_6 \end{bmatrix} \tag{4}$$

where x, y, are the local set of axes, x being the fibre direction and y transverse to the fibre and in the lamina. m and n are $\cos\theta$ and $\sin\theta$ respectively.

Taking an element of the kth layer, of unit width and length and distance z from the midplane, the strain energy associated with σ_x, σ_y and σ_{xy} in this element can be separated into three components, viz.

$$\delta U = \delta U_x + \delta U_y + \delta U_{xy},$$

where the strain energy stored in tension/compression in the fibre axis is

$$\delta U_x = \frac{1}{2}\ \sigma_x(S_{11}\ \sigma_x + S_{12}\ \sigma_y)\ \delta z,$$

the strain energy stored in tension/compression transverse to the fibre axis is

$$\delta U_y = \frac{1}{2}\ \sigma_y\,(S_{12}\ \sigma_x + S_{22}\ \sigma_y)\ \delta z,$$

and the strain energy stored in longitudinal shear is

$$\delta U_{xy} = \frac{1}{2}\ \sigma_{xy}{}^2\ S_{66}\ \delta z.$$

It will be noted that as the stresses are related to the fibre axes, $S_{16} = S_{26} = 0$ and

$$S_{11} = \frac{1}{E_L}, \qquad S_{22} = \frac{1}{E_T},$$

$$S_{66} = \frac{1}{G_{LT}}, \qquad S_{12} = \frac{-\nu_{LT}}{E_L}$$

By carrying out dynamic flexural tests on a 0° unidirectional beam, where only longitudinal tension/compression stresses exist (within the limitations of the simple theory, shear effects can be accommodated), the variation of the longitudinal S.D.C., $\psi_L(\sigma_x)$, with peak homogeneous stress σ_x can be determined. (The method of obtaining the homogeneous damping/stress relationship from the flexural results, which involve a non-homogeneous state of stress, has been described in Reference [7]: the precautions that must be taken to eliminate shear effects are elaborated in Reference [12].) Then the energy dissipated in the element in longitudinal tension/compression (from the definition of S.D.C.) is

$$\delta(\Delta U_x) = \frac{1}{2}\ \psi_L(\sigma_x)\sigma_x(S_{11}\ \sigma_x + S_{12}\ \sigma_y)\ \delta z\,.$$

Similarly, if a 90° specimen is tested in flexure, the variation of the transverse S.D.C., $\psi_T(\sigma_y)$ with peak homogeneous stress σ_y can be determined and the energy dissipated in the element is

$$\delta(\Delta U_y) = \frac{1}{2}\ \psi_T(\sigma_y)\sigma_y(S_{12}\ \sigma_x + S_{22}\ \sigma_y)\ \delta z\ .$$

By testing a 0° specimen in longitudinal shear, the variation of shear S.D.C., $\psi_{LT}(\sigma_{xy})$, with cyclic homogeneous shear stress σ_{xy} can be found. The energy dissipated in the element in this mode is then

$$\delta(\Delta U_{xy}) = \frac{1}{2}\ \psi_{LT}(\sigma_{xy})\ \sigma_{xy}{}^2\ S_{66}\ \delta z\,.$$

Hence the total energy dissipated in the element can be written

$$\delta(\Delta U) = \delta(\Delta U_x) + \delta(\Delta U_y) + \delta(\Delta U_{xy})$$

This expression can now be integrated for the whole plate section to yield the total energy dissipation,

$$\Delta U = \sum_{k=1}^{n} \int_{h_{k-1}}^{h_k} \delta\,(\Delta U) \tag{5}$$

where h_k and h_{k-1} are the ordinates of the upper and lower surfaces of the kth layer and n is the number of layers.

In general, the stress is a linear function of z, while the damping is a non-linear function of stress. For these reasons, the integral must be evaluated numerically.

The maximum strain energy stored by the plate under the peak force and moment resultants, $\{N\}$ and $\{M\}$, is

$$U = \frac{1}{2}\ [N]\ \{\epsilon\} + \frac{1}{2}\ [M]\ \{\kappa\}. \tag{6}$$

The specific damping capacity of the laminate in this mode is then

$$\psi = \frac{\Delta U}{U} = \frac{2\sum_{k=1}^{n} \int_{h_{k-1}}^{h_k} \delta(\Delta U)}{[N]\,\{\epsilon\} + [M]\,\{\kappa\}} \quad . \tag{7}$$

For idealised laboratory testing the investigator aims to reduce the complexity of loads applied to a specimen by controlling the mode of vibration. Hence the general equation simplifies as components of $\{N\}$ and $\{M\}$ are made zero.

Flexural Young's Modulus

For the case where a beam is subjected to a pure bending moment and any resulting twist, owing to the coupling term D_{16}, is allowed to occur the beam is said to be in "free flexure." If the twisting is constrained to zero then the beam is said to be in "pure flexure" [9, 10]. The case of free flexure will be analysed in detail here as it is more relevant to the conditions imposed by the flexural rig.

With reference to the plate constitutive equations, the following conditions are employed in free flexure:

$$\{N\} = 0 \qquad \text{(no in plane forces), and}$$

$$M_2 = M_6 = 0 \qquad \text{(no constraining moments).}$$

As M_1 is the only applied moment,

$$\kappa_1 = D'_{11} M_1$$

and the effective Young's modulus in free flexure is given by

$$E_{FF} = \frac{12}{h^3 D'_{11}} \quad .$$

In pure flexure, $\kappa_6 = 0$, so $M_6 = -D'_{16}/M_1/D'_{66}$, which yields

$$E_{PF} = \frac{12 D'_{66}}{h^3 [D'_{11} D'_{66} - (D'_{16})^2]}$$

Prediction of Damping in Free Flexure

With the same assumptions used for determining the flexural modulus, the stresses acting on an element in the kth layer, at height z from the midplane are given by Equation (3)

$$\sigma_i^k = \bar{C}_{ij}^k (B'_{j1} + zD'_{j1})M_1$$

The total energy dissipated in the plate, ΔU_F, can be determined as in Equation (5).

The strain energy simplifies, since M_1 is the only traction, to

$$U_F = \frac{1}{2} M_1 \kappa_1 = \frac{1}{2} M_1{}^2 D'_{11}$$

Then the specific damping capacity in free flexure is given by

$$\psi_{FF} = \frac{\Delta U_F}{U_F} .$$

For this plate analysis it is not necessary to make any assumptions for the components of the compliance coupling matrix $[B']$ for the manner in which they influence the stress distributions. However, the dynamic coupling effects in vibration may cause significant perturbation of resonant frequencies: these are discussed in a later section.

An Approximate Method to Find the Variation of S.D.C. with Fibre Orientation in Free Flexure

With most of the theoretical damping predictions given in this work, the solution to the equations is usually only feasible using numerical techniques on a digital computer, and it has only been possible to outline the method. However, the following example is amenable to a simple solution if the assumption is made that the damping coefficients ψ_L, ψ_T and ψ_{LT} are independent of stress. This means that there is no need to take into account the variation of stress across the section and any unit cube of the material will represent the whole.

Thus, in free flexure, $M_2 = M_6 = 0$, $\{N\} = 0$, and for a unidirectional lamina $[B] = 0$; σ_1 exists and it can be shown that $\sigma_2 = \sigma_6 = 0$. Transforming σ_1 to the fibre axes through angle θ gives

$$\sigma_x = m^2 \sigma_1 ,$$

$$\sigma_y = n^2 \sigma_1 ,$$

$$\text{and } \sigma_{xy} = -mn\sigma_1 .$$

The energy dissipated in an element of unit volume is

$$\Delta U_F = \frac{1}{2}\sigma_1{}^2 [m^2(S_{11}m^2 + S_{12}n^2)\psi_L + n^2(S_{12}m^2 + S_{22}n^2)\psi_T + m^2 n^2 S_{33} \psi_{LT}] .$$

The strain energy stored in the element is

$$U_F = \frac{1}{2}\ \sigma_1{}^2\ \bar{S}_{11}$$

where $\bar{S}_{11}$ is the compliance in the specimen axis.
Substituting for $[S]$, the S.D.C. in free flexure is then given by

$$\psi_{FF} = \frac{\Delta U_F}{U_F} = \frac{1}{\bar{S}_{11}} \left[\frac{\psi_L}{E_L} m^4 + (\psi_L + \psi_T)\frac{\nu_{LT}}{E_L}\ m^2 n^2 + \frac{\psi_T}{E_T}\ n^4 + \frac{\psi_{LT}}{G_{LT}}\ m^2 n^2\right]$$

where $\bar{S}_{11} = m^4\ S_{11} + m^2 n^2 (2S_{12} + S_{66}) + n^4\ S_{22}$.

Since for CFRP $E_L >> G_{LT}$, $E_L >> E_T$, $\psi_L << \psi_T$ and $\psi_L << \psi_{LT}$, then to a very good approximation, for $5° < \theta < 90°$,

$$\psi_{FF} = \frac{1}{\bar{S}_{11}} \left[n^4\ \frac{\psi_T}{E_T} + m^2 n^2\ \frac{\psi_{LT}}{G_{LT}}\right] \quad . \tag{8}$$

Torsion

In a similar manner to flexure it is possible to define "pure" torsion as the case where a torque is applied to a plate and any resulting bending curvature, κ_1, due to torsion/flexure coupling, D_{16}, is constrained to zero by an additional bending moment M_1. Free torsion occurs when any coupling effects are allowed to take place. The case of pure torsion will be dealt with here as it more closely represents the behaviour of a specimen in the torsion apparatus.

With reference to the plate constitutive Equations (1, 2) putting $\{N\} = 0$ and $M_2 = 0$.

$$\begin{aligned} \kappa_1 &= D'_{11}\, M_1\ + D'_{16}\, M_6 \\ \kappa_6 &= D'_{16}\, M_1\ + D'_{66}\, M_6 \end{aligned} \tag{9}$$

But since $\kappa_1 = 0$,

$$M_1 = \frac{-D'_{16}}{D'_{11}}\ M_6 \quad .$$

Substituting for M_1 in (9) yields

$$M_6 = \frac{D'_{11}}{D'_{66}\, D'_{11} - (D'_{16})^2}\ \kappa_6 \quad .$$

Then for $b>>h$ and making the assumption that the shear stresses in torsion are distributed over the cross-section in a similar way to tensile/compressive stresses in flexure [11], the effective shear modulus in pure torsion is

$$G_{PT} = \frac{12\,D'_{11}}{h^3\,[D'_{66}\,D'_{11} - (D'_{16})^2]}$$

Classical plate theory makes this assumption for the shear stress distribution *ab initio.* The validity of the simplification has been investigated in this report for a multilayer composite whose properties vary from layer to layer across the thickness.

In free torsion, $M_1 = M_2 = 0$ and $N = 0$, so the free torsion modulus is

$$G_{FT} = \frac{12}{h^3\,D'_{66}}$$

Prediction of S.D.C. in Pure Torsion

Using the same assumptions for obtaining the shear modulus in pure torsion, the stresses in the kth layer at a height z from the mid-plane are given by Equation (3).

$$\sigma_i^k = \bar{C}_{ij}^k\,(B'_{j\ell} + z\,D'_{j\ell})M_\ell$$

$$\text{where} \quad M_1 = \frac{-D'_{16}}{D'_{66}}\,M_6 \text{ and } M_2 = 0.$$

The energy dissipated in torsion, ΔU_T, can be evaluated from Equation (5).

The total strain energy of the plate in torsion is

$$U_T = \frac{1}{2}\,M_1\,\kappa_1 + \frac{1}{2}\,M_6\,\kappa_6 \quad .$$

However, $\kappa_1 = 0$, so

$$U_T = \frac{1}{2} M_6^{\,2} \left[\frac{-(D'_{16})^2}{D'_{11}} + D'_{66} \right] ,$$

and the specific damping coefficient in pure torsion is given by

$$\psi_{PT} = \frac{\Delta U_T}{U_T} \quad .$$

RESULTS AND DISCUSSION

Basic Unidirectional Composite Moduli and Damping

In order to make predictions for the variations of dynamic properties with lamination geometry it was necessary to determine the basic composite anisotropic moduli (E_L, E_T, G_{LT}, ν_{LT}) and specific damping coefficients (ψ_L, ψ_T, ψ_{LT}) as functions of stress.

E_L and ψ_L were determined from a flexural test *in vacuo* on specimens with a length/thickness ratio of 110 so as to reduce shear effects to a minimum [12]. The transverse properties, E_T and ψ_T were obtained by a similar test on a 90° specimen, but in this case lower length/thickness ratios could be used. The longitudinal shear modulus and damping were determined from cylindrical gauge length specimens in the torsion pendulum. It was found that for some composites, ψ_{LT} was amplitude dependent, and a least squares fit to the data was made (allowing also for the variation of shear stress with radius). The major Poisson's ratio, ν_{LT}, was determined using a well-proved static strain gauge technique, employing strain gauges of very low cross-sensitivity.

Apart from ψ_{LT} for the HM-S/DX209 system, none of the other dynamic moduli or damping coefficients of the materials investigated here was significantly stress dependent, and it was sufficiently accurate to use the average values given in Tables 1 and 2. The theoretical and experimental damping values of the laminated specimens made from HM-S/DX209 are all given for a known maximum cyclic bending or twisting moment to allow for the amplitude dependence of ψ_{LT}: for consistency, the other results are also quoted at a fixed cyclic moment.

Dynamic Flexure

Variation of Dynamic Flexural Properties with Fibre Orientation + θ. – Unidirectional plates were moulded with HM-S fibre and Shell Epikote DX209 resin, 10 X 10 in by 0.100 in thick and specimens cut at 0°, 5°, 10°, 20°, 30°, 35°, 40°, 45°, 60°, 70° and 90° to the lay of the fibre. Because the flexural apparatus provided no torsional constraint, the specimens vibrated in "free flexure." The D_{ij} matrix of the constitutive equation is fully populated for specimens with the fibre orientation at + θ and, thus, there is bending/torsion coupling. The torsional motion was observed to be small compared with the large transverse motion, resulting in only a small addition to the kinetic energy of the system. Theoretically, no strain energy is stored in torsion because there are no applied torques ($M_6 = 0$). Hearman [9] has shown that, for plywood, the longer is the beam, the more closely will the case of free flexure be approached. Brown [10] has shown more rigorously that there will only be strong interaction of flexure and torsion if the resonant frequencies of modes having the same number of moment loops are close together; higher modes have a very weak influence. Clary [5] comments that the resonant frequencies for boron/epoxy beams could be calculated using the effective modulus and normal beam theory and that these results agreed well with values obtained by finite element analysis.

Table 1. Basic Composite Moduli and Damping Coefficients

Parameter / Fibre/Resin	E_L 10^6 lbf/in²	E_T 10^6 lbf/in²	G_{LT} 10^6 lbf/in²	ν_{LT}	ψ_L %	ψ_T %	ψ_{LT} %
HM-S/ DX209	27.4	0.882	0.395	0.30	0.64	6.9	10 @ 250 lbf/in²
HT-S/DX210	15.01	1.10	0.555	0.30	0.49	5.48	6.75

Table 2. Basic Composite Moduli and Damping Coefficients of Materials Used for the General Plate

Layer	Fibre Type	E_L 10^6 lbf/in²	E_T 10^6 lbf/in²	G_{LT} 10^6 lbf/in²	ψ_L %	ψ_T %	ψ_{LT} %
1, 10	GR-A-S	14.8	1.08	0.463	0.38	4.57	12.8
2, 9	MOD-III-S	14.3	1.08	0.567	0.33	4.61	9.4
3, 8	GR-HM-S	27.2	0.829	0.415	0.38	4.92	9.5
4, 7	MOD-I-NS	25.8	0.948	0.613	0.50	5.0	9.2
5, 6	GR-A-NS	15.3	1.07	0.355	0.40	4.9	9.5

where GR-A-S is Grafil high strain treated fibre
MOD-III-S is Modmor high strain treated fibre
GR-HM-S is Grafil high modulus treated fibre
MOD-I-NS is Modmor high modulus untreated fibre
GR-A-NS is Grafil high strain untreated fibre

The matrix was Shell Epikote DX 210 in all cases.

As the fibre orientation deviated from 0° the line of the nodes became angled to the specimen axis, reaching a value of approximately 45° at a fibre angle of 10°. The nodal lines reverted to 90° for fibre orientations greater than about 45°. This behaviour was also reported by Clary and is associated with the torsional motion.

Plate theory predicted a minimum modulus at approximately 70° of 0.875×10^6 lbf/in² compared with the 90° value of 0.882×10^6 lbf/in². The experimental values (Figure 1) are very close to the theoretical ones in this region but do not go to a minimum, no doubt because there is only a fine difference in the values.

The experimental and theoretical results for the variation of damping with fibre orientation are also shown in Figure 1. There was extremely good correlation of

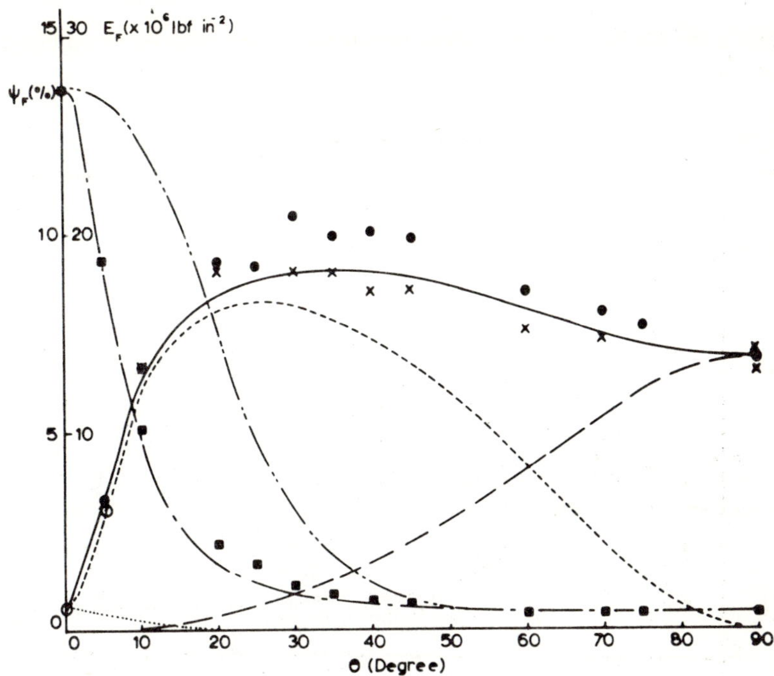

Figure 1. Variation of flexural Young's modulus E_F and damping ψ_F with fibre orientation θ for HM-S carbon fibre in DX209 epoxy resin, $v = 0.5$. ●, × ψ_F (plates 1 and 2), ○ ψ_F in vacuo; ■ E_F, average for plates 1 and 2 (values virtually coincident). Theoretical predictions; ——— – – ———E_{PF} (pure flexure); ——— – ———E_{FF} (free flexure); ———ψ_{FF} total; – – – –ψ_{xy}; —— —— ——ψ_y; . . . . ψ_x. Maximum cyclic bending moment: 2 lbf in.

theory with experiment for the whole range of θ. The S.D.C. peaks near 35° fibre angle and is associated with the C_{16} term of the transformed stiffness matrix which relates shear stress σ_6 to axial strain ϵ_1. The value of C_{16} is a maximum at 30° and, as the shear damping coefficients are large and the shear modulus is quite small, this leads to large energy dissipation in shear. The contributions of the three damping mechanisms are also displayed and it will be noted that the damping associated with stresses in the fibre direction ψ_x rapidly becomes insignificant at fibre angles > 10°.

As a comparison, the orientation series was repeated with GFRP, using E-glass fibres in Shell Epikote resin DX210. The damping showed a very similar relationship to fibre orientation as that of CFRP with good correlation between theoretical and experimental results.

Torsion tests on rectangular specimens (described later) showed a dependence of damping and modulus on specimen width. To investigate this effect in the flexure of + θ beams, two series of specimens were made at 0.5 and 1 in widths using HT-S fibre and Shell Epikote DX210 resin. The results are given in Figure 2 and good

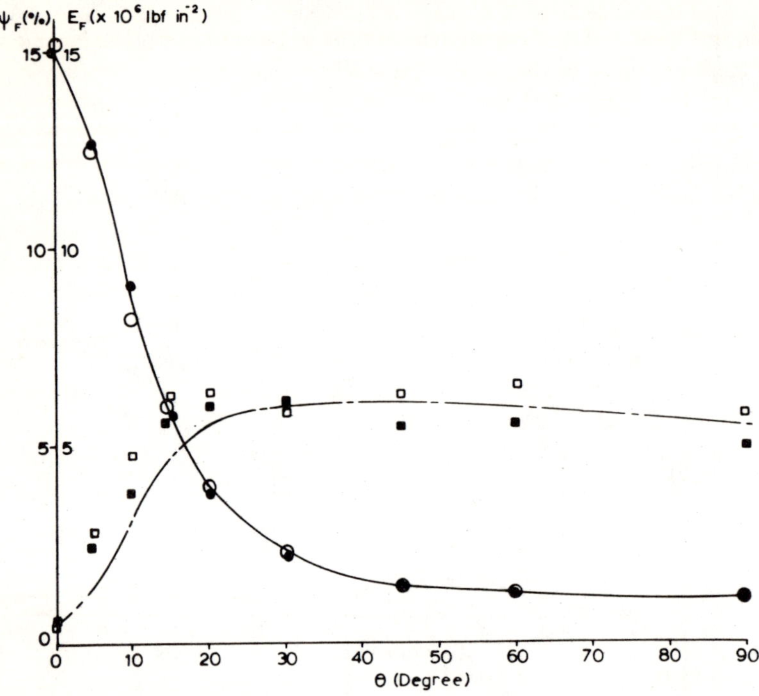

Figure 2. Variation of flexural Young's modulus E_F and damping ψ_F with fibre orientation θ for HT-S carbon fibre in DX210 epoxy resin, v = 0.5. ○ E_F, □ ψ_F, 1 in wide specimens; • E_F, ▪ ψ_F, 1/2 in. wide specimens. Prediction for free flexure; ——— E_{FF}, — – — ψ_{FF}. Maximum cyclic bending moment: 2 lbf in.

agreement was obtained with the plate theory prediction. No significant effect of beam width was noticed.

Angle Plies (± θ).– Angle plies are laminates with the orientations of layers at alternately plus and minus θ. A series of plates, 10 × 3 in by 0.100 in, were moulded from 10 layers of HM-S pre-preg (DX209 resin) with the following values of ply angle, ± 20°, ± 30°, ± 45°, and ± 60°. Specimens 0.5 in wide were cut from these plates.

For an even number of layers, the D_{16} and D_{26} terms in the D_{ij} matrix of the plate constitutive equation are zero and, therefore, there is no bending/twisting coupling. However, since the B_{ij} matrix, relating moments to in-plane strains, is partially populated ($B_{16}, B_{26} \neq 0$) there will be mid-plane strains during flexure. However, the effect of the coupling term B_{16}, B_{26} weakens very rapidly as the ply angle deviates from 45° and the terms are small for a reasonable number of layers [8].

The results for the variation of damping and modulus with ply angle ± θ are given in Figure 3. The damping tended to increase with cyclic bending moment for the angle ply range so the values of ψ in Figure 3 are given at a bending moment of 2 lbf in. The amplitude dependence could not entirely be explained by the increase in S.D.C. due to ψ_{LT} increasing with σ_{xy}. The non-linearity must be associated with a more complex state of stress than predicted by plate theory. It was observed that the node lines curved across the beam indicating that this was so.

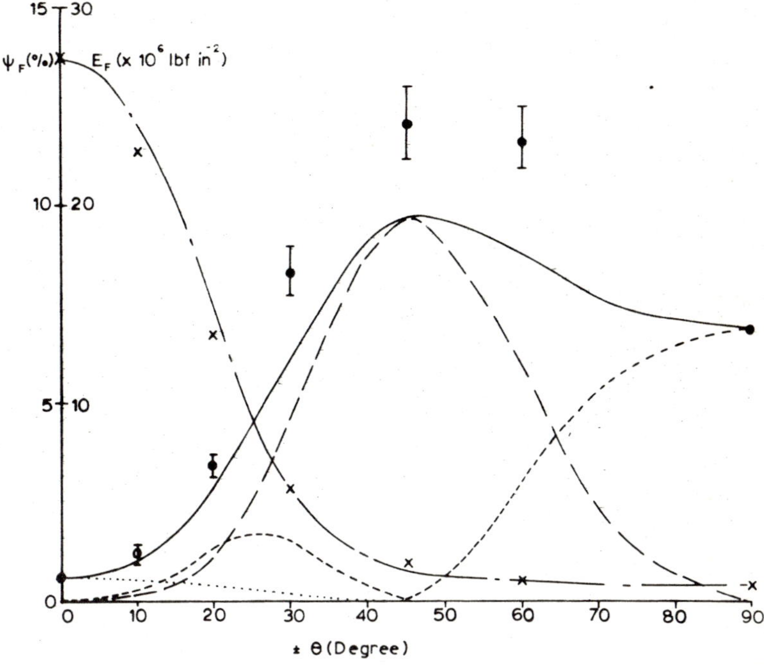

Figure 3. Variation of flexural Young's modulus E_F and damping ψ_F with ply angle ± θ for HM-S carbon fibre in DX209 epoxy resin, ν = 0.5. Experimental points; × *E_F;* • *ψ_F;* ○ *ψ_F* ***in vacuo.*** *Free flexure prediction: ——— – ———E_{FF}; ——— total ψ_{FF}; — — —ψ_{xy}; – – – –ψ_y; . . . . ψ_x. Maximum cyclic bending moment: 2 lbf in.*

There was good correlation of the theoretical effective Young's modulus with the experimental values. Plate theory indicates a minimum modulus of 0.874 × 10^6 lbf/in^2 near a ply angle of 70° but, practically, the difference would not be significant compared with the 90° value of 0.882 × 10^6 lbf/in^2. Comparing the variation of modulus for the fibre orientation series (Figure 1) it is noticeable that the modulus of angle plies does not decrease as rapidly for small values of θ because the constraint of each layer upon its neighbour essentially eliminates twist and makes each layer deflect in "pure flexure."

The S.D.C. reflects the increased constraint by remaining at a low value for $0 < \theta < 15°$. The contributions due to the three damping mechanisms (ψ_x, ψ_y, ψ_{xy}) have been summed for the plate at each ply angle, and it can be seen that where the total S.D.C. peaks this is almost entirely due to shear. The axial contribution, ψ_x, is significant for $0 < \theta < 30°$. The experimental values are all greater than the theoretical prediction but follow the theoretical trends closely. The fact that the damping for these angle plies was amplitude dependent could explain the discrepancy.

Further specimens were made from HT-S fibre in DX210 resin in two widths, 1/2 in and 1 in, to investigate the effect of shear stresses and the assumptions made by plate theory. The flexural results are presented in Figure 4 where it can be seen that the width has little effect on either the modulus or the damping. The greatest discrepancy in damping occurs at 45°, this being associated with a maximum of shear coupling.

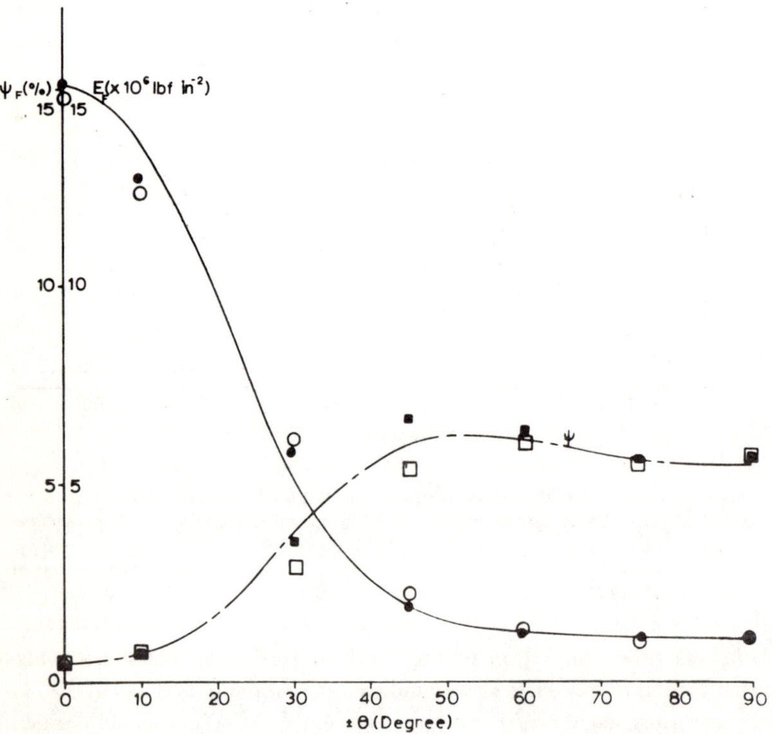

Figure 4. Variation of flexural Young's modulus E_F and damping ψ_F with ply angle $\pm\theta$ for HT-S carbon fibre in DX210 epoxy resin, $\nu = 0.5$. o E_F, □ ψ_F, 1 in wide specimens; • E_F, ▪ ψ_F, 1/2 in. wide specimens. Prediction for free flexure, ——— E_{FF}, – – – – ψ_{FF}. Maximum cyclic bending moment: 1 lbf in.

Cross Ply (0°/90°).– Cross plies are defined as having n layers, where all the odd layers are at 0° and of the same thickness and all the even layers are at 90°, and of the same thickness, which may be different from that of the odd layers. A cross ply ratio is defined as follows:

$$m = \frac{\text{total thickness of odd layers } (0^\circ)}{\text{total thickness of even layers } (90^\circ)} .$$

A series of 3- and 5-layered composites were moulded having the following lay-ups:

'1' indicates a 0° and '—' a 90° layer.

3-layered	1 – – – – – – – – – 1	$m = 0.25$
	11 – – – – – – – 11	$m = 0.666$
$n = 3$	111 – – – – – 111	$m = 1.5$
	1111 – – 1111	$m = 4$.
5-layered	1 – – – 1 – – – 1	$m = 0.5$
	11 – – 11 – – 11	$m = 1.5$
$n = 5$	111 – 111 – 111	$m = 4.5$

A nine-layered specimen was also fabricated having the following lay-up:

$n = 9$ 1 – 1 – 1 – 1 – 1 $m = 1.25$

The modulus and damping in flexure are dominated by the properties of the outer layers. For these cross plies, the outer layers are at 0° and it was found that in many cases the cross ply moduli and damping values did not differ greatly from those of a 0° unidirectional specimen. For this reason it was necessary to carry out all the flexural damping measurements *in vacuo*.

It was only at low values of the cross ply ratio that there was a reduction of the flexural modulus as the outer layers became progressively thinner (Figure 5). As n increases the modulus decreases rapidly at first but after $n = 9$ there is little change, and in the limit as $n \to \infty$, $m \to 1$ and the flexural modulus

$$E_F \to \frac{E_L + E_T}{2} \simeq \frac{E_L}{2} .$$

For $n = 3$, E_{FF} may be predicted by sandwich beam theory [13] and, as can be

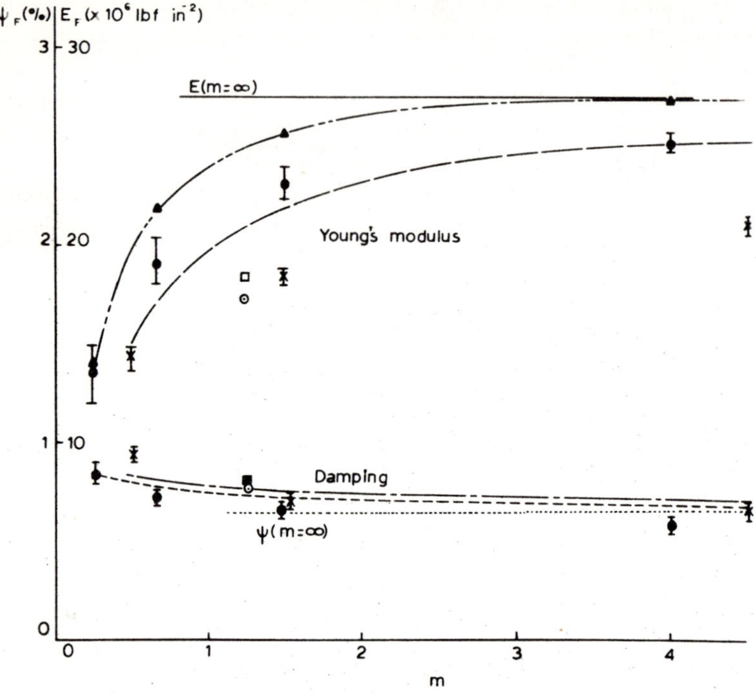

Figure 5. *Variation of flexural Young's modulus E_F and damping ψ_F with cross-ply ratio m for HM-S carbon fibre in DX209 epoxy resin, v = 0.5. Experimental values (E_F and ψ_F), • 3-ply, × 5-ply, ⊙ 9-ply. Theoretical predictions, ——— – – ——E_F, 3-ply; —— ——E_F, 5-ply; □ E_F, 9-ply; ——— – ———ψ_F, 3-ply; – – – –ψ_F, 5-ply; ■ ψ_F, 9-ply; ▲, 3-ply, sandwich beam theory. Maximum cyclic bending moment: 5 lbf in.*

seen from Figure 5, the moduli are almost exactly the same as those given by anisotropic plate analysis.

The theoretical prediction of damping follows closely the experimental values and increases with n and decreases with m. The S.D.C. is very low for cross plies and would not be sufficient for structural damping.

General Plate.– Several plates with lay-ups of the type used by Shultz and Tsai [4] were fabricated. These gave excellent correlation between experimentally and theoretically determined moduli and damping coefficients.

To investigate fully the effectiveness of the analytical techniques, a 10 × 10 in plate 0.100 in thick was moulded from several batches of "pre-preg" having the same resin system, Shell Epikote DX210, but different fibre types. The orientations of the 10 layers were as follows:

$$0°, 0°, 30°, 30°, 45°, 45°, 30°, 30°, 0°, 0°$$

The composite was symmetric about its mid-plane and the order of fibre types with their experimentally determined composite properties are given in Table 2. A value for ν_{LT} of 0.3 was used in each case.

The plate had essentially 5 layers (0°, 30°, 45°, 30°, 0°) and its properties, taken at $+\theta$ and $-\theta$, were designed to be asymmetrical. As the number of layers in a plate are reduced, the anisotropy is increased.

The experimental and theoretical results for the general plate are given in Figure 6. There was good correlation of theory with experiment for modulus and the

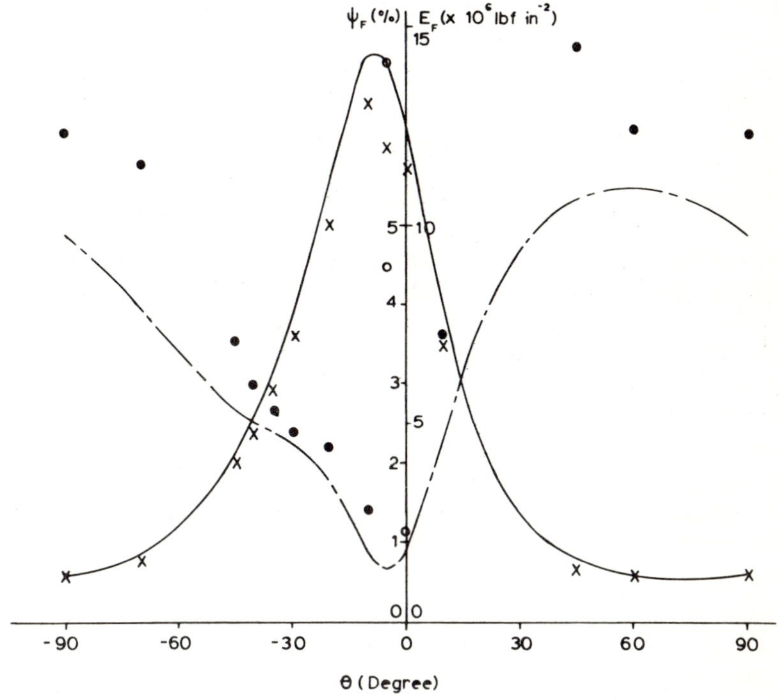

Figure 6. Variation of flexural Young's modulus E_F and damping ψ_F with angle θ for the general plate (for composition, see text). × E_F; • ψ_F *in air;* ○ ψ_F ***in vacuo****; free flexure prediction:* ——— E_{FF}; — — — ψ_{FF}. *Maximum cyclic bending moment: 2 lbf in.*

correlation for damping was reasonable except at 5° where there was a very large error. The 5° specimen was subsequently examined closely and found to have an interlaminar crack running for about 1 in. along its length. The damping of this specimen was 6 times higher than the theory predicted, whereas the modulus was only 14% down. This demonstrates the sensitivity of damping to damage and its potential in quality control. In general the damping was slightly higher than predicted which can be associated with combinations of bending/twisting coupling. The D_{ij} matrix is fully populated and all the specimens exhibited this coupling effect. However, it is evident that the prediction technique is quite adequate for normal purposes in estimating the damping and modulus values of laminated glass and carbon F.R.P.

Dynamic Torsion

Variation of Dynamic Torsional Properties with Fibre Orientation ($+\theta$).– Rectangular specimens from the flexural orientation series (HM-S fibre, DX209 resin) were tested in the torsion pendulum and the effective shear modulus and S.D.C. found (Figures 7 and 8). The *effective* shear modulus was calculated from the measured torsional stiffness of the specimen, K_s, using the expression [11]

$$G = \frac{K_s \, \ell}{c \, h^3 \, b}$$

where c is a geometrical factor dependent on the ratio b/h. Typically, for $b/h = 5, c = 0.2913$. This equation gives the shear modulus of an equivalent homogeneous specimen and is useful as a means of comparing the torsional properties of various lay-ups. The modulus given by plate theory will be a function of the anisotropy of the laminate but not a function of the b/h ratio. It is assumed that shear stresses follow the same distribution across the section as direct stresses and are constant across the width. However, in torsion, these stresses must form loops and if the specimen is not wide enough for the end effects to be insignificant then spurious results will occur. This system was tested with a flat Duralumin specimen 0.500 × 0.100 in. in section, with end pieces glued at a gauge length of 4 in. to increase the section to 0.500 × 0.500 in. The value of shear modulus obtained from this specimen was 3.91×10^6 lbf/in^2 which can be compared to the value of 3.94×10^6 lbf/in^2 obtained from a cylindrical specimen made from the same material. This demonstrated that the technique was accurate for isotropic materials.

As was mentioned in the flexural testing of the fibre orientation series, these specimens exhibit strong coupling between torsional and flexural motion. The torsional apparatus has a dummy specimen to restrain flexural motion so in theory the specimens are constrained to vibrate in "pure torsion." It was observed that specimens in the range 15–30° fibre orientation did not vibrate purely but that there was a certain amount of flexure present. The twisting/bending coupling is very strong in that range and even the dummy specimen began to flex. This tendency to vibrate in flexure brings the system towards the "free torsion" case which results in lower effective shear moduli (Figure 7). For the rest of the orientation range, flexure was not significant and good agreement was obtained between the theoretical and experimental results. In Figure 8 are displayed the experimental and theoretical relationships of torsional S.D.C. to fibre orientation. There was

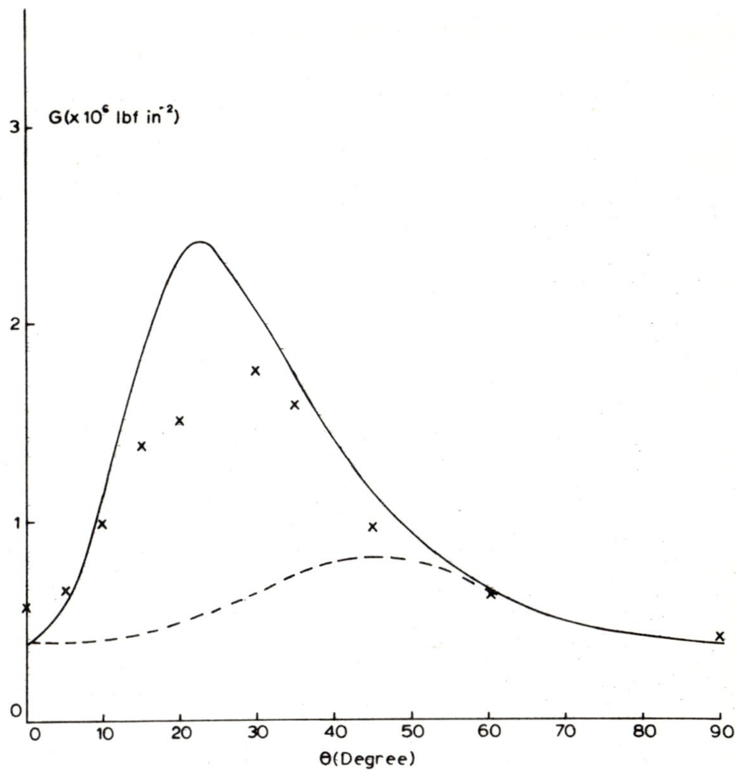

Figure 7. Variation of torsional modulus G with fibre orientation θ for HM-S carbon fibre in DX209 epoxy resin, ν = 0.5. Experimental values, ×. Predictions for G; ——— pure torsion;— — — free torsion.

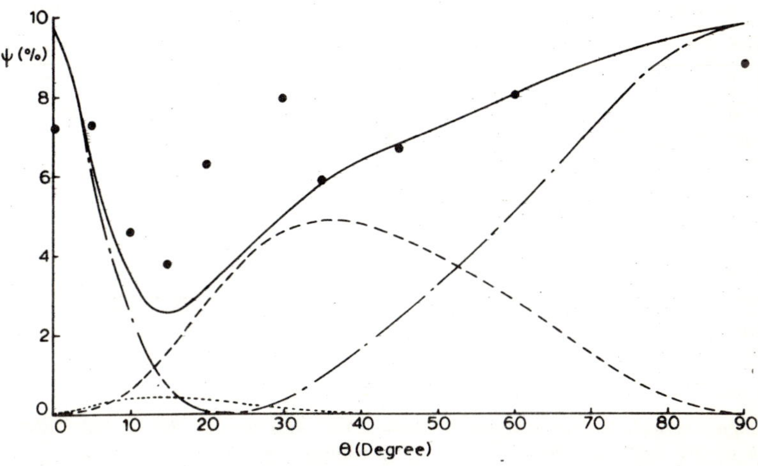

Figure 8. Variation of torsional damping ψ with fibre orientation θ for HM-S carbon fibre in DX209 epoxy resin, ν = 0.5. • experimental values. Prediction for pure torsion: ——— total damping; – – – – – – –ψ_{xy}, — — —ψ_y; ···········–ψ_x. Maximum cyclic torque: 1 lbf in.

good correlation between these values at most angles, except where the twisting/bending coupling is at a maximum (≃ 30°). The additional flexural motion dissipated further energy per cycle which resulted in higher damping. It is interesting to note that, in theory, the energy dissipated in the fibre orientation range 15–45° is predominantly due to transverse stresses and shear damping is a minimum near 25°. The axial contribution to the damping is small or negligible for the whole range of fibre orientation.

Anomalous results for the shear modulus of ± θ angle plies (see following section) led to an investigation of the effect of b/h ratio on the effective shear modulus and damping of rectangular sections: the results are given in Figure 9.

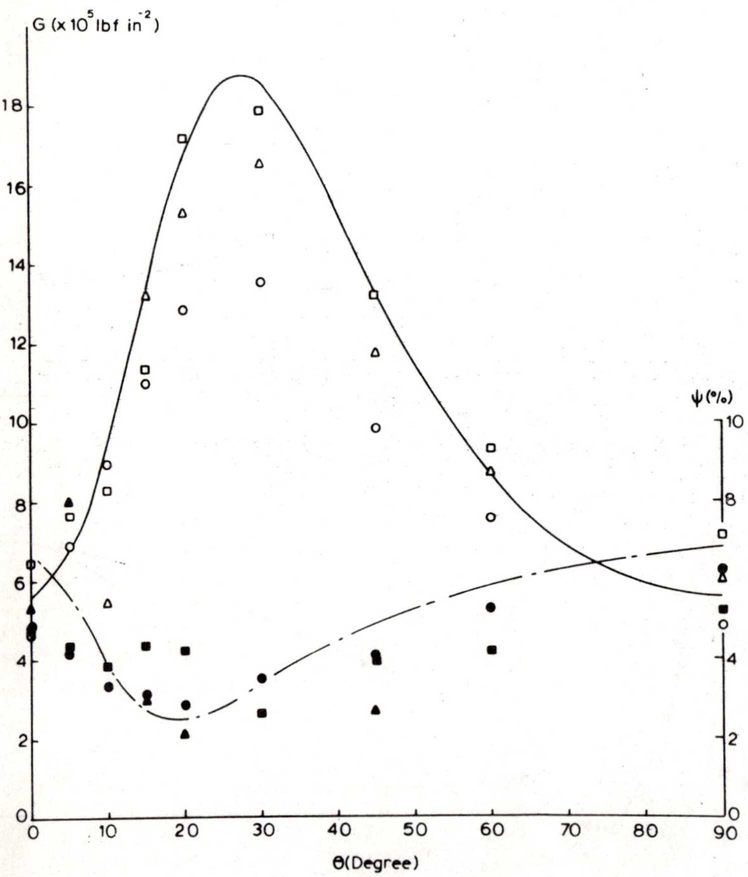

Figure 9. Variation of torsional modulus G and damping ψ with fibre orientation θ for HT-S carbon fibre in DX210 epoxy resin, $v = 0.5$. □ G, ■ ψ, $b/h = 17.5$; △ G, ▲ ψ, $b/h = 10$; ○ G, ● ψ, $b/h = 5$. Pure torsion prediction: ——— G_{PT}. — — — ψ_{PT}. Maximum cyclic torque: 1.2 lbf in.

Where the torsion/bending coupling is large, the effect of an increase in b/h is to increase the effective modulus and bring it up to the theoretical value. At 0° and 90° there is a discrepancy due to the anisotropy of the material. The torsion of 90° specimens is more complex than that of 0° specimens as the 2-dimensional plate analysis does not take into account the contribution of the transverse shear modulus (G_{TT}), and only gives the shear modulus of 90° specimens as G_{LT}. However, for volume fractions of 50%, G_{TT} is approximately the same value as G_{LT} as both are largely dependent on the modulus of the resin. There is an increase in effective G with increase in b/h ratio which would indicate that $G_{LT} > G_{TT}$ since the assumptions of the torsion of rectangular sections are more nearly followed.

Torsion of Angle Plies ($\pm\ \theta$). – In a similar manner to the fibre orientation torsion series, specimens were prepared from angle plies (HM-S fibre, DX209 epoxy matrix). These were then tested in dynamic torsion and the results of the effective shear modulus and damping are given in Figure 10. It was immediately obvious that the modulus values were nearly a factor of two too low, whereas the damping values were many times too high. After careful checking of the results and experimental techniques a possible explanation offered was that the discrepancy was a

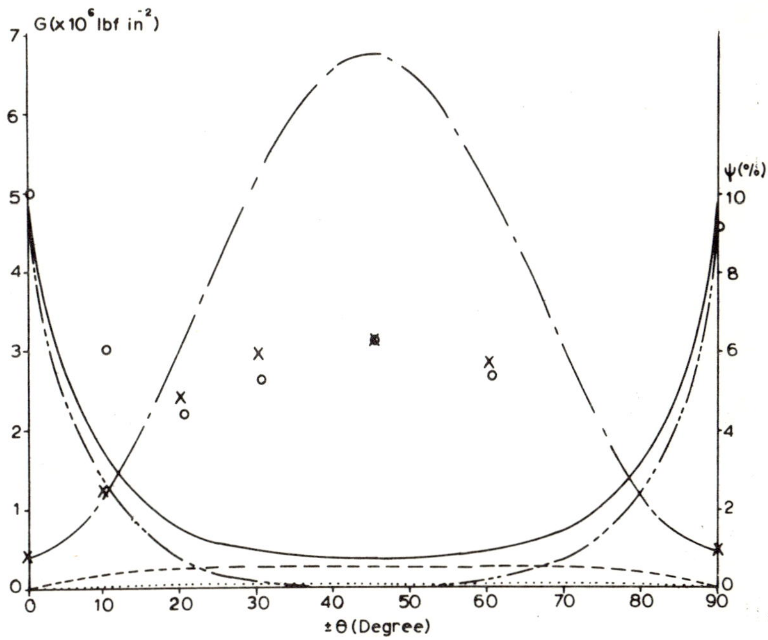

***Figure 10.** Variation of torsional modulus G and damping ψ with ply angle ± θ for HM-S carbon fibre in DX209 epoxy resin, v = 0.5. Experimental points: ×, ψ; ○, G. Pure torsion predictions: —— – —— G_{PT}; ——— total ψ_{PT}; —— – – —— ψ_{xy}, - - - - - - ψ_y; ·············· ψ_x. Maximum cyclic torque: 1 lbf in.*

direct result of the specimens not fulfilling the assumptions of plate theory. To check this hypothesis, a further series of angle plies was fabricated from HT-S fibre in DX210 resin, with b/h ratios ranging from 5 to 17.5. The results of torsion tests on these specimens are shown in Figure 11. It was evident that the ratio b/h was extremely critical in the determination of the torsional properties and accounted for the discrepancy between theoretical and experimental results for the HM-S 0.5 in specimens. Large changes in damping and effective modulus occurred with increase in b/h ratio and at $b/h = 17.5$ there was reasonable correspondence between theory and experiment. The results of this section made the previous work in flexure and torsion suspect which is why it was necessary to repeat much of the

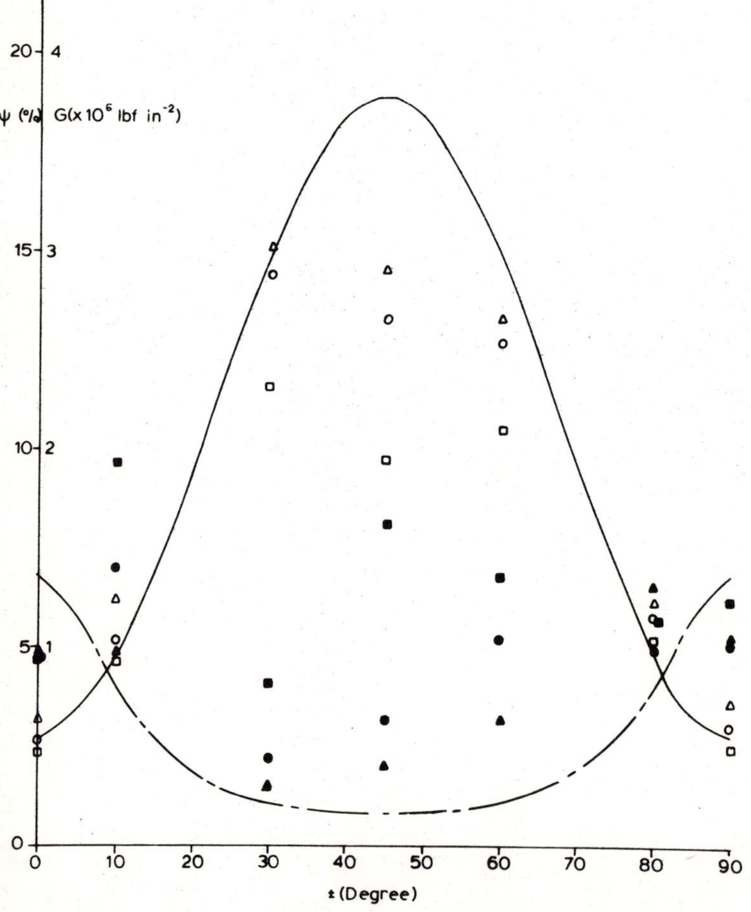

Figure 11. Variation of shear modulus G and damping ψ with ply angle ± θ for HT-S carbon fibre in DX210 epoxy resin, ν = 0.5. △ *G,* ▲ *ψ, $b/h = 17.5$;* ○ *G,* ● *ψ, $b/h = 10$;* □ *G,* ■ *ψ, $b/h = 5$. Predictions for pure torsion: ——— G_{PT}, — — — ψ_{PT}. Maximum cyclic torque: 1.3 lbf in.*

work with varying b/h ratios to determine the significance of this parameter. It is estimated that provided $b/h \geqslant 5$, in flexure its effect can be neglected, but in torsion b/h must be greater than 17.5 to approach the theoretical prediction. In Figure 11 the shear modulus of the 45° specimen is lower than that of the 30° specimen whereas plate theory predicts a maximum at 45°, and it could be that even larger b/h ratios are needed in this case. The concept of a larger b/h ratio with anisotropic laminated materials in torsion, is analogous to the larger aspect ratio ℓ/h required for the testing of these materials in flexure [12].

Cross-Plies.– In torsion, plate theory predicts that all cross-plies should have the same effective shear modulus and damping as a 0° layer. This is, of course, a direct result of the assumptions made in the 2-dimensional analysis that the properties in shear of 0° and 90° specimens are identical. In Figure 12 are presented the torsion results for 3, 5 and 9 layered cross plies. There was some scatter on the damping values, but these were grouped around the predicted S.D.C. All the shear modulus values were higher than predicted which is associated with the testing of rectangular 0° beams. However, all the modulus values were within 10% of a nominal average of 0.5×10^5 lbf/in^2 bearing out qualitatively, if not quantitatively, the results of plate theory.

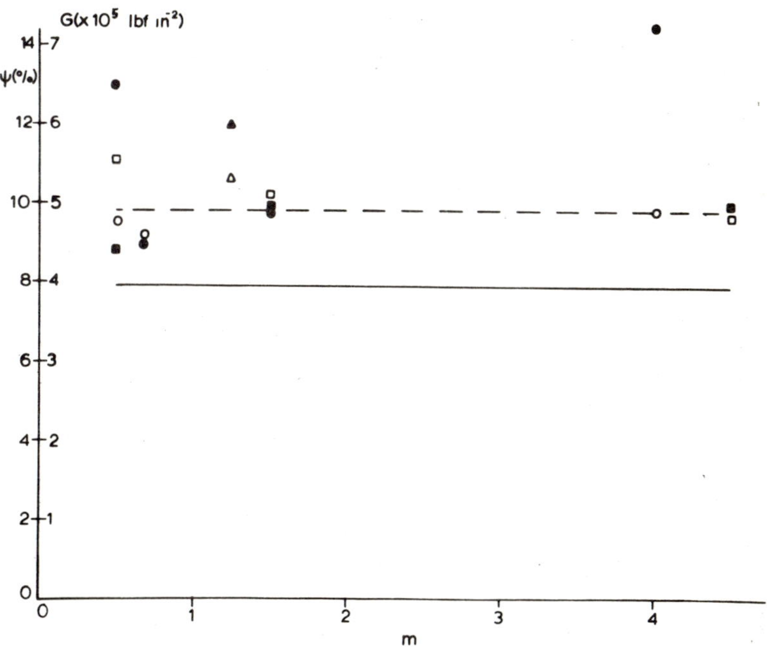

Figure 12. Variation of torsional modulus G and damping ψ with cross ply ratio m for HM-S carbon fibre in DX209 epoxy resin, ν = 0.5. ○ G, ● ψ, 3-ply; □ G, ■ ψ, 5-ply; △ G, ▲ ψ, 9-ply. ——— G for solid 0° specimen; — — — ψ for solid 0° specimen. Maximum cyclic torque: 1 lbf in.

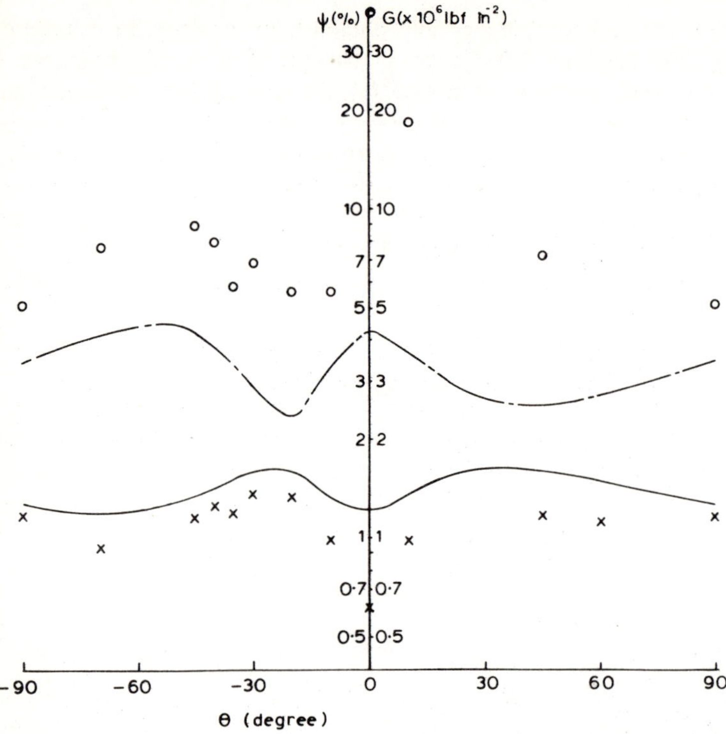

Figure 13. Variation of torsional modulus G and damping ψ with angle θ for the general plate (for composition, see text). × *G;* ○ *ψ: prediction for pure torsion:* ——— G_{PT}, ——–—— ψ_{PT}. *Maximum cyclic torque: 1 lbf in.*

General Plate – The flexural specimens of the general plate were tested in dynamic torsion and the results are presented in Figure 13. There are quite significant discrepancies in the predicted and experimental values but the trends were generally followed. The stress field induced in torsion becomes increasingly complex as the anisotropy of the plate increases and the precepts of plate theory are not upheld. The nature of the stress path near the free end across lamination boundaries is not fully understood and is considered to give rise to these spurious results. On the basis of the results for the torsion of angle plies ($\pm\ \theta$), it is reasonable to attribute a large proportion of the discrepancy between the predicted and measured values of modulus and damping to the b/h ratio being too small.

CONCLUSIONS

The combination of accurate experimental data and a new technique for predicting the dynamic properties of laminates has made it possible to investigate in detail the anisotropic properties of these materials in dynamic flexure and torsion. The variation of the properties with fibre geometry was examined and excellent correlation with theory was generally obtained. By dividing the contributions to damping into shear and direct stresses, the sources of energy dissipation can be examined in detail. It is evident that shear is the predominant factor in a lamination geometry that gives high damping. Transverse direct stresses can sometimes give high energy dissipation whilst longitudinal direct stresses result in minimal, if not negligible, damping for structural purposes.

ACKNOWLEDGMENTS

The authors wish to thank the Science Research Council for financial support for this work.

REFERENCES

1. R. D. Adams, "The damping characteristics of certain steels, cast irons and other metals," *J. Sound and Vibration*, Vol. *23* (1972), p. 199.
2. J. C. Snowdon, "Representation of the mechanical damping possessed by rubberlike materials and structures," *J. Acoustical Soc. America*, Vol. *35* (1963), p. 821.
3. R. D. Adams and D. G. C. Bacon, in preparation.
4. A. B. Schultz and S. W. Tsai, "Measurements of complex dynamic moduli for laminated fiber-reinforced composites," *J. Composite Matls.*, Vol. *3* (1969), p. 434.
5. R. R. Clary, "Vibration characteristics of unidirectional filamentary composite material panels," Composite Matls: Testing and Design (Second Conference), ASTM STP 497 (1972), p. 415.
6. R. D. Adams, M. A. O. Fox, R. J. L. Flood, R. J. Friend and R. L. Hewitt, "The dynamic properties of unidirectional carbon and glass fiber-reinforced plastics in torsion and flexure," *J. Composite Matls.*, Vol. *3* (1969), p. 594.
7. R. D. Adams and D. G. C. Bacon, "Measurement of the flexural damping capacity and dynamic Young's modulus of metals and reinforced plastics," *J. Phys. D: Appl. Phys.*, Vol. *6* (1973), p. 27.
8. S. W. Tsai, "Structural behaviour of composite materials," N.A.S.A. CR 71, 1964.
9. R. F. S. Hearmon, "The significance of coupling between shear and extension in the elastic behaviour of wood and plywood," *Proc. Phys. Soc.*, Vol. *55* (1943), p. 67.
10. W. F. Brown, Jr., "Interpretation of torsional frequencies of crystal specimens," *Phys. Rev.*, Vol. *58* (1940), p. 998.
11. S. P. Timoshenko, *Theory of plates and shells*, D. Van Nostrand, 1934.
12. R. D. Adams and D. G. C. Bacon, "The dynamic properties of unidirectional fibre-reinforced composites in flexure and torsion," *J. Composite Matls.*, Vol. 7 (1973), p. 56.
13. H. G. Allen, "Measurement of the shear stiffness of sandwich beams," *Trans. J. Plastics Inst.*, (1967), p. 359.

Improved Mechanical Properties of Composites Reinforced with Neutron-Irradiated Carbon Fibers

E. L. McKague, Jr., R. E. Bullock and J. W. Head

General Dynamics, Convair Aerospace Division
Fort Worth, Texas 76101

(Received March 12, 1973)

ABSTRACT

Mechanical property improvements of neutron-irradiated carbon fibers are reviewed, and an experiment is described which verifies that mechanical properties of carbon-epoxy laminates are improved when reinforced with such fibers. Increased flexural and shear strengths of composites are confirmed to result from the use of irradiated fibers. The future potential of radiation treatment of carbon fibers to improve mechanical properties of composites is discussed; it is concluded that the process offers commercial promise.

INTRODUCTION

The purpose of this paper is to report the enhanced structural properties resulting from reinforcing epoxy-matrix composites with neutron-irradiated carbon fibers. Irradiation-induced increases in tensile strengths and elastic moduli of carbon fibers themselves have previously been reported [1–4], and results herein show that these property improvements largely translate into composites reinforced with such fibers. Moreover, as an added benefit, fiber-to-matrix bonding is improved in irradiated-fiber composites; this is of particular importance because chemical surface treatments of fibers to improve their bondability with an epoxy matrix often result in strength losses [5], whereas irradiation treatment increases both the strength and bondability of carbon fibers.

PROPERTIES OF IRRADIATED CARBON FIBERS

Selected room-temperature mechanical properties of irradiated PAN-based HT-S (Type II) carbon fibers are listed in Table 1 in order of increasing properties, and irradiation temperature is obviously of importance since the ordering does not correspond to increasing neutron exposure. Increases in tensile strengths range up to 40%, while modulus increases are 15% or less. Thus, strain-to-fracture (ϵ) and work-to-fracture (*WF*) increase along with the strength and modulus of the fiber, ϵ by as much as 20% and *WF* by as much as 70%, and this is very important. Designers of structural composites in the aircraft industry usually prefer as stiff a reinforcement fiber as possible, but they require some minimum strain before fiber failure occurs. In particular, fibers are usually required to have a minimum failure strain of at least 6,000 μin./in. [6], and fibers with failure strains above 10,000 μin./in. are strongly preferred. Of the carbon fibers currently available, the requirement of a 1% failure strain dictates a modulus of less than 35 $\times$ 10^6 psi (35 Msi), because the failure strain and the strength fall off rapidly for higher moduli [6].

Table 1. Property Improvements of Irradiated HT-S Carbon Fibers

Test No. & [Ref.]	*Fiber Condition*	*Single-Fiber Mechanical Properties*				*Irradiation Conditions***
		σ*(ksi)*	*E(Msi)*	ϵ* *(μin/in)*	*WF* (in-lb/in³)*	
1	as received	387	37.1	10,400	2,020	~ 7 × 10^{17} n/cm²
[1]	irradiated	434	41.2	10,500	2,280	in air
	% increase	12.1	11.0	1.0	12.9	at ~ 120°F
2	as received	382	39.4	9,700	1,850	8.5 × 10^{17} n/cm²
[2,3]	irradiated	445	42.7	10,400	2,320	in air
	% increase	16.5	8.4	7.2	25.4	at ~ 175°F
3	as received	415	31.2	13,300	2,760	1.8 × 10^{20} n/cm²
[4]	irradiated	495	34.4	14,400	3,560	in vacuum
	% increase	19.3	9.3	8.3	29.0	at ~ 1000°F
4	as received	382	39.4	9,700	1,850	3.0 × 10^{18} n/cm²
[2,3]	irradiated	494	45.1	10,900	2,700	in LN_2
	% increase	29.4	14.4	12.4	46.0	at −320°F
5	as received	415	31.2	13,300	2,760	3.3 × 10^{20} n/cm²
[4]	irradiated	582	36.0	16,150	4,700	in vacuum
	% increase	40.2	15.4	21.4	70.4	at ~ 1000°F

*Strain-to-fracture (ϵ) and work-to-fracture (*WF*) were not given in the original references [1–4]; however, because of the perfectly linear stress-strain behavior of the fibers, these quantities could be calculated from reported values of strengths (σ) and moduli (E) as $\epsilon = \sigma/E$ and $WF = \sigma^2/2E$.

**All fast-neutron fluences (n/cm^2) are reported only for neutrons with energies above 1 MeV.

Obviously, if one could obtain a higher modulus fiber (45 Msi, say) of sufficient strength (> 450 ksi for the modulus in question) to give the required 1% failure strain, this would be a significant fiber advancement. This degree of improvement has been accomplished by irradiating as-received HT-S fibers in a nuclear reactor (Test 4 of Table 1). Moreover, the irradiation conditions of Table 1 do not seem to be optimized by any means, and there is reason to believe that the modulus can be raised considerably higher and the 1% strain still retained if the proper starting fiber (higher modulus) and irradiation exposure and temperature (higher doses at lower temperatures) are properly selected.

Finally, of equal importance, the work per unit volume required to fracture fibers (*WF*) increases very sharply with radiation exposure, with percentage increases in *WF* being roughly 3.5 times as great as percentage strain increases. Consequently, since fibers are the principal energy absorbers in epoxy-matrix composites [7, 8], an irradiated-fiber composite should be considerably more impact resistant than a composite reinforced with as-received fibers. For a typical unidirectional carbon-epoxy composite in which about 70% of the impact energy is absorbed by the fibers [7], the increased impact resistance obtained by using irradiated fibers should range up to 50% (fibers from Test 5 of Table 1), and again it should be possible to improve on this through more carefully controlled irradiation conditions.

Other data reported in References [1–3] and [9], which are omitted in Table 1, indicate that radiation effects on PAN-based HM-S (Type I) carbon fibers are considerably less in magnitude than are those for HT-S fibers. This was probably a poor choice of a high-modulus-type fiber for radiation strengthening, however, for cool-down defects appear to be particularly severe for these higher heat-treated, circular-shaped fibers [10]. Strengths of such fibers are apparently controlled by these cool-down flaws, which do not respond to radiation treatment, and this appears to account for the different type of behavior of irradiated HT-S and HM-S fibers [11].

It might well be possible to strengthen other types of high-modulus fibers in which the fiber strength does not fall off with heat-treatment temperature over the process range required to attain the higher modulus. Included in this category, for example, would be the dog-bone-shaped Fortafil family of fibers from Great Lakes Carbon and the rayon-based Thornel family of stress-graphitized fibers from Union Carbide. More radiation-effects work on a variety of fiber types is needed before the full potential of the strengthening process can be determined, but certainly the results of Table 1 show that radiation treatment has potential for certain types of fibers.

IRRADIATED-FIBER COMPOSITES

The objective of the present work was to verify that property improvements of irradiated fibers translate into epoxy composite laminates which are reinforced with such fibers and then cured under heat and pressure. Fibers selected for this purpose had been strengthened by 16.5% (Test 2 of Table 1) through exposure to a fast-neutron fluence of 8.5×10^{17} n/cm^2 ($E > 1$ MeV) at 175°F in General Dynamics' Ground Test Reactor (*GTR*). These high-strength, 48-in.-length fibers were manufactured by Courtaulds Limited from a polyacrylonitrile (PAN) precursor. They had been surface treated to improve bondability with an epoxy matrix and were identified by their manufacturer as HT-S fibers. Strength translations of these fibers into laminates were judged by comparing mechanical properties of specimens from sets of two composite panels that were laid up one after another from irradiated and as-received fibers (from the same batch) and then were given identical cure treatments at the same time.

Fabrication and Testing

Fibers were impregnated with Union Carbide's ERLA 4617 resin that had been reacted with a metaphenylenediamine (MPDA) catalyst. A uniform 4-mil film of properly staged resin (4 in. wide and 48 in. long) was first applied to Patapar release paper. Tows of HT-S fibers (~ 10,000 individual 8.5μ filaments) were then laid down parallel and uniformly spaced (23 tows to the 4 in. width) by means of an indexing alignment guide. Standard manual handling methods were used since the radioactivity of the high-purity carbon fibers was very slight ($<0.3 \times 10^{-6}$ Curie per tow). After the fibers were positioned on the tacky resin and covered with release paper, a heatable hydraulic press was used to apply 40 psi and 150°F to the layup for one-half hour to thoroughly impregnate the fibers with resin.

This one-ply layup was cooled under pressure and then was cut into sections (4 in. by 4 in.); these sections were stacked one on another to make up a 11-ply unidirectional laminate. This laminate was cured as follows: (1) A pressure of 85 psi was applied, the temperature was raised to 250°F, and the pressure and temperature were held constant for one hour. (2) The temperature was raised to 350°F and held there for another hour at the same pressure. (3) The laminate was cooled under pressure, removed from the press, and then was postcured in a circulating-air oven for two hours at 350°F. The cure cycle used has been documented in some detail because of the possibility that it could have detrimental effects on enhanced properties of the irradiated fibers, but this did not occur, as will be discussed presently.

Two different resin batches (*A* and *B*) were used in making sets of panels. One panel reinforced with irradiated fibers and one panel reinforced with control fibers (unirradiated, but otherwise identical) were made up from each batch of resin. The cured 11-ply panels had a thickness of about 0.080 in. and a density of about 1.51 g/cm^3. Their fiber content was 56% by volume and 64% by weight.

Flexural specimens (3 in. by 0.5 in.) were cut from each of these panels with their major axes both parallel to the fiber direction (longitudinal specimens) and perpendicular to it (transverse specimens). Horizontal shear specimens (0.6 in. by 0.25 in.) were also cut from each panel with their major axes lying along the fiber direction. The longitudinal (0°) flex specimens and short-beam shear specimens were loaded at their mid-planes while being supported over spans of 2.5 and 0.4 in., respectively. The transverse (90°) flex specimens were loaded at two points, 0.5 in. on each side of their mid-planes, while being supported over a 2.0 in. span. Instron crosshead speed was 0.05 in./min. for all tests, and all tests were performed at room temperature. Ultimate flexural and shear strengths were calculated from standard beam formulas, as in Reference [12].

Mechanical Properties and their Discussion

Test results for the two laminated panels made up from each of the resin batches *A* and *B* are given in Table 2. The longitudinal flex strength of the irradiated-fiber composite was about 11% greater than that of the unirradiated-fiber composite for each of the two resin batches. Increases of about 8% in horizontal shear strengths of both irradiated-fiber composites were also found. Thus, not only did the 16% strength enhancement of the irradiated fibers largely translate into the composite, but better bondability between fiber and matrix was also obtained. In addition, the longitudinal modulus of the irradiated-fiber composite was increased by about 5% for each set of test panels. The transverse flex strength was considerably larger for the irradiated-fiber composite than for the control with resin batch *A*, but the transverse strength of the control composite was somewhat larger for resin batch *B*. These differences in transverse flex strengths apparently have very little influence on the translation of fiber properties into composites, since longitudinal flex strengths were very nearly the same for a given fiber (irradiated or control) laid up in both batches of resin.

Percentage translations of fiber properties into composite flexural properties are shown in Table 3. The translations for irradiated fibers are only a few percent below those for control fibers, and property comparisons between fiber and composite are in excellent agreement with rule-of-mixture predictions. Therefore, considering the many variables involved, it is assumed that radiation-enhanced fiber properties were fully translated into the cured composites. The percentage translations would not likely have been in such good agreement with rule-of-mixture estimates if composite specimens had been tested in tension, however, for in our testing of 0° specimens it has generally been found that (Figure 1, for example)

$$\text{flex strength} = 1.35 \times \text{tension strength},$$

in good agreement with Weibull's theoretical estimate of a 1.41 flex-to-tension ratio for rectangular, homogeneous beams [13].

Therefore, in tension, the strength translations of Table 3 would probably have been in the neighborhood of 70 to 75%. Although of uniform fiber translation, and therefore adequate for the purpose at hand, our laminated layups were apparently of poorer quality than those made up using tapes prepared by commercial "pre-preggers." For composites made from good quality "prepreg," strength translations of fibers into flexural and tension specimens are typically 125 and 95% (Figure 1), respectively, while modulus translations of fibers are about 100% for each test.

Table 2. Property Comparisons of Laminates Reinforced with Irradiated and Unirradiated HT-S Carbon Fibers

Panel No.	Reinforcement Fiber	0° Flex Strength (10^3 psi)	0° Flex Modulus (10^6 psi)	Horizontal Shear (10^3 psi)	90° Flex Strength (10^3 psi)
A1	Control HT-S	219	24.1	10.3	14.8
		206	21.8	10.6	10.7
		236	25.3	12.6	11.6
		209	22.2	13.0	
	Average	218	23.4	11.6	12.4
A2	Irradiated HT-S	244	24.0	11.9	17.4
	(8.5×10^{17} n/cm^2)	236	23.3	12.0	17.7
		245	25.7	12.5	14.5
		243	25.1	13.0	
	Average	242	24.5	12.4	16.5
B1	Control HT-S	215	22.5	13.2	13.1
		213	22.0	14.1	16.1
		221	24.1	13.3	13.0
	Average	216	22.9	13.5	14.1
B2	Irradiated HT-S	235	23.5	14.6	13.0
	(8.5×10^{17} n/cm^2)	242	24.3	14.3	14.5
		241	23.9	15.3	13.6
	Average	239	23.9	14.7	13.7

Table 3. Translation of Fiber Tensile Properties into Flexural Properties of Composite Specimens

Fiber Condition	Ultimate Stress (10^3 psi)			Modulus (10^6 psi)			Ultimate Strain (μin/in)		
	Fiber	Composite	% Trans.*	Fiber	Composite	% Trans.*	Fiber	Composite	% Trans.
As received	382	218	102	39.4	23.4	104	9,700	9,300	96
		216	101		22.9	102		9,400	97
Irradiated	445	242	97	42.7	24.5	101	10,400	9,900	95
		239	96		23.9	98		10,000	96

*These percentage translations are based on rule-of-mixture calculations for composites having 56% fibers by volume, where the flexural strength and modulus of the ERLA 4617 matrix were taken as 14 ksi and 0.8 Msi, respectively.

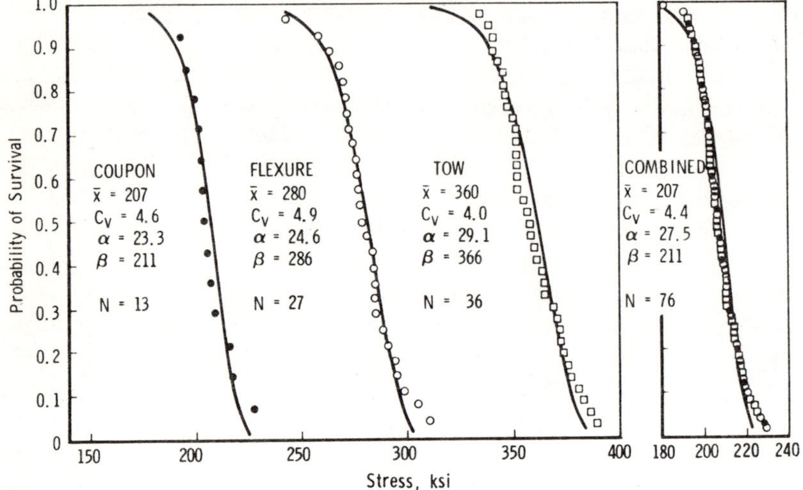

Figure 1. The translation of fiber tensile strengths into tensile and flexural strengths of composite laminates.

The cure cycle, then, which involved holding the layup at a temperature of 350°F for 3 hours, did not cause any appreciable decrease in properties of irradiated fibers. Property degradations were not expected, since irradiated fibers had been given this same heat treatment in vacuum without loss of strength before the layup was attempted. Tensile strengths of such fibers are presumably increased because of carbon atoms that are displaced (through neutron collisions) into interstitial positions between planes of lattice atoms that are widely spaced (3.44Å) and weakly bound from plane to plane, but closely spaced (1.42Å) and tightly bound within a given plane.

These displaced atoms in the interlayer spaces within crystallites of carbon fibers apparently act as an "atomic adhesive" to bind planes of atoms together in somewhat the same way that epoxy binds together the strong carbon fibers in the composite itself. These interstitial atoms can be made to diffuse back into vacant lattice sites in basal planes by raising the temperature to various threshold levels, the higher levels being needed to permit diffusion of interstitial atoms which have bonded together to form higher-order complexes of carbon atoms, but no loss of strength occurs for carbon fibers heated to 350°F. It is not yet known how much higher the temperature can be carried before strength losses begin to occur, and this will likely be a function of irradiation temperature.

As for the increased horizontal shear strength of the irradiated-fiber composites, this might occur as a result of either or both of two processes. The carbon-fiber/matrix interface divides into three regions [14]: (1) the epoxy surface adjoining the fiber, (2) the chemical interface between fiber and matrix, and (3) the first several atomic planes within the surface of the carbon fiber itself. Shear failures will occur in the weakest of these regions, of course, and the first region can be eliminated as a source of failure in good resin systems, since electron micrographs do not reveal an epoxy layer left on fibers after shear failures [14]. The second region is certainly the weakest in carbon fibers that are not surface treated, but there is evidence that the chemical interface can be sufficiently strengthened through surface treatment to shift the zone of weakness into the surface layers of the carbon fiber itself [15]. Neutron irradiation should provide strengthening in both of the critical interface regions (2 and 3).

As to the chemical interface (2), the surfaces of high-modulus fibers are covered largely with basal planes of carbon atoms in which valences of interior atoms are fully saturated [5, 14], so that such fibers do not bond well with epoxies. A greater number of the much more chemically active "edge-type" atoms (having unsaturated valences) are exposed at vacant lattice sites left by the displacement of surface atoms, and this should enhance fiber-to-matrix bonding [16]. At the same time, the weak shear regions between the atomic planes within the carbon fiber itself (3) are being strengthened by the displaced atoms that come to rest within the interplanar regions. Thus, neutron irradiation should be effective in strengthening both zones of interface shear weakness, and, indeed, irradiated-fiber composites had almost 10% higher shear strengths than did composites reinforced with chemically surface-treated HT-S fibers. Moreover, strengths and moduli of the composites reinforced with irradiation-treated fibers increased along with the shear strength, whereas these properties often decrease when the shear strength is increased through chemical treatments [5]. It might also be mentioned here that shear strengths of carbon/carbon composites have been increased by about 25% through irradiation of the finished laminate [17].

FUTURE POTENTIAL

The practical usefulness of nuclear-radiation treatment of carbon fibers, as a means for improving mechanical properties of composite structures, depends upon at least two factors. One of these is the extent to which composite properties can be improved, and the other is the cost of achieving the improvements. The far-from-optimum results already obtained would certainly seem to justify the treatment on a purely technical basis. Increases in fiber strengths of as much as 40% have previously been reported [4], and the results of this paper indicate that these increases should largely translate into epoxy-matrix composites reinforced with such fibers. More uncertainty is involved in the second factor, and, of course, the costs that could be justified would depend on what eventual technical improvements might be realized in optimized irradiation treatments.

Only very preliminary estimates have been made on the cost of irradiation treatment of carbon fibers, and this would vary greatly according to the type and power rating of the nuclear reactor used. For reactors of the *GTR* type in which materials are irradiated outside the reactor core for long periods of time in order to achieve a given exposure dose (e.g., a 600-hour operation at a power of 10 MW is required to produce a fast-neutron fluence of 1×10^{19} n/cm^2 outside the contained reactor core), the irradiation of small research quantities of fibers would be exceedingly expensive. Consequently, all fiber irradiations at General Dynamics have been done in conjunction with other experiments in which large quantities of materials were irradiated, and resulting irradiation conditions have not been ideally suited for fiber treatments. Research-type irradiations to optimize exposure conditions could be done much less expensively with reactors in which small quantities of materials can be introduced directly into the reactor core to achieve the desired exposure dose in a much shorter period of time.

However, for the irradiation of large quantities of materials, the advantage swings back to a *GTR*-type reactor, for it costs no more to irradiate several thousand pounds of fibers than to irradiate only a small research quantity. Based on the construction of a specially designed *GTR*-type reactor with a power of 20 megawatts, which would be amortized over a ten-year period, the cost for radiation strengthening of carbon fibers would be roughly \$15 per pound for the treatment of 100,000 pounds of fibers per year. This is expensive, but probably not prohibitively so if fiber properties were well in excess of those obtained by other methods. The ideal situation, under which the costs could be considerably lowered, would be to obtain irradiated carbon fibers as a by-product of the generation of electricity in a larger power reactor. Thus, while still some time away, the commercial radiation strengthening of carbon fibers[1] is not without promise if the technical potential is as great as now seems likely.[2]

CONCLUSIONS

Neutron irradiation causes changes in the crystalline structure of carbon fibers [18], and these changes have been found to result in significant improvements of strength and strain capacities of the fibers [1–4]. Now it has been demonstrated that these enhanced properties largely translate into epoxy laminates reinforced with such fibers. Moreover, the irradiation improves the strength of the fiber/matrix interface, as indicated by increased short-beam shear strengths. Because further improvements in fiber properties can reasonably be expected when irradiation conditions are optimized, the cost surcharge that is projected for the irradiation treat-

---

[1] U.S. Patent pending.

[2] Logarithms of the percentage strength increases of Table 1 fall within a linearly increasing band when plotted against the square root of the neutron exposure dose divided by the absolute irradiation temperature. Consequently, strengths of 800 and 1000 ksi are projected for nominal 400-ksi HT-S fibers that are irradiated at 175°F to an exposure dose of 3.3×10^{20} n/cm^2, this being the dose that gave a strength of 580 ksi for fibers irradiated at 1000°F (Table 1). These strengths are in the range that Whitney and Kimmel suggest as a target for the next generation of graphite fibers [*Nature Physical Science,* Vol. 237 (1972), p. 93].

ments makes the process appear quite feasible for practical structural applications in which strength-to-weight ratios are of importance.

ACKNOWLEDGMENT

The authors are grateful to Mr. C. W. Rogers for encouragement and support during the course of this investigation.

REFERENCES

1. S. Allen, G. A. Cooper and R. M. Mayer, "Carbon Fibers of High Young's Modulus," *Nature*, Vol. 224 (1969), p. 684.
2. R. E. Bullock, "The Effects of Fast-Neutron Irradiation on Tensile Strengths of Carbon Fibers," *Radiation Effects*, Vol. 11 (1971), p. 107.
3. R. E. Bullock, "The Effects of Fast-Neutron Irradiation on Elastic Moduli of Carbon Fibers," *Radiation Effects*, Vol. 14 (1972), p. 263.
4. B. F. Jones and I. D. Peggs, "Strengthening of Carbon Fibres by Fast Neutron Irradiation," *Nature*, Vol. 239 (1972), p. 95.
5. L. A. Joo', R. J. Diefendorf and J. C. Goan, "Graphite Fiber Surface Treatments," AFML-TR-70-302, May, 1971.
6. J. C. Halpin, "Structure-Property Relations and Reliability Concepts," *J. Composite Materials*, Vol. 6 (1972), p. 208.
7. R. C. Novak and M. A. DeCrescente, "Impact Behavior of Unidirectional Resin Matrix Composites," ASTM, STP 497, 1972, p. 311.
8. C. C. Chamis, M. P. Hanson and T. T. Serafini, "Impact Resistance of Unidirectional Composites," ASTM, STP 497, 1972, p. 324.
9. B. F. Jones and I. D. Peggs, "The Effect of Fast Neutron Irradiation on the Structure and Mechanical Properties of a High Modulus Graphite Fiber," *J. Nuclear Materials*, Vol. 40 (1971), p. 141.
10. C. W. LeMaistre and R. J. Diefendorf, "The Effect of Thermal Expansion Anisotropy on the Strength of Carbon Fibers," 10th. Biennial Conference on Carbon, Lehigh University, 1971, Summary of Papers, p. 163.
11. R. E. Bullock, "An Interpretation of Radiation Effects on Mechanical Properties of Carbon Fibres Based on a 'Sheath' and 'Core' Model of Fibre Structure," *J. Mat. Sci.*, Vol. 7 (1972), p. 964.
12. R. E. Bullock, "Radiation Effects on Strengths of Boron-Epoxy Composites in Various Environments," *J. Nuclear Materials*, Vol. 44 (1972), p. 175.
13. W. Weibull, "A Statistical Theory of the Strength of Materials," Ing. Vetenskaps Akad. Handl. NR 151, 1939.
14. B. L. Butler and R. J. Diefendorf, "Graphite Filament Structure," Carbon Composite Technology Symposium, Albuquerque, New Mexico, 1970, Proceedings, p. 109.
15. F. Tuinstra and J. L. Koenig, "Characterization of Graphite Fiber Surfaces with Raman Spectroscopy," *J. Composite Materials*, Vol. 4 (1970), p. 492.
16. R. J. Dauksys and J. D. Ray, "Properties of Graphite Fiber Nonmetallic Matrix Composites," *J. Composite Materials*, Vol. 3 (1969), p. 684.
17. R. E. Bullock and E. L. McKague, Jr., "Radiation Effects on Mechanical Properties of Carbon/Carbon Composites," *Carbon*, Vol. 11 (1973), in press.
18. R. E. Bullock, D. E. Gordon and B. C. Deaton, "Irradiation-Induced Structural Changes in Carbon Fibers," *Carbon*, Vol. 11 (1973), in press.

Macroscopic Fracture Mechanics of Advanced Composite Materials

M. E. WADDOUPS, J. R. EISENMANN, AND B. E. KAMINSKI

General Dynamics Corporation
Fort Worth, Texas 76101

(Received May 13, 1971)

ABSTRACT

The application of classical fracture mechanics to laminated composites is discussed. A convenient method is presented for predicting the static strength of a flawed specimen. Theoretical predictions are compared with experimental data for specimens containing two types of flaws.

INTRODUCTION

THE INVESTIGATION of fracture behavior of advanced composite laminates has produced a paradox. Static and fatigue data were obtained for a $[0/90]_S$ graphite-epoxy laminate with various stress concentrations. The results have been summarized in Table I. Note that the notched survivors of the fatigue experiment showed a residual strength equal to, or greater than their corresponding static controls. Also note that the residual strengths of the notched specimens are approximately equal to that of the unnotched specimens. Finally, introduction of the stress concentrations did not nucleate a finite through crack after 10^6 cycles at 80 percent of static ultimate. Thus, the paradox is that although the material is statically brittle, it shows no propensity for the nucleation and growth of a through crack as exhibited in metals.

Table 1.* $[0/90]_s$ *Graphite-Epoxy Fatigue, RT, R = 0.1.

*Specimen***	*Average Static Strength (psi)*	*Average Residual Strength After 5 × 10^6 Cycles (psi)*
No stress concentration	83,500	77,200
.063 dia. hole	68,900	77,700
.063 dia. hole + .018 × .004 notch	72,200	74,300
.063 dia. hole + .078 × .004 notch	58,300	79,400

*This study was conducted under an inhouse experimental program at General Dynamics, Convair Aerospace Division, Fort Worth Operation by E. L. McKague and R. J. Stout.
**Coupon specimens 1.0 inch in width and 9.0 inches in length.

Wu, Reference 1, found that under certain conditions the techniques of isotropic fracture mechanics can be directly applied to composite materials. The conditions were:

1) The orientation of the flaw with respect to the principal axis of symmetry must be fixed.
2) The stress intensity factors defined for the anisotropic cases must be consistent with the isotropic case in stress distribution and in crack displacement modes.
3) The critical orientation coincides with one of the principal directions of elastic symmetry.

Unidirectional materials satisfy these restrictive conditions, and the experiments confirmed the applicability of fracture mechanics. Unidirectional materials, however, are of little design interest in structures, and hence, fracture criteria remained unobtainable for practical laminates.

A carefully conducted stress concentration experiment performed at General Dynamics has led to data which lends clarification to the apparent paradox. Tensile coupons for a $[0/\pm45]_{2S}$ graphite-epoxy laminate were tested over a range of hole sizes. The data obtained is tabulated in Tables 2 and 3

Table 2. Small Hole Data Summary.*

Specimen**	Static Strength (psi)
Control	67240
.062 in. Dia. Hole	45000
.031 in. Dia. Hole	51500
.015 in. Dia. Hole	60900

*This work was sponsored by the Air Force Materials Laboratory under Contract F33615-69-C-1494.
**Coupon specimens were 1.0 inch in width and 9.0 inches in length.

Table 3. Large Hole Data Summary.*

Specimen**	Static Strength (psi)	
	Actual	Corrected for Finite Width
Control	76000	76000
1.0 in. Dia. Hole	26600	27900
2.5 in. Dia. Hole	15900	22800
3.0 in. Dia. Hole	13250	23000

*This work was sponsored by the Air Force Materials Laboratory under Contract F33615-69-C-1494.
**Coupon specimens were 5.0 inches in width and 38.0 inches in length. Isotropic width correction taken from Reference 5.

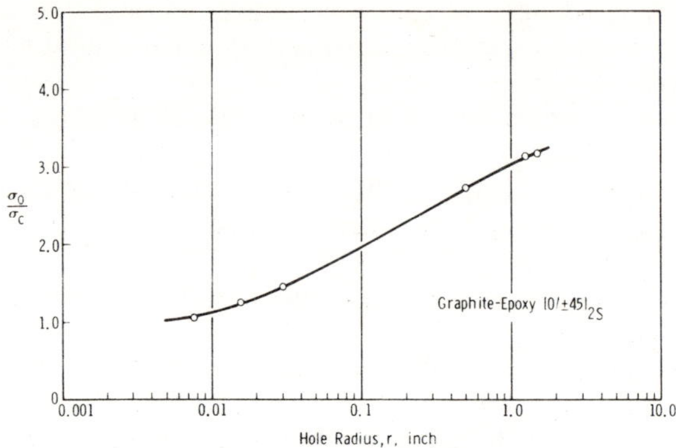

Figure 1. *Stress concentration study.*

and plotted in Figure 1 in terms of the ratio of σ carried without a flaw (σ_o) to σ carried with the flaw (σ_c). Figure 1 shows that σ_c varies with hole radius instead of remaining constant. The observed behavior may be explained within the theory of fracture mechanics.

ANALYSIS

Irwin's (Reference 2) development of the equivalence of the energy-rate and stress intensity approach to fracture is central to the interpretation of the data. In the consideration of a plane strain Mode I crack, the energy available for crack extension in isotropic materials is given by (Reference 1):

$$\mathcal{G}_I = \frac{(1 - \nu)\pi}{2G} K_I^2 = \frac{(1 - \nu^2)\pi}{E} K_I^2. \tag{1}$$

Admit as a basic model of the system the geometry shown in Figure 2 having a region of intense energy defined in terms of a characteristic length (a) based upon a $1/\sqrt{r}$ type of singularity. Obviously, the generating macroscopic stress function may be different; but it is only the total available energy that is of interest. Hence, the exact shape of the stress function is not important. The characteristic length (a) is to be considered small, but finite. Bowie (Reference 3) developed a solution for the stress intensity factor, K_I, for the problem of cracks emanating from a circular hole for an isotropic, homogeneous material. With respect to stress concentration, the $[0/\pm45]_{2S}$ graphite-epoxy laminate has a K_T of 2.98 versus the isotropic value of 3.00. The symmetric crack analysis was chosen since the intense energy regions are symmetric.

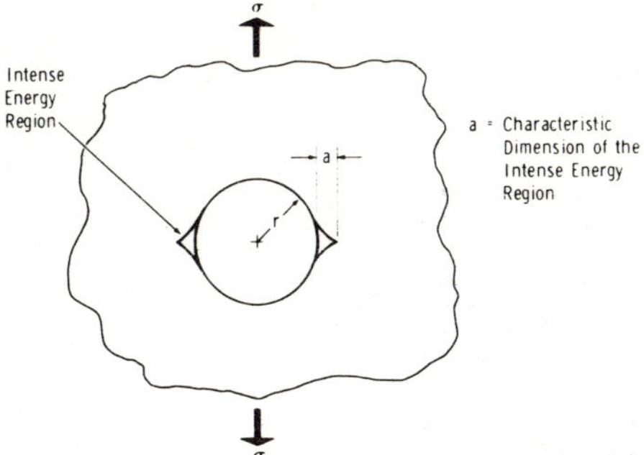

Figure 2. *Fracture model.*

From Reference 4,

$$K_I = \sigma\sqrt{\pi a}\, f(a/r). \tag{2}$$

Combining equations (1) and (2) yields:

$$\mathcal{G}_I = \left[\pi\sqrt{\frac{a(1-\nu^2)}{E}}\right]\sigma f(a/r). \tag{3}$$

Should the material be an ideal Griffith solid, $\mathcal{G}_I$ will be a material constant.

Assuming that changes in (a) are of second order relative to changes in r or $f(a/r)$, the bracketed quantity in equation (3) will be constant. This leads to the conclusion that the quantity

$$\sqrt{\mathcal{G}_I}\left[\pi\sqrt{\frac{a(1-\nu^2)}{E}}\right] = \sigma f(a/r) \cong \text{constant}. \tag{4}$$

Solving equation (2) for σ_c for the Bowie specimen,

$$\sigma_c = \frac{K_{I_c}}{\sqrt{\pi a}\, f(a/r)}. \tag{5}$$

For a specimen with no hole (control),

$$\sigma_o = \sigma_c \big|_{a/r=\infty} = \frac{K_{I_c}}{\sqrt{\pi a}(1.00)} \tag{6}$$

Thus,

$$\frac{\sigma_o}{\sigma_c} = f(a/r) \tag{7}$$

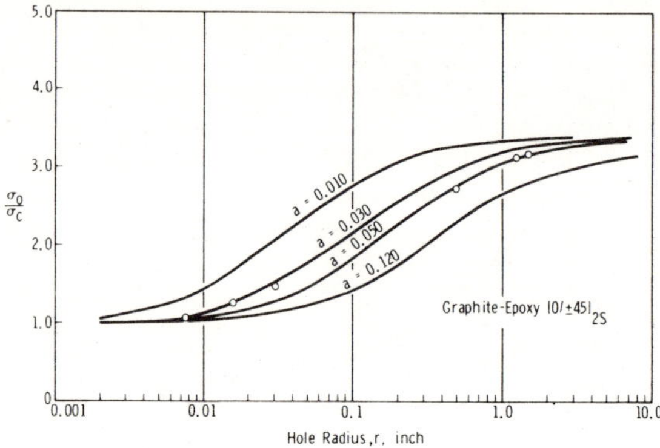

Figure 3. *Parametric study.*

By parametrically evaluating (σ_o/σ_c) as a function of r for given values of (a) the curves shown in Figure 3 were obtained. Superimposed on these curves are the data from the stress concentration study. The experimental behavior is consistent with the behavior as predicted by the fracture mechanics analysis for a constant value of a = 0.040 inch. No crack, that is, no macroscopically measurable crack was present. This fact is attributable to the internal structure of the composite. The material reacts to the stress intensity and ruptures catastrophically as the critical energy level is reached.

Another series of tests performed on the same laminate provided strain measurements at the net section. Strains measured by a gauge bonded to the edge of the laminate inside the hole and by gauges on the surface of the laminate near the hole were in excess of the ultimate strain of a pristine, unnotched control specimen. The strain data is plotted in Figure 4 as a function of the distance from the edge of the hole. These exceedingly high strains may be a manifestation of the intense energy region.

The correlation between the experimental behavior and fracture analysis suggests the important result that $\mathcal{G}$, a measure of the energy available to resist a crack, remains constant as the stress intensity factor K_I is changed as a function of r. This supports the thesis that the material behaves in fracture as a near perfect, brittle Griffith solid.

APPLICATIONS

Several examples are presented below which illustrate the application of fracture mechanics to advanced composite materials. The general procedure is to define values of (a/r) and (a) using two specimens, one having an induced stress concentration and the other with none. Having established (a/r) and

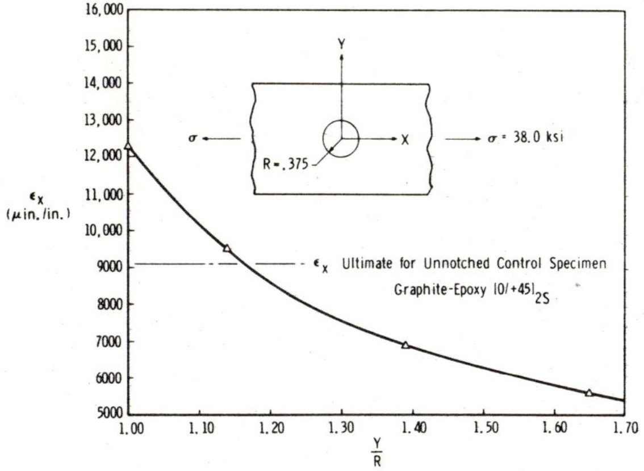

Figure 4. *Strain profile at net section.*

(a), it is postulated that σ_c can be predicted for other stress concentrations of similar geometry.

Consider the experimental data shown in Figure 3 for radii less than 0.5 inch. The static strengths for the three specimens containing a hole and the control specimen are presented in Table 2. Substituting the static strengths of the first two entries of the table into equation (7),

$$f(a/r) = \frac{\sigma_o}{\sigma_c} = \frac{67240}{45000} = 1.495.$$

From Reference 6,

$$a/r = 0.89$$

and, therefore,

$$a = 0.0276 \text{ inch.}$$

The stress intensity factor, K_{I_c}, can now be calculated from equation (2).

$$K_{I_c} = \sigma_c \sqrt{\pi a}\, f(a/r) = 19800 \text{ psi} \sqrt{\text{in.}} \tag{8}$$

Knowing K_{I_c}, the static strengths of the remaining two specimens can now be predicted. Solving equation (8) for σ_c,

$$\sigma_c = K_{I_c} / \sqrt{\pi a}\, f(a/r). \tag{9}$$

For the specimen having the 0.031 inch diameter hole,

$$r = 0.0155,$$

$$a/r = 1.78,$$

and

$$f(a/r) = 1.22.$$

Thus,

$$\sigma_c = 52{,}200 \text{ psi}$$

which agrees well with the measured value of 51,500 psi. Similarly, for the specimen with an 0.015 inch diameter hole, σ_c was found to be 60,600 psi versus the measured value of 60,900 psi.

Next, consider the data shown in Figure 3 for radii greater than or equal to 0.5 inch. The static strengths for the specimens containing a hole are presented in Table 3.

Using σ_o equal to 76,000 psi and the second entry in Table 3,

$$f(a/r) = \frac{\sigma_o}{\sigma_c} = \frac{76000}{27900} = 2.72.$$

From Reference 6,

$$a/r = 0.100$$

and, therefore,

$$a = 0.0500 \text{ inch}.$$

Then, from equation (2),

$$K_{I_c} = 30{,}100 \text{ psi} \sqrt{\text{inch}}.$$

For the specimen having the 2.50 inch diameter hole,

$$a/r = 0.040,$$

and

$$f(a/r) = 3.13.$$

Thus, from equation (9)

$$\sigma_c = 24{,}200 \text{ psi}$$

which agrees well with the measured value of 22,800 psi. Similarly for the specimen having a 3.00 inch diameter hole, σ_c was found to be 23,900 psi versus the measured value of 23,000 psi.

Note that the characteristic lengths (a) for the small and large hole data, 0.0276 and 0.050 inch respectively, bracket the nominal value of 0.04 inch (Figure 3) which was obtained from the parametric study.

Thus far, fracture mechanics analysis has been shown to be applicable to discontinuities of one similar geometry, i.e., circular holes. Now consider the case of a narrow slit in a composite specimen. The stress intensity for a Griffith flaw in isotropic material is given by, Reference (6),

$$K_I = \sigma\sqrt{\pi L} \tag{10}$$

where L is defined as the half-crack length. Again, an intense energy region is assumed to exist at the tip of the slit for a composite material. This assumption results in an effective half-crack length of $(L + a)$. Thus, equation (10) becomes

$$K_I = \sigma\sqrt{\pi(L + a)} \tag{11}$$

Here,

$$\frac{\sigma_o}{\sigma_c} = \sqrt{\frac{L + a}{a}} \tag{12}$$

The general procedure is now to determine (a) from two test points which provide values for σ_o and σ_c. For example, consider the data in Table 1. Assume that the stress distribution around the flaw is dominated by the presence of the slit. That is, the size of the hole is small compared to that of the slit. Using the first and last entries of Table 1, equation (12) can be solved for the characteristic length (a).

$$a = \frac{L}{(\sigma_o/\sigma_c)^2 - 1}$$

Now,

$$L = \frac{\text{hole dia.} + \text{slit}}{2}$$

$$= \frac{0.063 + 0.078}{2}$$

$$= 0.0705 \text{ inch},$$

$$\sigma_o = 83{,}500 \text{ psi},$$

and

$$\sigma_c = 58{,}300 \text{ psi}.$$

Therefore,

$$a = 0.0671 \text{ inch}$$

The critical stress intensity factor, K_{I_c}, can be found from

$$K_{I_c} = \sigma_o\sqrt{\pi a}$$

Thus,

$$K_{I_c} = 38{,}400 \text{ psi} \sqrt{\text{inch}}.$$

Knowledge of K_{I_c} allows prediction of σ_c for the third entry in Table 1.

$$\sigma_c = K_{I_c}/\sqrt{\pi(L + a)}$$

Now,

$$2L = 0.063 + 0.018 = 0.081 \text{ inch}$$

$$a = 0.0671 \text{ inch}.$$

Then, σ_c is equal to 66,800 psi versus the measured value of 72,200 psi.

CONCLUSION

The critical stress (σ_c) may be computed in terms of a characteristic dimension (a) and K_{I_c} given the geometry and stress intensity factor. Unlike metals, the geometry of the specimen does not substantially alter with repetitive loads and, in fact, massive material damage at the point of peak K_I may improve the effective geometry. Although the apparent critical flaw length is small, the lack of flaw geometry change with cyclic loads renders the long term structural capacity predictable. Design criteria with respect to anticipated ply-plane flaw induced failure may reduce to, for aircraft structure,

1) prediction of K_I (and the distribution of K_I) for significant design discontinuities
2) anticipation of the character of service usage induced flaws
3) prediction of the probability of failure under structural overload.

Implementation of a quantitative reliability plan, as suggested in Reference 7, becomes increasingly feasible for laminates similar to the one characterized in this discussion. This is because a brittle material with a small coefficient of variation with respect to strength, combined with the property that repeated loads do not generally reduce K_I by changing the flaw geometry renders the material exceedingly deterministic. Unlike previous investigations, the laminate tested and failure modes are representative of an efficient structural laminate. Because of material heterogeneity, the physical condition of the material in the region of intense macroscopic stresses is not known. It is feasible, however, that the material provides a resistance to cracking which is constant for the orientation tested and may be modeled as a surface energy $\mathcal{G}$. It is suggested that these results be investigated for different geometries and orientations. Since failure modes change with filament orientation, $\mathcal{G}$ is anticipated to be a function of laminate configuration.

REFERENCES

1. E. M. Wu, "Fracture Mechanics of Anisotropic Plates," *Composite Materials Workshop*, Technomic Publishing Co., (1968).
2. G. R. Irwin, "Fracture Dynamics," *Fracturing of Metals*, A.S.M., Cleveland, (1948).
3. O. L. Bowie, "Analysis of an Infinite Plate Containing Radial Cracks Originating from the Boundary of an Internal Circular Hole," *Journal of Mathematics and Physics*, Vol. 35 (1956), p. 60.
4. P. C. Paris and G. C. Sih, "Stress Analysis of Cracks," ASTM STP 381, (1970), p. 70.
5. R. E. Peterson, *Stress Concentration Design Factors*, J. Wiley and Sons, (1953).
6. A. A. Griffith, "The Phenomena of Rupture and Flow in Solids," *Philosophical Transactions of the Royal Society*, Vol. 221A (1920), p. 163.
7. J. C. Halpin, J. R. Kopf, and W. Goldberg, "Time Dependent Static Strength and Reliability for Composites," *Journal of Composite Materials*, Vol. 4 (1970), p. 462.

Experimental Investigation of Fracture in an Advanced Fiber Composite

H. J. KONISH, JR, J. L. SWEDLOW, AND T. A. CRUSE

Department of Mechanical Engineering
Carnegie Institute of Technology
Carnegie-Mellon University
Pittsburgh, Pennsylvania 15213

(Received November 21, 1971)

ABSTRACT

Pilot tests were run to determine whether the concepts of linear elastic fracture mechanics might be used to describe behavior of initially cracked specimens of advanced fiber composite laminates. The results, presented in some detail, show a positive finding.

INTRODUCTION

LINEAR ELASTIC fracture mechanics (LEFM) is now accepted as the rationale for characterizing crack toughness of materials that are ostensibly homogeneous and isotropic, the outstanding examples being a wide range of metallic alloys. The basic experience that supports this approach is that presence of a macro crack dominates a structure's response to remote loading. With the advent of advanced fiber composites, however, there arises the question of whether the applicability of LEFM to date is a consequence of the apparent uniformity of the structure surrounding the crack. In particular, there is concern over whether heterogeneity and anisotropy will preclude practical use of LEFM in composites.

Vigorous discussion of this issue is important, but the interchanges so far have tended to be theoretical and even speculative. In an effort to supply some physically based information, we have undertaken a pilot series of experiments. Our objective was to answer two specific questions:

1. If a cracked, composite specimen is loaded to failure, is the path of crack prolongation determined by the geometry of the initial crack and the loading, or by material orientation?
2. Can LEFM, suitably modified to account for material anisotropy, be usefully applied to composites?

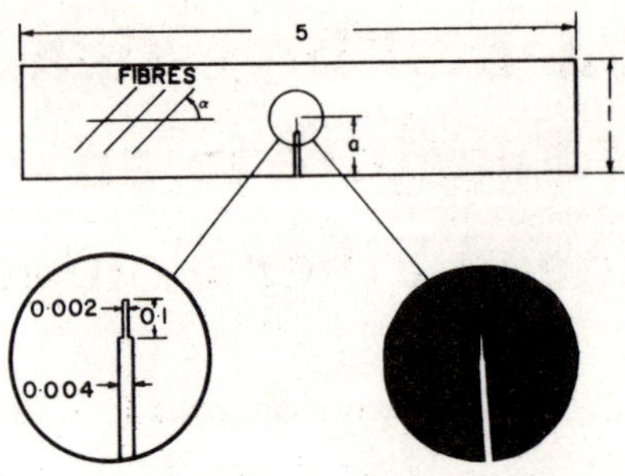

Figure 1. Three-point bend specimen geometry, with crack shape shown in inserts, both schematic (left) and actual (right). Fiber direction given by α, crack length by a. Specimen thickness 0.5 in (nom); all dimensions given in inches.

The data now in hand, although limited, indicated that a crack in a composite is at least influential in determining failure patterns and, in many cases, the crack is dominant; and that LEFM provides useful procedures for evaluating crack toughness of composites.

This paper gives a brief review of the test procedures, methods of data reduction, and experimental results. Observations made during the course of the tests are reported, and failure surfaces are shown. Analytical work stimulated by these results is underway and will be reported subsequently.

TEST PROCEDURES; PROGRAM

It was obvious from our objective that our test procedures should follow those developed within the framework of conventional fracture mechanics. There is, in fact, a wealth of literature on this subject including an ASTM Tentative Method [1] and extensive interpretation of it (see, e.g., [2,3]). Departures from the specifications in [1] were minimal and were dictated either by the special nature of the material under test or by simple practicality.

We chose the three-point bend specimen prescribed in [1] largely to bypass problems associated with gripping the test piece. (See Figure 1.) In the extensive data base that now exists for metals testing, results for this configuration compare well to those for other geometries so that, among other matters, we had no reason to expect that the bearing load opposite the crack front should influence unduly the processes of crack prolongation. In fact, the data reduction scheme in [1] accounts for such details of specimen geometry and load arrangement.

The specimen proportions shown in Figure 1 follow the recommendations in [1] except that the crack front was not sharpened under fatigue loading. Instead, the notch was produced by a sawcut followed by a final lengthening and sharpening using an ultrasonic cutter.

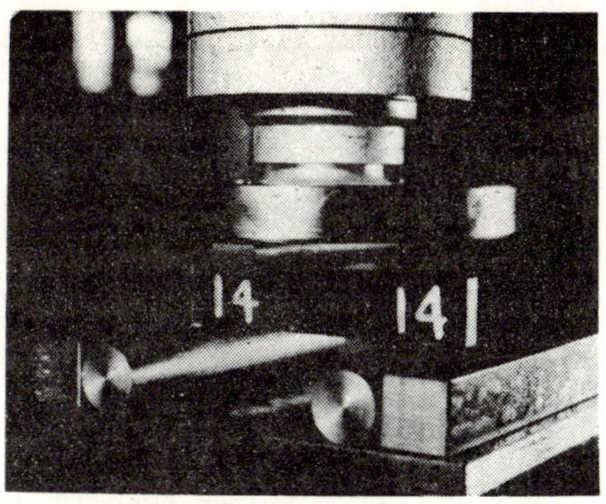

Figure 2. Test jig at beginning of loading.

As shown in Figure 2, each specimen was centered on two parallel rollers (1 in dia) whose centerlines were 4 in apart. A third parallel roller was then located directly above the crack, and the specimen was loaded vertically downward. Testing was performed in an Instron machine of 10,000 lb capacity, and cross-head motion was set at 10^{-2} in/min to minimize dynamic effects. Load and cross-head motion were monitored during each test and then cross-plotted to give the basic data for later reduction. While the requirement of [1] is to record crack-mouth opening by means of a special clip gauge, both the basic linearity of material response and the machine's rigidity seemed to make this degree of fidelity to [1] unnecessary for the pilot test series.

The program involved twenty-three specimens, thus allowing for two reproducibility tests, and for the testing of both uni- and multi-directional laminates having a range of starter crack lengths. The material used was a NARMCO graphite-epoxy with Morganite II fibers in 5206 resin.*

Reproducibility was evaluated by testing two sets of five specimens, each set of the same lay-up and geometry. The first set was a uni-directional laminate ($\alpha = 0°$) and had an initial crack length of 0.4 in. The second set was multi-directional ($\alpha = (0°/\pm 45°/90°)_s$) and had the same starter crack length. Single tests were run for $\alpha =$ 0°, 45°, 90°; $(\pm 45°)_s$; and $(0°/\pm 45°/90°)_s$. Starter crack lengths were 0.2, 0.4, and 0.6 in, the shortest of which are less than the requirements in [1]. We included such specimens to permit evidence of material dominance to develop.

DATA REDUCTION; RESULTS

A typical load-cross-head displacement trace is reproduced in Figure 3. There is an initial region of increasing slope during which slack in the load train is taken up, and bearing surfaces under the loading rollers develop. This is followed by a linear

*We are most grateful to the Advanced Composites Group, Convair Aerospace Division, General Dynamics Corporation, Fort Worth, for fabricating and supplying the test specimens.

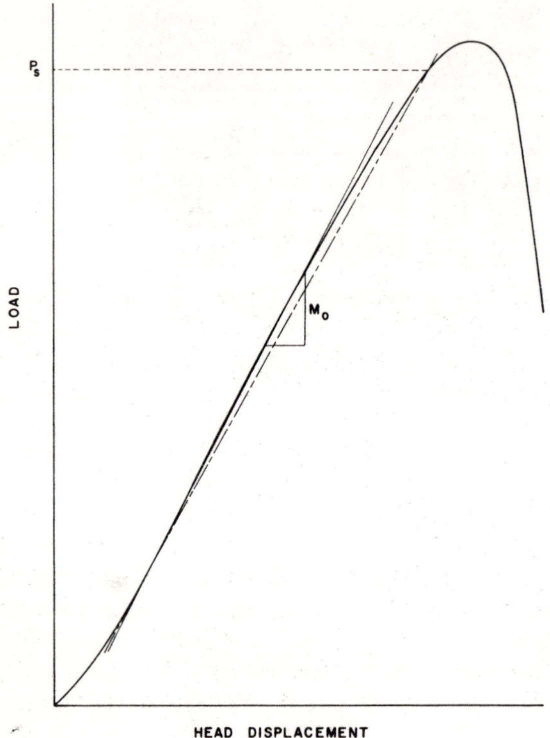

Figure 3. Typical trace of load applied to specimen vs. cross-head displacement, showing method used to determine P_S.

region in which the specimen deforms elastically. A third region of decreasing slope then begins as a result both of nonlinear load-displacement behavior and of damage initiation. Finally the load peaks and falls off as the test piece breaks in two.

In order to differentiate the nonlinear effects from those ascribable to damage, the Tentative Method prescribes the following data reduction scheme.* The slope M_o of the linear portion of the curve is identified, and a line of slope 5% less than M_o is drawn as shown in Figure 3. This line intersects the curve at a load termed P_S. If P_S is the greatest load withstood by the specimen to that point in the test, P_S is set equal to P_Q. If any load maximum precedes P_S, then P_Q is equated to that maximum value. In either case, the experience in metals testing has shown P_Q to correspond reasonably well to the point of failure initiation. In the absence of a suitable data base for composites, we used this procedure to find P_Q and thereby regard the data obtained as surely consistent and probably conservative. Together with specimen

*It should be borne in mind that the present discussion is but an abstract of a most explicitly defined procedure; the interested reader is urged to consult [1] for complete details.

geometry, P_Q is then used to compute K_Q, the critical stress intensity or candidate fracture toughness.** See [2,3].

For each laminate, the K_Q values were averaged to give $\bar{K}_Q$ which, in turn, was used to find a critical strain energy release rate G_Q – see [3,4]. The results are shown in Table 1. Also of interest are the failure surfaces, depicted in Figures 4-8; a specimen that did not part fully is shown in Figure 9.

DISCUSSION

At the outset, two questions were posed regarding the utility of LEFM in characterizing fracture of composites. The first concerns paths of crack prolongation, and we infer our answer from inspection of the failure surface. The second involves use of LEFM as a data reduction scheme, and we look to physical measurements for a response.

Table 1. Experimental Results

fiber orientation angle, α	K_Q, $lb/in^2\sqrt{in} \times 10^{-3}$ a = 0.2 in	 a = 0.4 in	 a = 0.6 in	$\bar{K}_Q$, $lb/in^2\sqrt{in} \times 10^{-3}$		G_Q, in lb/in^2
0°	—[1]	28.8	36.3	32.6	+11% −11%	117.
90°	1.66	1.46	—[2]	1.56	+ 6.3% − 6.3%	0.943
45°	0.690[3]	2.22	2.39	2.30	+ 3.9% − 3.8%	—[4]
$(\pm 45°)_s$	18.5	18.5	16.3	17.7	+ 4.8% − 9.4%	45.0
$(0°/\pm 45°/90°)_s$	23.5	21.7	20.5	21.9	+ 7.3% − 8.6%	55.1

Notes:

1. **Specimen was crushed before crack propagation occurred.**
2. **Instrumentation failure.**
3. **This value omitted when calculating $\bar{K}_Q$.**
4. **No G_Q available because the crack propagated in a mixed mode, which could not be directly uncoupled.**

****In metals testing, certain additional steps are taken to establish the validity of an individual test result. Since these steps necessitate use of the yield stress, we were unable to follow them in our work. Thus we report only *candidate* values of fracture toughness, or K_Q. We cannot presume our data to give K_{Ic} for these materials because compliance with the strict requirements of [1] are definitionally impossible.**

Figure 4. Failure surfaces for $\alpha = 0°$ specimens of three starter crack lengths (a = 0.6, 0.4, 0.2 in).

Figure 5. Failure surfaces for $\alpha = 45°$ specimens of three starter crack lengths (a = 0.6, 0.4, 0.2 in).

The appearance of the failure surfaces suggests that, in the main, the crack and loading dominate fracture. In Figure 4 (specimens for which $\alpha = 0°$), the path of crack growth is observed to be roughly coplanar with the starter crack. Note that in the case of the longest crack ($a = 0.6$ in), where a longitudinal secondary crack formed, the path is generally forward. Indeed, the crack seems to have made a series of sharp turns to regain its coplanar path.

It is not surprising, on the other hand, to see that, in the $\alpha = 45°$ specimens, the crack grew along a plane containing no fibers. This is clear in Figure 5 and, although fracture occurred as the result of crack propagation (in the matrix), the mode is a mixture of opening and sliding [3]. More sophisticated instrumentation would have permitted articulation of the relative presence of each mode, but the data we collected precluded such resolution.

Forward crack growth is evident for the $\alpha = 90°$ specimens as depicted in Figure 6. Growth again was along a plane containing no fibers which, in this case, is coplanar with the starter crack.

Figure 6. Failure surfaces for $\alpha = 90^\circ$ specimens of three starter crack lengths (a = 0.6, 0.4, 0.2 in).

Figure 7. Failure surfaces for $\alpha = (\pm 45^\circ)_s$ specimens of three starter crack lengths (a = 0.6, 0.4, 0.2 in).

During testing the uni-directional specimens described above emitted popping noises prior to failure. Because the fracture process also involved matrix breaking of one sort or another, we believe the two phenomena are related. Even in the $\alpha = 0^\circ$ specimens, the crack appears at the outset to have operated on virtually independent fiber bundles as they pulled out from the matrix. The resulting failure surfaces are very rough for the early stages of growth but then become more nearly uniform. The noise levels for the remaining specimens were much lower, and their failure surfaces are less suggestive of matrix cracking.

Figure 7 is instructive in that it shows for the $\alpha = (\pm 45^\circ)_s$ test pieces an increasing crack dominance as the starter crack is made longer. For $a = 0.2$ in, the crack path almost immediately turns 45° from its initial orientation, there being but a slight indication of forward growth. A greater tendency toward coplanar growth is apparent when $a = 0.4$ in, and crack dominance is manifest when $a = 0.6$ in. Crack growth is not possible on a plane containing no fibers – there being none by virtue of the lay-up – and some zig-zagging is apparent. This group of specimens thus shows a transition from some material dominance where the starter crack is shorter than required by the Tentative Method to a fracture pattern fully dominated by the starter crack.

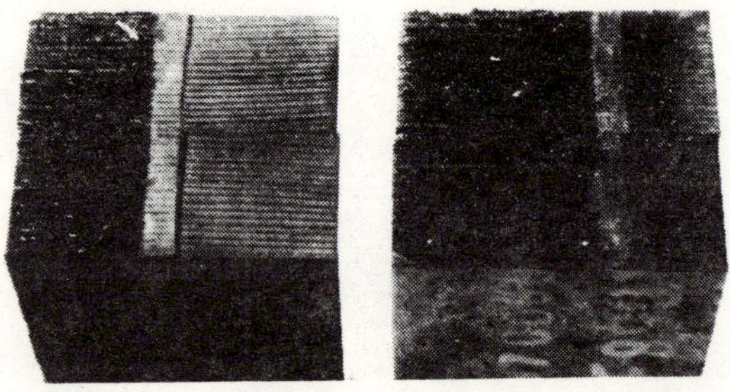

Figure 8. Failure surfaces for $\alpha = (0°/\pm 45°/90°)_s$ specimens of two starter crack lengths (a = 0.6, 0.4 in).

Figure 9. Failed but unbroken specimen ($\alpha = (0°/\pm 45°/90°)_s$, a = 0.2 in).

Crack dominance is also clear in Figure 8, where we show failure surfaces for $\alpha = (0°/\pm 45°/90°)_s$. In these specimens, the crack moved in its own plane but apparently grew further in the interior of the test piece than on its surface. An indication of this behavior, not uncommon in metals testing, is shown in Figure 9.

The use of K_Q to characterize behavior of these specimens appears, on the whole, to be warranted. The reproducibility tests on the $\alpha = 0°$ specimens* and the $\alpha = (0°/\pm 45°/90°)_s$ specimens were satisfactory. Load-displacement traces are shown in Figures 10 and 11, and the average K_Q values found are

$$\alpha = 0° \quad : \quad K_Q = 28.8 \times 10^3 \text{ lb/in}^2\sqrt{\text{in}} \quad {}^{+0.4\%}_{-5.5\%}$$

$$\alpha = (0°/\pm 45°/90°)_s \quad : \quad K_Q = 21.7 \times 10^3 \text{ lb/in}^2\sqrt{\text{in}} \quad {}^{+1.7\%}_{-3.2\%}$$

The scatter is not unlike that found in metals testing. For three laminates, our data are fairly consistent with values obtained independently by Halpin [5] (25-28 × 10^3

*One exception occurred for the $\alpha = 0°$ specimen set; because it was the first specimen of the entire series tested, we are inclined to blame our own inexperience more than material variation.

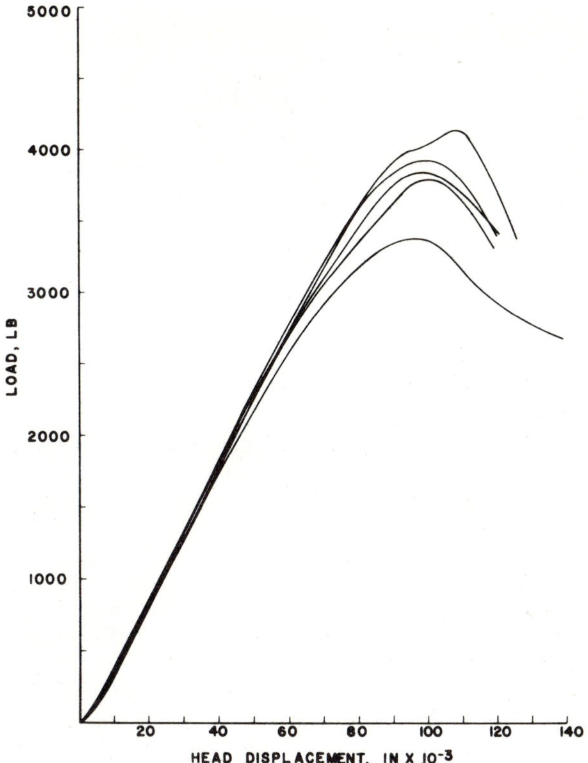

Figure 10. Traces of load vs. cross-head displacement for five specimens used in reproducibility tests for a uni-directional laminate ($\alpha = 0°$, $a = 0.4$ in).

lb/in$^2\sqrt{\text{in}}$, $\alpha = (0°/\pm45°/90°)_s$) and by Weiss [6] ($31 \times 10^3$ lb/in$^2\sqrt{\text{in}}$, $\alpha = 0°$; 19×10^3 lb/in$^2\sqrt{\text{in}}$, $\alpha = (\pm45°)_s$) using other specimen geometries (shape and thickness) and load arrangements.

Inspection of Table 1 will show further that the K_Q values for various starter crack sizes are within a reasonable range of the average $\bar{K}_Q$ for each laminate. It should also be noted that the majority of largest deviations occur for subsize starter cracks, and none of these is serious.

CONCLUDING REMARKS

This pilot test series has been successful, for it has answered the questions posed at the outset. The failure mechanism of the specimen tested is crack dominated in most cases, and the procedures of LEFM can be applied even where the overt failure mechanism is not so obviously dominated by the starter crack.

There remains, however, a variety of questions about cracks in an advanced fiber composite material. Some concern the effects of specimen geometry and load arrangement, and can be answered only by further testing. Such work is needed, first, to define and delineate more fully the respective influence of cracks and material.

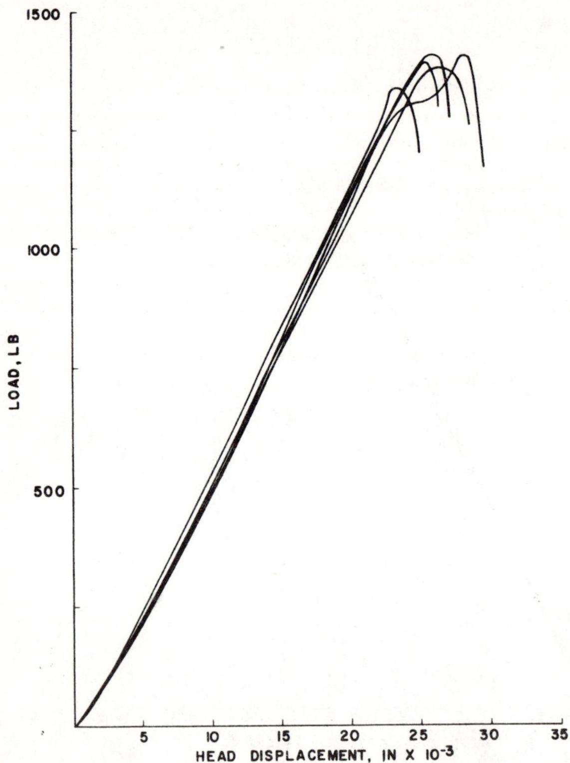

Figure 11. Traces of load vs. cross-head displacement for five specimens used in reproducibility tests for a multi-directional laminate ($\alpha = (0°/\pm45°/90°)_s$, $a = 0.4$ in).

Further, the entire matter of fracture in composites needs for its resolution an extensive data base similar to that which has evolved for metals. The building of this kind of experience is important not only to determine what constitutes meaningful laboratory work, but also to provide guidance in treating service situations. We need also to learn how experimentally determined K_Q values for given laminates might be related to the properties of individual laminae within other laminates. Ultimately, the designer should be in a position to use fracture toughness as he would other material properties.

It would now appear that efforts to address these questions are warranted, for the present test series indicates that, when suitably modified to account for anisotropy, linear elastic fracture mechanics may usefully be applied to advanced fiber composite materials.

ACKNOWLEDGMENTS

This work was supported jointly by NASA Research Grant NGR-39-002-023 and U. S. Air Force Contract F33615-70-C-1146.

REFERENCES

1. Tentative Method of Test E 399 for Plane-Strain Fracture Toughness of Metallic Materials, 1971 *Annual Book of ASTM Standards,* Part 31, American Society for Testing and Materials, Philadelphia (1971).
2. J. E. Srawley and W. F. Brown, Jr., *Plane Strain Crack Toughness Testing of High Strength Metallic Materials,* ASTM STP 410, American Society for Testing and Materials, Philadelphia (1967).
3. *Fracture Toughness Testing and its Applications,* ASTM STP 381, American Society for Testing and Materials, Philadelphia (1965). (n.b. p. 52 et seq.)
4. G. C. Sih, P. C. Paris, and G. R. Irwin, "On Cracks in Rectilinearly Anisotropic Bodies," *International Journal of Fracture Mechanics* Vol. 1 (1965) 189-203.
5. J. C. Halpin, U. S. Air Force Materials Laboratory, private communication, 1971.
6. O. E. Weiss, Convair Aerospace Division, General Dynamics Corporation, Fort Worth, private communication, 1971.

Mixed-Mode Fracture of Unidirectional Graphite/Epoxy Composites

J. M. McKinney, *Lockheed-Georgia Company, Marietta, Georgia 30060*

(Received October 29, 1971)

Unidirectional graphite/epoxy laminates containing crack-like defects parallel to the fibers were tested. The results of these tests were used to construct an interaction diagram for mixed-mode fracture of graphite/epoxy. This interaction diagram, when plotted in normalized form, is closely described by the equation $k_1/k_{1c} = 1 - k_2/k_{2c})^2 - 1$ just as Wu [1] found for two other unidirectionally orthropic materials, Scotchply and balsa wood.

SPECIMEN PREPARATION

Specimens used in this investigation were fabricated of Rolls Royce low modulus graphite fiber prepregged with 3M's PR-279 resin system and were of two different configurations. Type I specimens were used to obtain critical stress-intensity-factor data for opening and combined opening/forward-sliding modes of crack extension. Type II specimens were used to obtain critical stress-intensity-factor data for the forward sliding mode of crack extension. A strip of white "dye-chem developer" was sprayed on the surface of each specimen to allow visual detection of dye penetrant (see "Testing Procedure" section) which was used to define the extremities of the crack. All panels were 4-plies thick. Specimen load introduction was achieved by means of fiberglass tabs bonded to the specimen ends.

A crack was introduced into the center of each specimen via the following sequence of operations.

1. A 0.060" diameter hole was drilled in the geometric center of the specimen.
2. Starting from this hole, a jeweler's saw (nominal width, 0.010") blade was used to extend a slit along the fiber direction and symmetrically located about the hole.
3. Each end of the slit was then wedged open using a razor blade. The resulting crack tip configurations were thought to more realistically simulate a fatigue crack.

The resulting initial crack lengths were measured with the aid of an optical comparator.

TESTING PROCEDURE

All specimens were tested in a 30,000 lb_f capacity Reihle U.T.M. Both 1500 lb_f and 6000 lb_f load ranges were used during the test program. Both ranges had accuracies of ±0.5% of indicated load. Loading was accomplished through spherically seated end fittings at approximately 0.025 inch/minute head travel rate for Type I specimens. Type II specimens were loaded through spherically seated end fittings at approximately 0.050 inch/minute head travel rate.

Because the graphite/epoxy specimens tested were characteristically black and

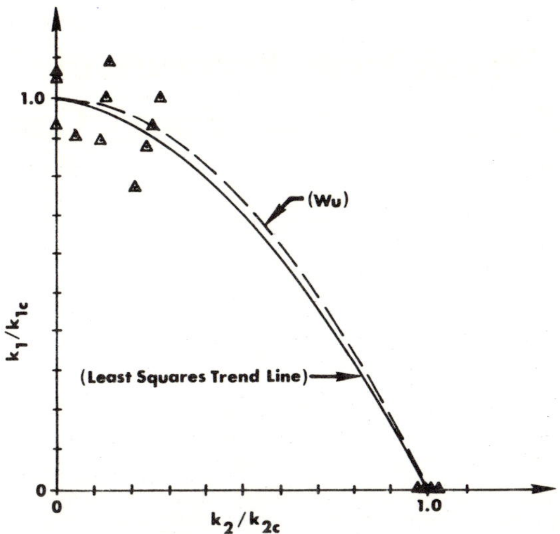

Figure 1. Normalized interaction diagram for mixed-mode fracture of unidirectional graphite/epoxy.

because the cracks under observation were extremely narrow, unaided visible detection of any slow crack growth taking place prior to catastrophic fracture was almost impossible. The procedure finally adopted for detecting crack lengths is outlined below.

1. A strip of (white) developer was sprayed on the front surface of each specimen.
2. An acoustic emissions amplifier was located in the vicinity of the crack.
3. Whenever either end of the crack extended stably during load application, the acoustic emission device amplified the sound accompanying this extension so that the test machine operator could hear it. Further loading was then stopped until the crack length could be measured.
4. At the instant a noise was heard, loading was stopped and dye penetrant was sprayed onto the back side of the specimen at the crack tips. The penetrant quickly traveled through to the front side of the specimen where, upon reacting with the developer, it clearly revealed the extremities of the crack tip. The effect of dye penetrant on propagation of cracks in graphite/epoxy is not known. Therefore, in interpreting these test results, it was assumed that there is no effect.
5. A microscope and steel scale were then used to measure the total crack length. This value was recorded and loading of the specimen resumed until the next noise was heard.

RESULTS

Both Types I and II specimens were tested under monotonically increasing load conditions (see *Testing Procedure*). During this loading process, stable crack lengths were measured and recorded. This crack growth information was then recorded in graphical form and used to determine both the critical crack length and the critical value of applied load. This information was then used to determine critical stress-intensity-factors, k_{1c} and k_{2c} [1, 2]. These values are used to plot the normalized interaction diagram shown in Figure 1. Although the curve is based on only 15 data

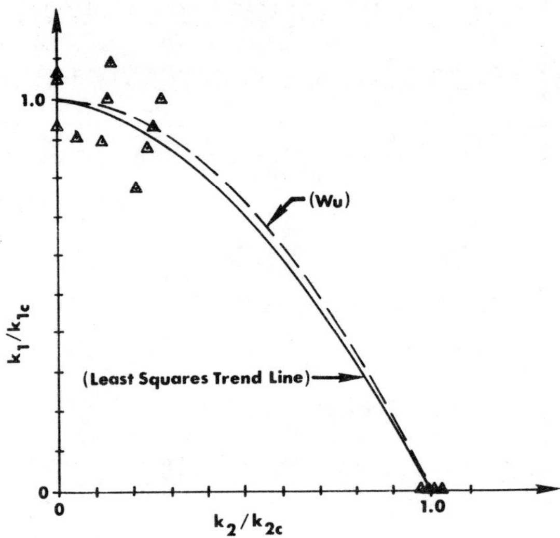

Figure 1. Normalized interaction diagram for mixed-mode fracture of unidirectional graphite/epoxy.

because the cracks under observation were extremely narrow, unaided visible detection of any slow crack growth taking place prior to catastrophic fracture was almost impossible. The procedure finally adopted for detecting crack lengths is outlined below.

1. A strip of (white) developer was sprayed on the front surface of each specimen.
2. An acoustic emissions amplifier was located in the vicinity of the crack.
3. Whenever either end of the crack extended stably during load application, the acoustic emission device amplified the sound accompanying this extension so that the test machine operator could hear it. Further loading was then stopped until the crack length could be measured.
4. At the instant a noise was heard, loading was stopped and dye penetrant was sprayed onto the back side of the specimen at the crack tips. The penetrant quickly traveled through to the front side of the specimen where, upon reacting with the developer, it clearly revealed the extremities of the crack tip. The effect of dye penetrant on propagation of cracks in graphite/epoxy is not known. Therefore, in interpreting these test results, it was assumed that there is no effect.
5. A microscope and steel scale were then used to measure the total crack length. This value was recorded and loading of the specimen resumed until the next noise was heard.

RESULTS

Both Types I and II specimens were tested under monotonically increasing load conditions (see *Testing Procedure*). During this loading process, stable crack lengths were measured and recorded. This crack growth information was then recorded in graphical form and used to determine both the critical crack length and the critical value of applied load. This information was then used to determine critical stress-intensity-factors, k_{1c} and k_{2c} [1, 2]. These values are used to plot the normalized interaction diagram shown in Figure 1. Although the curve is based on only 15 data

The Effect of Specimen and Testing Variables on the Fracture of some Fibre Reinforced Epoxy Resins

C. D. ELLIS AND B. HARRIS

School of Applied Sciences
University of Sussex
Brighton BN1 9QT, England

(Received November 13, 1972)

ABSTRACT

The work of fracture of composite materials has often been measured without sufficient attention, for example, to specimen geometry and test conditions. In this work a range of testing techniques has been used to investigate how the measured work of fracture of unidirectionally-reinforced carbon fibre/epoxy composites, containing fibres of types 1 and 2, changes with various specimen and test variables. The measured work of fracture is seen to be independent of specimen width, and shows but slight dependence on notch root radius and test rate. More significant variations occur with orientation, specimen height and notch depth, however.

Attempts are described to measure other fracture parameters which, for this class of materials, might prove to be of more value than the total work of fracture. In particular, an effort has been made to adapt the methods of Linear Elastic Fracture Mechanics, and some degree of success has been achieved in relating K_{1c}, the critical stress intensity factor, measured by double edge notched plate techniques, and G_{1c}, the critical strain energy release rate, with those measurements discussed previously.

INTRODUCTION

THE WORK OF fracture (γ) of fibre reinforced materials has been measured by a number of investigators (see for example Cooper [1], Hancox [2], Harris et al [3], Sidey and Bradshaw [4] and Beaumont and Phillips [5]). A variety of test methods and specimen designs has been employed and it has not always been clear to what extent the measured values of work of fracture so obtained have been a function of the test method or specimen design. The first part of this work describes an attempt to isolate the effect of some test and specimen design variables on the measured work of fracture of a number of fibre reinforced epoxy materials.

The work of fracture, however, is not necessarily the most useful fracture parameter that can be determined for composites, either from a materials characterisation or from a design viewpoint, and this paper also describes attempts to measure other parameters which could prove to be of more value. In particular an effort has been made to adapt the test methods of Linear Elastic Fracture Mechanics (LEFM) to this class of materials, and to relate the parameters so measured to the work of fracture.

MATERIALS

The composites used most in this work have consisted of Types 1 and 2 untreated carbon fibres (Morganite Modmor Ltd.) incorporated in an epoxy resin (Epikote DX-209 cured with BF_3400 in the ratio 32:1). They were made by a pre-impregnation and hot-pressing technique to a standard fibre volume fraction of 0.4. A certain amount of comparative work has also been performed using a silica fibre reinforced epoxy composite supplied to participants in the AGARD (NATO Advisory Group on Aerospace Research and Development) cooperative research programme on composite materials. This material was of 0.7 V_f and was manufactured by the Company Technique du Verre Tissé by a pre-impregnation, filament winding and hot pressing technique.

The mechanical properties of these test materials are presented in Table 1.

Table 1. Mechanical Properties of Test Materials

Test	*Tensile*	*Long Beam Bend Test Span:depth = 16:1*		*Short Beam Bend Test; 5:1*
Property measured	U.T.S.	Flexural strength	Flexural modulus	Interlaminar shear strength
Units	$MN.m^{-2}$	$MN.m^{-2}$	$GN.m^{-2}$	$MN.m^{-2}$
Type 1 C.F./Epoxy ($V_f = 0.4$)	660	580	87.5	25.5
Type 2 C.F./Epoxy ($V_f = 0.4$)	821	743	61.1	38.4
Silica/Epoxy ($V_f = 0.7$)	1350	1070	41.8	68.5

WORK OF FRACTURE TESTING

Test Method

Four test methods have been used to determine work of fracture values: Charpy impact testing using a Hounsfield Plastics Impact Testing Machine of capacity 2.7 *J*

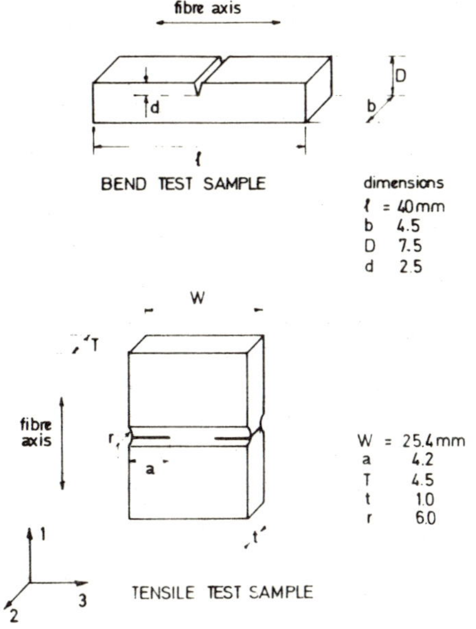

***Figure 1.** Details of samples used in notched bend tests and tensile tests.*

and striker velocity 2.4 m.sec^{-1}; instrumented drop-weight impact testing using a machine of capacity 27 *J* and maximum striker velocity 4.6 m.sec^{-1}; medium rate bend testing using a Cape Engineering servo-hydraulic fatigue testing machine; and slow bend testing using an Instron Universal Testing Machine.

The specimens employed were edge-notched bend specimens cut with a diamond wheel and notched on a milling machine using a frequently reground tool (Figure 1). All specimens except those used for investigating orientation-dependence were cut with their axis parallel to the fibre axis and were notched perpendicular to the fibre direction; in all the tests except those where the effect of notch root radius was being investigated the notch was sharpened with a scalpel.

A Charpy test machine gives a direct read-out of the energy lost by the pendulum in fracturing the sample. This is divided by twice the cross-sectional area in the plane of the notch (by analogy with surface energy measurements) to obtain the work of fracture (γ). The other test methods yield a load/deflection plot, either on a chart recorder in the case of the Instron tests, or on a storage oscilloscope in tests at higher deformation rates. This load/deflection plot is integrated to give the total energy required to break the sample completely, and this energy is similarly divided by twice the cross-sectional area in the plane of the notch to obtain γ.

The Charpy test, which furnishes only a measure of the total energy involved in the fracture process, is less useful than those methods which also provide load-deflection data, and the results can sometimes be misleading. For example, the Charpy work of fracture of standard sized silica/epoxy samples as received and after exposure to steam are quite similar, being measured as 1.4 and 1.2 kJ m^{-2} respectively. The load-deflection traces for the two materials are, however, very different as shown in Figure 2. The Charpy test can be placed on a par with the other test methods by instrumentation as described by Wullaert [6].

This particular example also illustrates the point that the total work of fracture (γ) is not necessarily the most useful fracture parameter that can be measured since it masks such effects. If γ_{total} is considered to be the sum of two components γ_i, the work of fracture prior to crack initiation, and γ_p, the work of fracture associated with crack propagation, it can be seen that the energy the material is

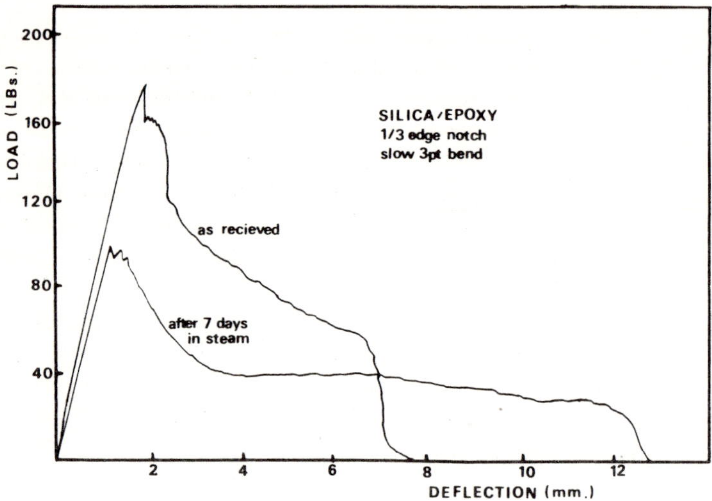

Figure 2. Load/deflexion traces obtained from three point bend tests on notched samples of SiO_2/epoxy composites, showing the effect of exposure to moisture.

capable of absorbing before initial crack propagation (γ_i) is possibly a more useful parameter than γ_{total} since, provided γ_p is sufficiently large to avoid catastrophic brittle failure, its actual magnitude is relatively unimportant. For this reason γ_i values are included in many of the results presented.

Results and Discussion

γ (and in some cases γ_i) have been measured as a function of the bending mode (i.e., three or four point bending) (Table 2), rate of loading (Figure 3), specimen notch root radius (Figure 4), specimen width (Figure 5), specimen height and notch depth ratio (Figure 6), and specimen orientation (Figure 7).

Table 2. Work of Fracture of C.F./Epoxy in 3 & 4 Point Bending.

BENDING MODE	TYPE 1		TYPE 2	
	Charpy Test	Slow Bend Test	Charpy Test	Slow Bend Test
3 Point	24.8	18.8	50.9	38.5
4 Point	26.0	23.2	54.7	43.7

All results are in kJ/m^2 and are averages of between 3 and 6 tests
All material 40% fibre volume fraction
Notch depth = 1/3 specimen depth (single edge notched)

From Table 2 the four point bending test can be seen to give consistently higher values of γ, the effect being more pronounced at lower testing rates. The main difference in the stress system between the two tests is that in four point loading the bending beam shear component is eliminated between the central loading supports. As untreated carbon fibre composites typically have a tensile/shear ratio of 30:1 initial failure in a mixed stress situation is likely to be in shear. Elimination of the bending beam shear component in the region of the notch could be the reason for the higher failure loads and fracture energies recorded in four point bending.

In some of these tests, however, a certain amount of compressive damage can be seen to be associated with the central loading support. Compressive strengths measured for materials of this type are lower than the tensile strengths and the lower values obtained in three point bending could be partly a result of compressive damage under the single loading support.

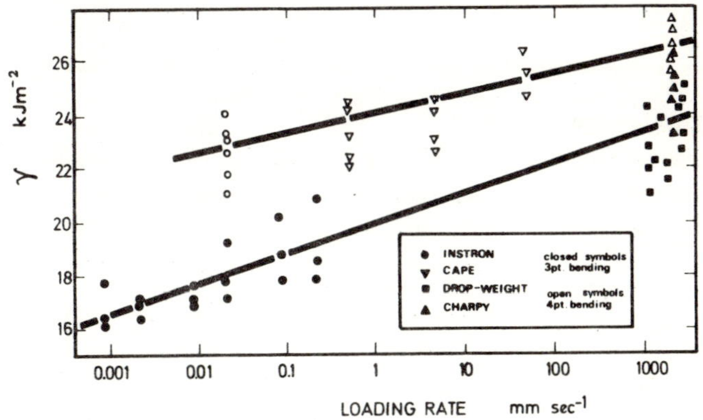

***Figure 3.** Dependence on rate of loading of the measured value of the work of fracture of carbon fibre composites (type I fibre). The loading rate is expressed as the rate of midpoint deflexion in the bend test sample shown in Figure 1.*

Figure 3 shows the results for three and four point bend testing of type 1 C.F.R.P., the work of fracture showing an increase with increasing loading rate. However this is not a large effect, amounting to only a 30% increase over seven decades. Comparing the load-deflection traces at various test rates (Figure 8) the increase in γ can be seen to be the result of an increase in the contribution to the total energy of the unloading portion of the test. This is the region in which fibre pull-out will be occurring and the pull-out contribution will be dependent on τ_i', the frictional interfacial shear stress. This frictional interfacial shear stress will probably be strain-rate dependent because of the viscoelastic nature of the resin and will increase with strain-rate as pointed out by Beaumont and Phillips [5]. An increase in τ_i' over the seven decades of about 1 MNm^{-2} would be sufficient to account for the observed strain-rate dependence of γ.

Figure 4 shows that the measured work of fracture of both C.F.R.P. and silica/epoxy decreases with decreasing notch root radius; again the effect is a relatively minor one. The results are as one would expect from a knowledge of notch stress theory; the sharper the notch, the greater the stress concentration and the lower the energy input necessary to propagate the crack. Also, predictably, the effect is greater in the silica/epoxy composite which has much higher interlaminar shear stress (68.5 $MN.m^{-2}$) than the untreated type 1 C.F.R.P. (25.5 $MN.m^{-2}$). Thus because delamination, leading to crack blunting in the manner described by Cook and Gordon [7], is easier in the C.F.R.P. the sharper crack is less effective than it is in the better bonded silica/epoxy.

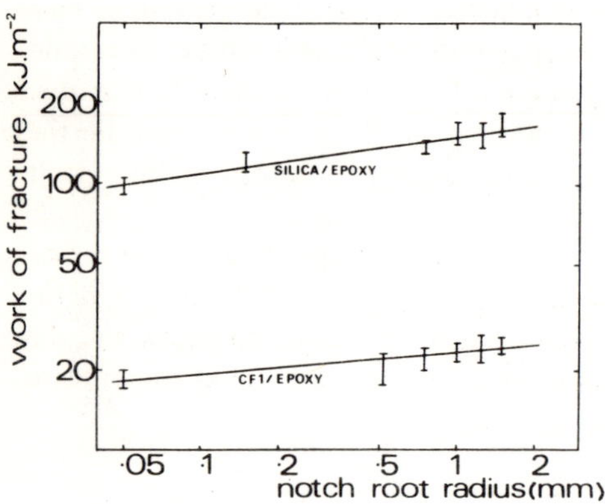

Figure 4. The effect of specimen notch tip radius on measured work of fracture values for SiO_2/epoxy and type I carbon fibre/epoxy.

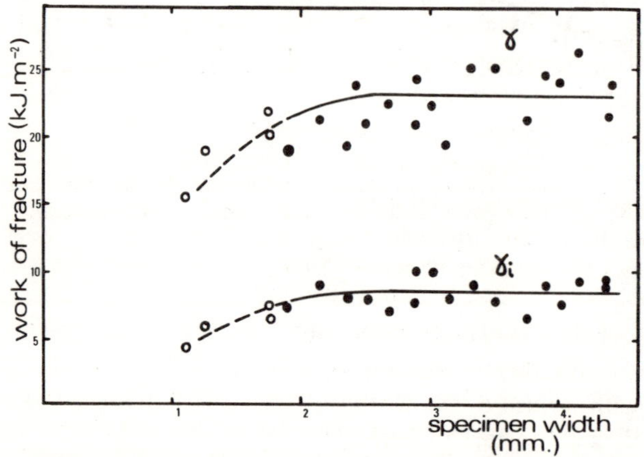

Figure 5. Dependence of work of fracture on sample width (type I carbon fibre/epoxy – 3 point notched bend tests).

Figure 5 shows that there is no effect of specimen width on measured γ until very low widths are reached. With these very thin samples it is not certain that what is being measured is a genuine material property since the specimens are prone to

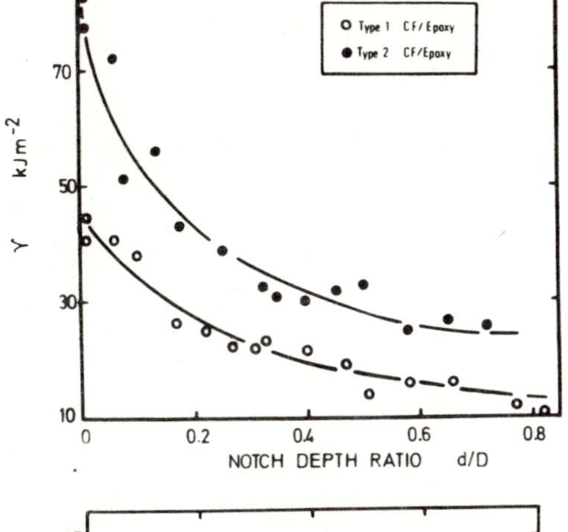

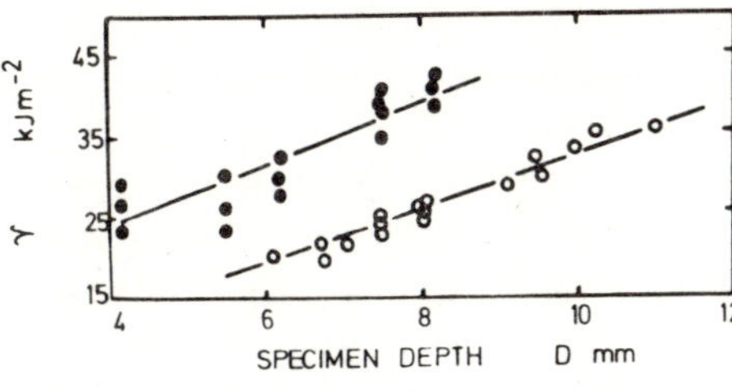

Figure 6. The effects of varying the notch depth ratio, d/D, and the sample height, D, on the measured work of fracture values for type I and type II carbon fibre/epoxy composites.

failure by buckling in this region. An attempt has been made to overcome this buckling by using samples in which the width is preferentially reduced in the cross-section of the notch. The results from these samples confirm the fall off in γ at low widths, but the complex stress situation introduced in these samples may also be affecting the fracture behaviour.

Figure 6a shows the variation in measured work of fracture using a standard sized sample and varying the notch depth ratio (d/D). Figure 6b shows the effect of taking samples with a fixed notch depth ratio, $d/D = 1/3$, but varying the height of the sample (D). Plotting these two sets of results on the same graph as a function of the uncracked height of the sample (D–d) gives Figure 9 where the work of fracture shows an almost linear dependence on the uncracked height. Samples of very different heights and notch depths can thus exhibit similar works of fracture if their uncracked heights are the same.

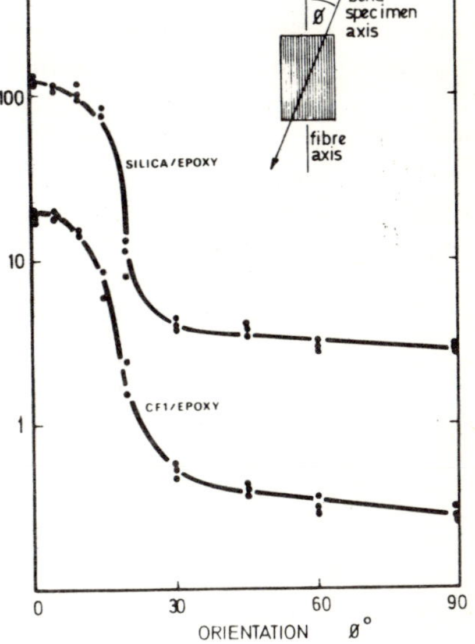

Figure 7. Variation of the work of fracture of SiO_2/epoxy and type I carbon fibre/epoxy composites with orientation. ϕ is the angle between the fibre axis and the normal to the crack face, i.e., for $\phi = 0$ the crack is travelling perpendicular to the fibres.

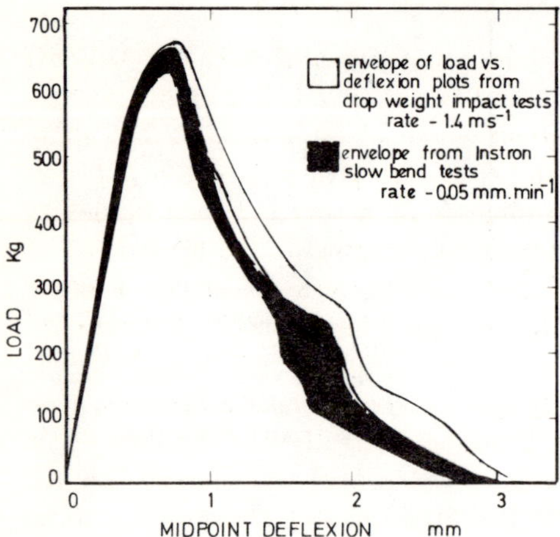

Figure 8. Load/deflexion traces obtained from type I carbon fibre/epoxy composites at various loading rates.

Hancox [2] found a similar dependence of measured work of fracture on sample height in Izod impact testing and although an expression similar to the one he derived does not explain too well the observed behaviour it does suggest that the stored shear strain energy in the beam, which of course increases with the height of the beam, is making an important contribution to the fracture energy.

The results of a series of tests in which the work of fracture was measured as a function of the angle between the specimen and fibre axes (ϕ) are shown in Figure 7. In both materials there is a rapid drop in the work of fracture once ϕ exceeds about 10°, the silica/epoxy retaining its fracture toughness at slightly greater values of ϕ, once again as a result of the better fibre/resin bond. These angles are about the same as those at which strength and modulus also begin to fall rapidly [8] which additionally emphasizes the extremely anisotropic behaviour of unidirectionally reinforced materials.

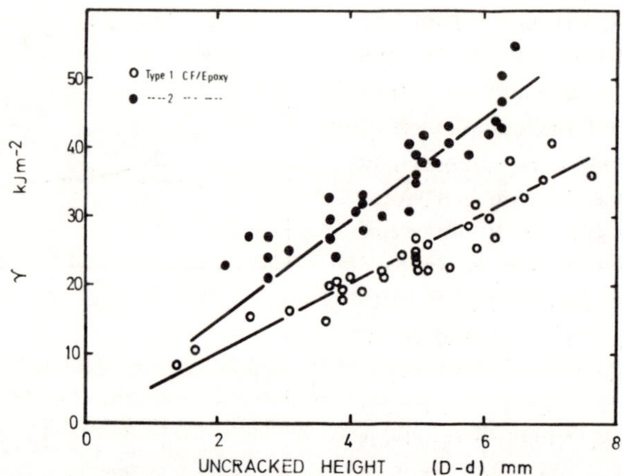

Figure 9. Data replotted from Figure 6(a) and (b) to show how the measured fracture work of type I and type II carbon fibre/epoxy composites varies with the sample depth under the notch or the "uncracked height," (D-d).

APPLICATION OF FRACTURE MECHANICS

In designing with high strength alloys, linear elastic fracture mechanics (LEFM) has proved to be invaluable in predicting the flaw sensitive failure of components on the basis of laboratory scale tests. Various workers have already tried to apply LEFM to fibre reinforced materials; Corten [9] and Wu [10] mostly from a theoretical point of view and Hardy [11], Sanford and Stonesifer [12] and Beaumont and Harris [13] principally from an experimental approach. The application of LEFM to fibre-reinforced materials obviously involves more than just a simple extension of the techniques used for the isotropic, homogeneous materials for which LEFM were developed. The materials being considered here are orthotropic and consist of two phases, one of which exhibits a degree of viscoelasticity.

The first departure of these materials from the assumptions of the theory, the orthotropy, has been studied by Sih, Paris and Irwin [14] who showed how the normal Irwin-Westergaard expressions for the stress about a crack tip has to be modified for a crack in a homogeneous, orthotropic solid. They showed that the expression for the critical stress intensity factor in the opening mode, K_{1c}, remained the same, but that the expression for the critical strain energy release rate, G_{1c} in plane strain, for crack propagation along a plane of elastic symmetry perpendicular to the orthotropic axis became:

$$G_{1c} = K_{1c}^2 \sqrt{\frac{a_{11}a_{22}}{2}} \left[\sqrt{\frac{a_{11}}{a_{22}}} + \frac{2a_{12} + a_{66}}{2a_{22}}\right]^{1/2} \tag{1}$$

with the notation as defined in Figure 1.

This can be written in terms of conventional engineering moduli and greatly simplified for a material with a reasonable degree of orthotropy (i.e., where $E_{11} E_{22} \gg G_{12}^2$). Expression (1) then becomes

$$G_{1c} \cong K_{1c}^2 / 2\sqrt{E_{11}G_{12}} \tag{2}$$

where the denominator can be considered as an effective modulus (E_{eff}) by analogy with the simple expression for an isotropic material in plane stress [15] $G_{1c} = K_{1c}^2/E$.

The usual method of dealing with the other two ways in which fibre-reinforced materials depart from the assumptions of LEFM is to assume that they do not; i.e , to treat them as being homogeneous and linearly elastic up to failure.

Experimental

The apparent critical stress intensity factor, K_{1c} has been measured using edge-notched bend samples identical to those used in the work of fracture testing, and edge-notched tensile samples also shown in Figure 1. K_{1c} values have been

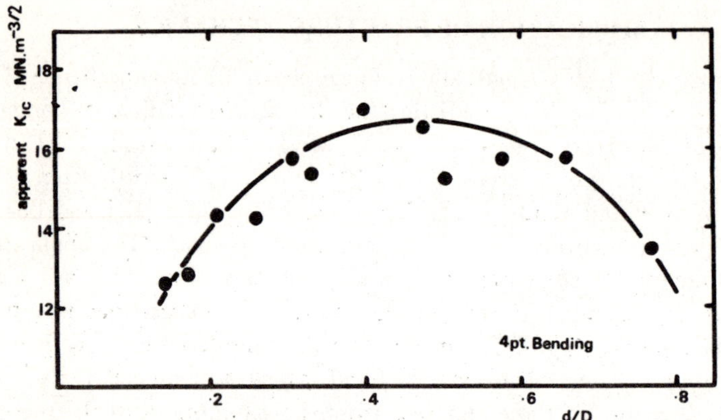

Figure 10. The variation of apparent K_{1c} for type I carbon fibre/ epoxy composites with notch depth ratio, d/D, in 4 point bending, and with the equivalent ratio, a/W, for edge notched plates tested in tension.

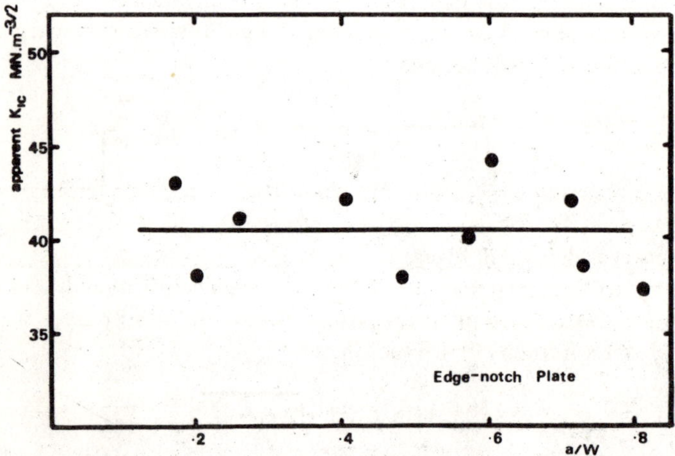

calculated using the analyses of Srawley and Brown [15]. Figure 10a shows the measured variation of apparent K_{1c} with d/D ratio for the bend samples and Figure 10b shows the equivalent variation with a/W ratio for the tensile plate samples. It can be seen that only in the case of the plate data do we appear to be measuring a possible material parameter.

The critical strain energy G_{1c} has also been measured for bend and tensile samples using a series of samples with increasing notch depths and plotting compliance/crack area curves. G_{1c} is then obtained as described by Corten [9] from the expression

$$G_{1c} = 1/2\ P^2\ (dC/dA)_A \qquad (3)$$

where P is the load to propagate a crack of area A: this method of evaluation does not involve prior consideration of anisotropy.

For a linear elastic material $G_{1c} = 2\gamma$, so it is possible, making the assumption of linear elastic behaviour, to compare G values obtained from the work of fracture tests, G values obtained from compliance testing and G values calculated from apparent K_{1c} values using Equation (2). The results are shown in Table 3.

Table 3. Comparison of Work of Fracture & Fracture Mechanics Data

Parameter measured	Type 1 C.F./Epoxy	Type 2 C.F./Epoxy
2γ (slow bend test; edge-notched sample)	36	78
$2\gamma_i$ (slow bend test; edge-notched sample)	17	36
$G_1 = \frac{K_{1c}^2}{E_{eff}}$ (tensile test; double edge-notched plate)	34	85
$G_1 = \frac{K_{1c}^2}{E_{eff}}$ (slow bend test; edge-notched sample)	5	11
G_1 (compliance analysis; bend test on edge-notched sample)	14	48

All results are in kJ/m^2 and are averages of 3 to 6 tests
All results obtained at a testing rate of 1.25 $mm.min^{-1}$
All bend samples $d/D = 1/3$, all plate samples $^{2a}/w = 1/3$

Discussion

The K_{1c} measurements and G_{1c} values obtained by the compliance technique both consider only the initial load to propagate the crack. It might be expected therefore that they could be most easily related to γ_i, the work of fracture prior to crack propagation, rather than to γ which involves the energy to propagate the crack over the whole sample. From Table 3 it can be seen that a good degree of correlation is achieved between the G_{1c} values obtained by the compliance technique and $2\gamma_i$. The correlation is less satisfactory in the case of the G_{1c} values obtained from the apparent K_{1c} results; the bend sample data gives G_{1c} values considerably less than $2\gamma_i$ and the values obtained from the tensile plate data are greater than $2\gamma_i$ and in fact much closer to 2γ.

The compliance results could be expected to show better agreement with the work of fracture data as they involve fewer assumptions and are, so to speak, self calibrating, whereas the G_{1c} values obtained from K_{1c} results involve not only the initial theoretical expression for K_{1c} in terms of the load to propagate the crack and specimen dimensions, but also Equation (1) relating K and G for orthotropic material.

Although to some extent the degree of correlation obtained might be claimed to be fortuitous the fact that all the values obtained are very much the same order of magnitude does suggest that the application of fracture mechanics to composite materials is a fruitful line of research.

CONCLUSIONS

The measured work of fracture values of several fibre reinforced materials has been shown to be very dependent on the dimensions of the sample and the type of test. Great care must be exercised if results are to be scaled up or down for design purposes. It is also vital that if valid comparisons are to be made between any fracture data complete test conditions must be specified.

An attempt has been made to compare work of fracture data with fracture parameters measured by a linear elastic fracture mechanics approach, but more theoretical and experimental work is required before fracture mechanics tests can be used as confidently with this class of materials as they are with high strength metals.

ACKNOWLEDGMENTS

The work described is part of a continuing programme of research on the fatigue and fracture of composites sponsored jointly by the Science Research Council, I.C.I. Ltd., G.K.N. Ltd., and by the Ministry of Defence. We gratefully acknowledge the support of these organisations. In addition C. D. Ellis was in receipt of an S.R.C. studentship whilst this work was carried out.

REFERENCES

1. G. A. Cooper, "The Fracture Toughness of Composites Reinforced with Weakened Fibres," *J. Mat. Sci.*, Vol. 5 (1970), p. 645.
2. N. L. Hancox, "Izod Impact Testing of Carbon Fibre Reinforced Plastic," *Composites*, Vol. 2 (1971), p. 41.
3. B. Harris, P. W. R. Beaumont and E. Moncunill de Ferran, "Strength and Fracture Toughness of Carbon Fibre Polyester Composites," *J. Mat. Sci.*, Vol. 6 (1971), p. 238.
4. G. R. Sidey and F. J. Bradshaw, "Some Investigations on Carbon Fibre Reinforced Plastics under Impact Loading and Measurement of Fracture Energies," International Conference on Carbon Fibres their Composites and Applications, London (1971), paper 25.
5. P. W. R. Beaumont and D. C. Phillips, "The Fracture Energy of a Glass Fibre Composite," *J. Mat. Sci.*, Vol. 7 (1972), p. 682.
6. R. A. Wullaert, "Applications of the Instrumented Charpy Impact Test," Impact Testing of Metals, ASTM STP 466 (1970), p. 148.
7. J. Cook and J. E. Gordon, "Mechanism for the Control of Crack Propagation in all Brittle Systems," *Proc. Roy. Soc.*, Vol. A282 (1964), p. 508.
8. J. Dimmock and M. Abrahams, "Prediction of Composite Properties from Fibre and Matrix Properties," *Composites*, Vol. 1 (1969), p. 87.

9. H. T. Corten, "Influence of Fracture Toughness and Flaws on the I.L.S.S. of Fibrous Composites," Fundamental Aspects of Fibre Reinforced Plastics Composites ed. Shwartz and Shwartz, New York Interscience (1966), p. 89.
10. E. M. Wu, "Some Unique Crack Propagation Phenomena in Unidirectional Composites and their Mathematical Characterization," Conference on Structure, Solid Mechanics and Engineering Design ed. M. Te'eni, New York Wiley-Interscience (1969).
11. G. F. Hardy, "Fracture Toughness of Glass Fibre-Reinforced Acetal Polymer," *J. Appl. Polym. Sci.*, Vol. 15 (1971), p. 853.
12. R. J. Sanford and F. R. Stonesifer, "Fracture Toughness Measurements in Unidirectional G.R.P.," *J. Composite Materials*, Vol. 5 (1971), p. 241.
13. P. W. R. Beaumont and B. Harris, "The Energy of Crack Propagation in Carbon-Fibre Reinforced Resin Systems," *J. Mat. Sci.*, in press.
14. G. C. Sih, P. C. Paris and G. R. Irwin, "On Cracks in Rectilinearly Anisotropic Bodies," *Int. J. Fracture Mechanics*, Vol. 1 (1965), p. 189.
15. J. E. Srawley and W. F. Brown, "Fracture Toughness Testing Methods," Fracture Toughness Testing and its Applications, ASTM STP 381 (1964), p. 133.

Fractographic Study of Graphite-Epoxy Laminated Fracture Specimens

THOMAS A. CRUSE AND MICHAEL G. STOUT
Department of Mechanical Engineering, Carnegie Institute of Technology, Carnegie-Mellon University, Pittsburgh, Pennsylvania 15213

(Received February 2, 1973)

INTRODUCTION

The purpose of this study was to determine what, if any, information can be deduced from the fracture surfaces of a series of graphite/epoxy fracture specimens. The fracture specimens were fabricated from NARMCO graphite/epoxy with Morganite II fibers in 5206 resin and were tested in three point bending. The results of the fracture test program have been reported previously [1].[1]

In the earlier report [1] the tentative conclusion was reached that the data reduction scheme of linear elastic fracture mechanics (LEFM) could be used to correlate the graphite/epoxy fracture strength data. Some of the important questions concerning the applicability of the LEFM model to composites center on the importance of the material microstructure as it affects the continuum model used in LEFM. In particular, we will address three major points in this report:

i. How much matrix cracking precedes fracture of the specimen?
ii. To what extent do the fibers fail independently?
iii. To what extent do the plies fail independently?

One of the tools of fractography is a post-mortem optical study of the fracture surface. In metal fractography this study is accomplished at very high magnification such that the path of fracture can be traced to its origin. In composites fractography this does not seem to be possible due to the very high variability of the local fracture processes. In this study magnifications on the order of 90*X* proved to be the most useful, making use of the scanning electron microscope (SEM).

FRACTURE SURFACE OBSERVATIONS

One of the typical fracture specimens from the (0/±45/90) laminate family from [1] was examined with the SEM. The SEM photographs are taken at 90*X* with the axis of the camera 20° from the plane of the original crack. Figures 1 and 2 are photographs taken along the plane of crack propagation, following the same set of

[1] Figures showing the test specimens and their fracture surfaces are given in [1]; for completeness the reader is encouraged to see these figures.

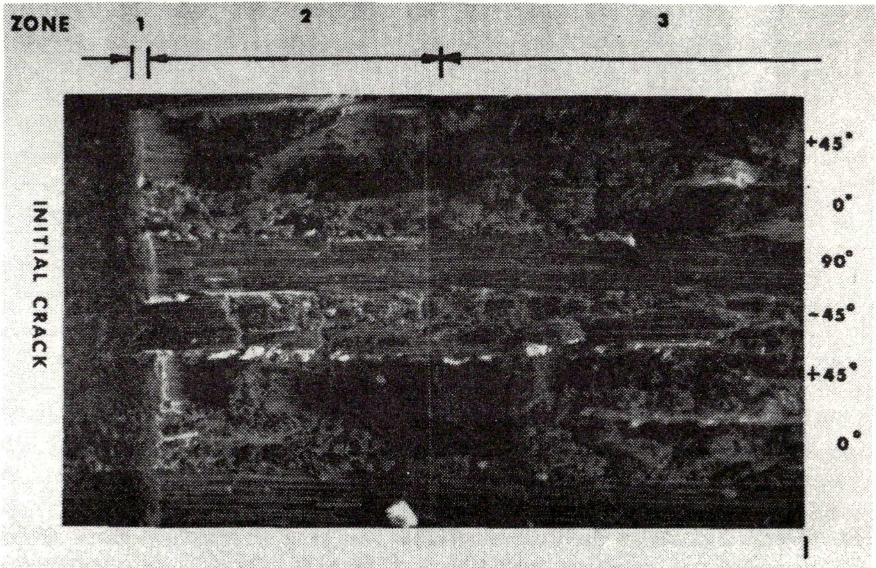

Figure 1. Fracture surface for (0/±45/90) laminate; plate 1.

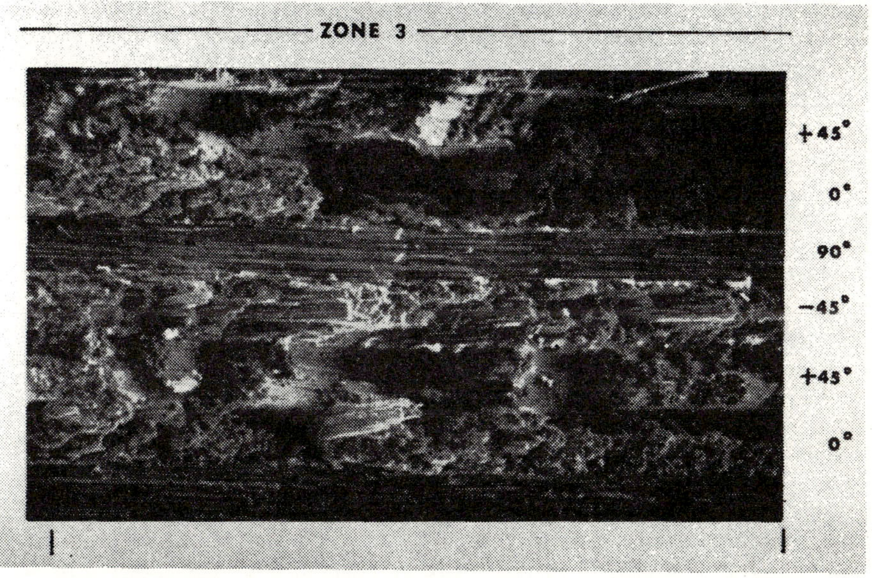

Figure 2. Fracture surface for (0/±45/90) laminate; plate 2.

plies. The small horizontal line at the right side of the figures indicates how the photographs overlap.

Three zones are deduced from the series of photographs. The first zone is the very small region of fiber ends and irregular damage from the razor cut made to sharpen the original crack. Figure 1 shows this small zone as a light area where the 90° layer and the original crack plane join.

The second zone is called the zone of initial damage. This zone is best seen in Figure 1 extending from zone 1 half-way to the top of the figure. While the effect is more dramatic in stereo, the plies are seen to have fractured along planes at significantly different elevations in adjacent plies. This is particularly visible in Figure 4 for the 45° plies. The length of Zone 2 (roughly 0.010 inches) correlates very well with the size of the region of material that exceeds fiber and matrix allowables, at the fracture load [2]. It is for this reason that we refer to Zone 2 as the initial damage zone.[2]

In Zone 3 (to about 0.05 inches from the original crack tip) the fracture surface elevation is random compared to Zone 2 and exhibits a very *dependent* nature of fracture in all of the plies. That is, fracture surfaces in adjacent plies are at the same elevation for most locations. This zone initially appears to be one of general instability along the crack plane, probably with simultaneous ply fracture in the portion of the zone shown in Figure 2. Zone 3 will be called the instability zone. It is the zone which is subject to a stress field governed by a level of stress intensity factor.

Figure 2 shows some inter-fiber surfaces in the (±45) plies which result from the saw-tooth pattern of failure in these plies. However, note the very strong dependence of the failure surfaces between the 0° ply and the adjacent 90° and 45° plies. This contrasts strongly with Zone 2 where the plies are failing in independent modes, as if isolated by matrix cracking.

Since Zone 2 is small compared to the original crack length, it may be concluded that the change in the nominal stress field due to initial damage is slight. The highly dependent nature of the fracture surface in the near crack-tip portion of Zone 3 indicates that the level of inhomogeneity associated with both fiber size and ply size is not important to the fracture instability of graphite/epoxy. This observation supports the success of the application of LEFM reported in [1]. As discussed in [4] the fibers appear to fail in groups, rather than as single fibers.

The photographs in Figure 3 and 4 follow the path of crack propagation for a (±45) laminate. As discussed in [1] this family of laminates failed along different surfaces depending on the original crack length. The specimen discussed herein is for the case $a = 0.4$, where the fracture generally ran along a plane at 45° to the original crack. However, Zone 3 is essentially coplaner with the original crack.

Zone 1 and Zone 2 (the zone of independent failure of the plies) show clearly in Figure 3 and 4. While the crack started to grow along the +45 plane, Zone 3 shows a strongly dependent series of fractures failing fiber bundles in both sets of plies. Zone 3 continues into Figure 4 showing fiber bundle failures in both plies following a typical saw-tooth pattern.

Again the zone of initial damage is very small and the instability area at the start of Zone 3 involves failure of both plies at a location near the plane of the original crack. This feature of the Zone 3 fracture again supports the conclusion of the applicability of LEFM to this specimen, as reported in [1].

---

[2] As described in [3], the standard data reduction method [4] makes allowance for the growth of this damage zone.

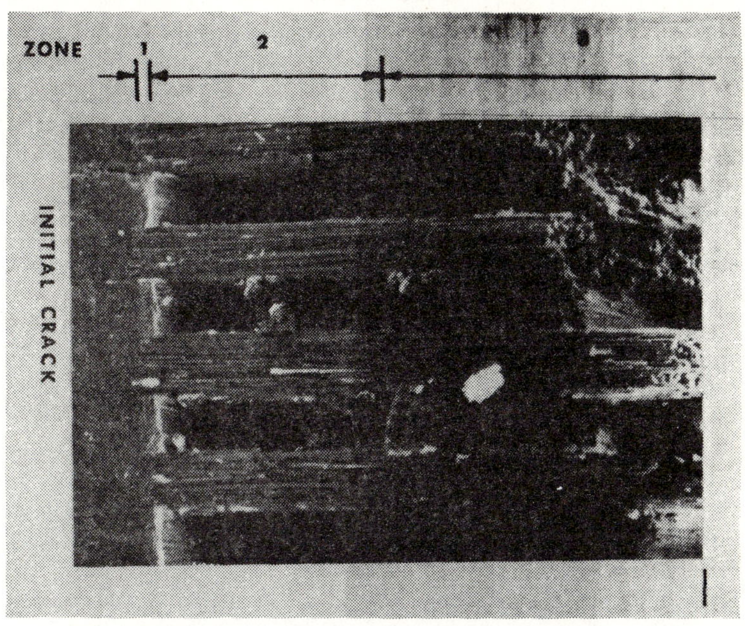

Figure 3. Fracture surface for (±45) laminate; plate 1.

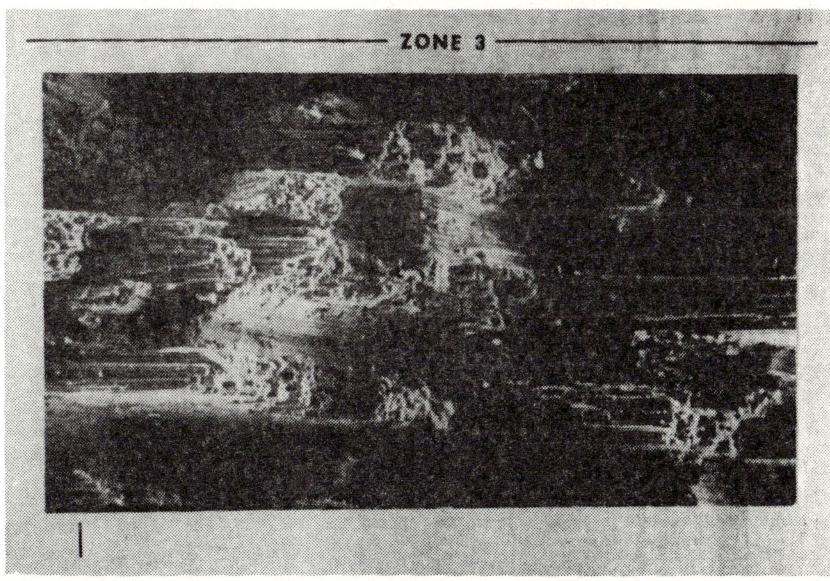

Figure 4. Fracture surface for (±45) laminate; plate 2.

CONCLUSIONS

Several possible conclusions may be drawn by observing the fracture surface at low magnifications. The fracture surface is characterized by a series of distinct zones. The first important zone is the zone of initial damage where fiber and matrix appear to have failed prior to specimen fracture. This conclusion derives from the independent fracture surfaces for the plies seen in this zone. The size of this zone is very small compared to the initial flaw size such that the stress and strain field ahead of the crack may be assumed to be unaffected by this damage due to the geometric constraint of the undamaged material.

The second important zone is in the region of material which is subjected a stress field characterized by a stress intensity factor. In this zone the fracture surface is seen to exhibit highly dependent failures of the plies. This observation supports the conclusion that the fracture instability involves roughly simultaneous ply fracture, especially at the start of Zone 3, supporting the use of a continuum model such as LEFM for graphite/epoxy. Thus the physical observations tend to confirm the conclusions of [1] as to the applicability of LEFM for these specimens.

ACKNOWLEDGMENTS

The authors gratefully acknowledge the significant support given this study by Professor J. R. Low of the Department of Metallurgy and Materials Science, Carnegie-Mellon University; and the financial support of the Air Force Materials Laboratory, Contract F33615-72-C-1214.

REFERENCES

1. H. J. Konish, Jr., J. L. Swedlow and T. A. Cruse, "Experimental Investigation of Fracture in an Advanced Fiber Composite," *J. Composite Materials*, Vol. 6 (1972), p. 114.
2. H. J. Konish, Jr., T. A. Cruse and J. L. Swedlow, "A Method for Estimating Fracture Strength of Specially Orthotropic Composite Laminates," in *Analysis of Test Methods for High Modulus Fibers and Composites, ASTM STP 521*, American Society for Testing and Materials, Philadelphia, Pennsylvania (1973).
3. T. A. Cruse and M. G. Stout, Fractographic Study of Graphite-Epoxy Laminated Fracture Specimens, Report SM-72-24(A), Department of Mechanical Engineering, Carnegie-Mellon University (December 1972).
4. Tentative Method of Test E 399 for Plane-Strain Fracture Toughness of Metallic Materials, *1971 Annual Book of ASTM Standards*, Part 31, American Society for Testing and Materials, Philadelphia, Pennsylvania (1971).

The Application of the Principles of Linear Elastic Fracture Mechanics to Unidirectional Fiber Reinforced Composite Materials

M. A. WRIGHT AND F. A. IANNUZZI

The University of Tennessee Space Institute
Tullahoma, Tennessee

(Received July 2, 1973)

ABSTRACT

The strength of individual carbon fibers was measured, and the results were used to calculate the lower bound of strength expected from carbon-fiber reinforced epoxy matrix materials. Using the value of the mean stress in the fibers at fracture of a composite and the principles of linear elastic fracture mechanics, the strength of coupon specimens containing edge slots could be predicted. However, the results of some initial compliance experiments indicated that some slight modifications of the expressions generated for isotropic pin-loaded specimens were necessary to account for the specific experimental techniques used.

INTRODUCTION

FOLLOWING THE EARLY work of Wu [1] numerous papers have been published dealing with the application of linear elastic fracture mechanics (LEFM) to predict failure of reinforced plastics containing slots or notches [2, 3, 4, 5, 6]. Most of the conclusions can be summed up by noting the recent comments of Beaumont and Tetelman [7], "The Fracture Stress of Aligned Fibrous Composites can be accurately predicted using the concepts of linear elastic fracture mechanics providing the concentration of longitudinal tensile stress reaches a maximum at the crack tip before transverse splitting occurs."

In this paper we suggest that the strength of these carbon reinforced epoxy specimens is best described in terms of a lower bound. The stress level necessary to cause failure is then simply the strength of a bundle of fibers. We show that a critical stress must exist at the crack tip for either matrix splitting or crack propagation to occur. Thus, the failure of notched specimens, either by matrix splitting or by propagation of a premachined notch can be described within the framework of fracture mechanics.

EXPERIMENTAL PROCEDURES

Panels of epoxy containing 35 v/o or 50 v/o carbon fibers designated Hy-E-1311C were fabricated by the Fiberite Corporation using resin X-904 as the matrix material. Fiberglass reinforcement doublers, 2 inches long, were bonded along the top and bottom edges of the 9-inch long panels before they were cut to give specimens 0.5, 1.0 and 1.5 in. wide. The thickness of each specimen varied along the specimen length particularly for the 35 v/o material. It was assumed that this difference resulted from loss of matrix during fabrication; thus, since the number of carbon fibers presumably remained constant and the load bearing capacity of the matrix was small, the loads applied to the composite were converted into stress values by dividing by the appropriate width and the constant thickness values of 0.022 inches.

Individual carbon fibers representative of those used to reinforce the epoxy were supplied with the specimens. Tensile tests were then carried out on those fibers and on fibers extracted from specimens by dissolving away the matrix in fuming nitric acid.

Fiber bundles 1, 2, and 4 inches long were prepared by dissolving the appropriate length of matrix from a specimen in fuming nitric acid. This procedure was accomplished using a slotted aluminum crucible to contain the specimens as shown in Figure 1. During the actual dissolving process excess acid was prevented from running out of the crucible by sealing the slots with silicone rubber. As can be observed in the figure the fibers were twisted together, and some of them were broken.

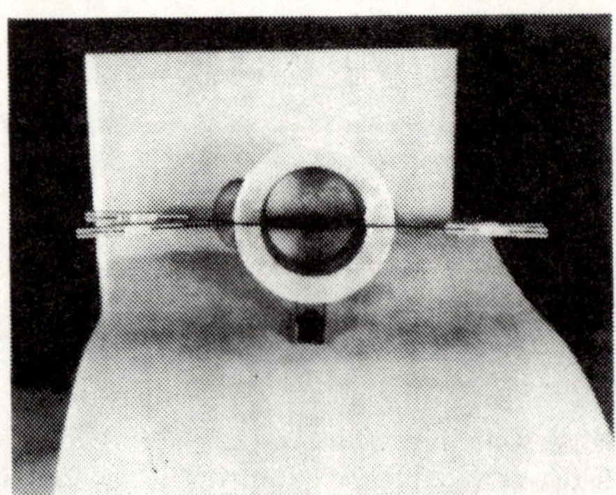

Figure 1. Aluminum crucible and carbon-epoxy specimen as used to obtain fiber bundles, by fuming nitric acid.

Double edge notched fracture specimens were prepared from the reinforced epoxy material by hand cutting slots with a diamond impregnated wire. The radius of the wire was 0.004 in. and this determined the notch root radius. Slot lengths ranged from 0.05 in. to 0.60 in. reflecting the differences in individual specimen widths.

Loads were applied to composite specimens by an Instron Tensile Test Machine operating at a crosshead speed of 0.02 inches per minute, which was the slowest speed of the machine. The normal, large capacity gripping arrangements were utilized; however, a spirit level was used to insure that the load axis of the machine was parallel to the longitudinal edges of the specimen.

The data used to calculate the crack extension force, G, was obtained using the experimental arrangements shown in Figure 2. Compliance gauges fabricated by bonding 2.5 in. long single wire strain gauges to 0.008 in. thick, age-hardened Cu-Be sheet were connected to the specimen through the notch plates as shown. Smaller notch plates were also bonded across one of the cut slots and the crack opening displacement was obtained by monitoring their movement with a clip gauge similar in design to that described by Fisher, et al [8].

Tensile tests on the reinforcements were carried out by first bonding individual fibers across a hole precut into a cardboard support. Then, fiber and support were gripped in the small pressure jaws of the Instron, the cardboard on each side of the hole was cut with scissors before the fiber was pulled to failure. The length of the rectangular hole, which was identical to the gauge length of the fiber, was adjusted to allow testing of different fiber lengths. The average diameter of the fibers, $d = 3.1 \times 10^{-4}$ in. was computed from optical measurements obtained from a photomicrograph taken of a composite specimen cross section.

RESULTS

The mean failure loads, $\bar{P}$, failure stress, σ, standard deviation, S, and coefficient of variation, c, computed from data obtained from as received fibers, 1 in., 2 in., 4 in. and 8 in. long and from extracted fibers 1.0 in. long is shown in Table 1. The mean strength of the as received fibers was identical to the strength values exhibited by the extracted fibers at the 5% significance level. It was therefore assumed that the nitric acid leaching treatment did not damage the fibers. It is quite apparent that the shorter fibers exhibited the larger mean strengths in agreement with the accepted probability failure criteria for brittle materials. The strengths of bundles of fibers fabricated by the process described previously are also shown in the table. It is interesting to note that the strengths of these bundles also increased as the fiber length decreased; in addition, the bundle strengths were always less than the mean strength of the individual fibers. Specifically, the bundle efficiency factor, ϵ, defined as the ratio of the bundle strength, σ_B, to mean strength, $\bar{\sigma}$, was appreciably less than one, i.e.:

$$\epsilon = \frac{\sigma_B}{\bar{\sigma}} < 1 \, .$$

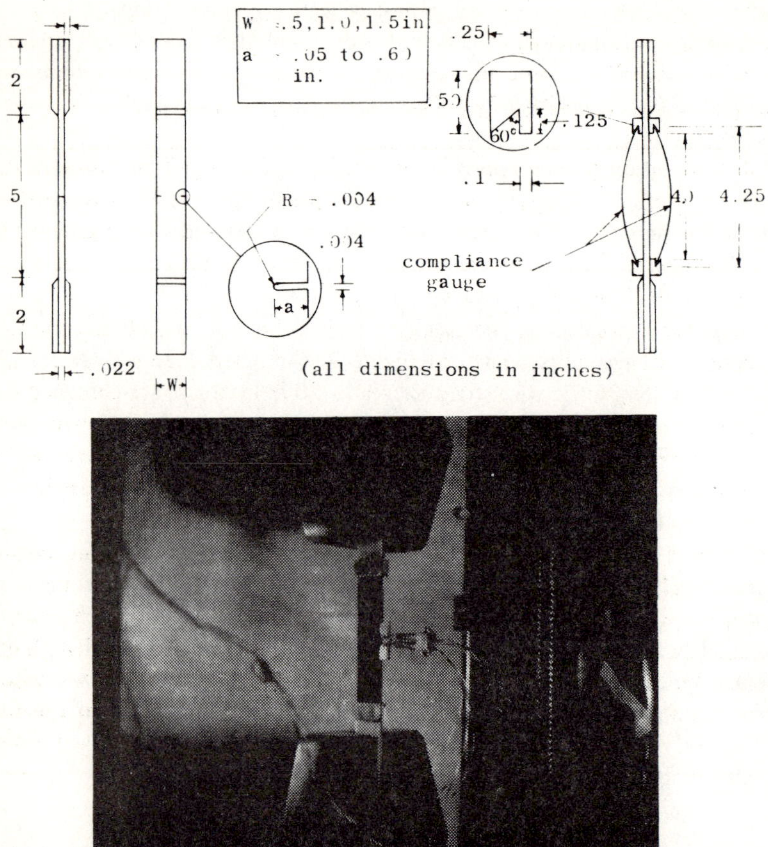

Figure 2. Specimen dimensions and experimental set up for fracture toughness measurements.

The strength and moduli of the unidirectionally reinforced materials are shown in Table 2. The scatter in the data is typical of the strength values obtained from most brittle materials. However, the mean modulus values 10.95×10^6 psi for the 50 v/o and 7.47×10^6 psi for the 35 v/o material compare favorably with those obtained by substituting into the "Rule of Mixtures" relationship typical values of 21.5×10^6 psi and 0.52×10^6 psi for the modulus values of the carbon fibers and the matrix respectively.

The tensile strength of a composite was greater than the strength that could be expected by assuming that the fibers functioned as a simple load carrying bundle. In effect, a synergistic strengthening effect was observed when the matrix surrounded the fibers. This effect is best appreciated by comparing the failure loads of a bundle of fibers, as shown in the table with the failure loads exhibited by a composite. In every case the composite material withstood the higher load; and, that load was much greater than that expected by treating the matrix as a simple tensile load bearing component.

Table 1. The Mean Strength of Individual Fibers and Bundles of Fibers

	Experimental Values								Calculated Values	
L (in)	$\bar{P}$ (lb × 10^{3})	S (lb × 10^{3})	C (lb × 10^{3})	$\bar{\sigma}$ Msi	P^* (lb)	P^{**} (lb)	σ^*_B Msi	σ_B^{**} Msi	$\bar{\sigma}$ Msi	σ_B Msi
1	32.20	6.23	17.64	0.424	629	1123	0.163	0.207	0.475	0.322
1′	31.50	9.58	41.67	0.414	–	–	–	–	0.475	–
2	32.38	6.58	19.69	0.426	546	984	0.142	0.180	0.424	0.288
4	29.49	6.36	18.39	0.388	401	841	0.104	0.153	0.379	0.258
8	28.48	6.45	18.90	0.375	–	–	–	–	0.338	–

$\bar{P}$ = mean load; S = standard deviation; c = coefficient of variation; A = area of fibers; $\bar{\sigma}$ = mean strength; P = failure load of bundle; σ_B = bundle strength; * = 35 v/o; ** = 50 v/o; ′ = extracted fibers.

Table 2. The Tensile Strength and Modulus of Composite Specimens .5 in. Wide Containing 50 v/o and 35 v/o Carbon Fibers

Specimen Identifi-cations	W (in)	t (in)	U.T. Load (lb)	Mean U.T.L. (lb)	S. D. (lb)	$\bar{\sigma}$ (ksi)	Modulus (Msi)	Mean Modulus (Msi)
50–1	.4965	.0232	1304				10.92	
50–2	.4970	.0227	1480				10.80	
50–3	.4985	.0198	1256	1283	188.18	118.77	10.98	10.95
50–4	.4965	.0235	1476				11.10	
50–5	.4965	.0225	1190				11.02	
50–6	.4974	.0187	981				10.90	
35–1	.4965	.0232	820				7.21	
35–2	.4970	.0205	892				8.40	
35–3	.4966	.0215	862	889	77.4	82.25	7.02	7.47
35–4	.4978	.0220	1040				7.05	
35–5	.4980	.0215	830				7.67	

composite. In every case the composite material withstood the higher load; and, that load was much greater than that expected by treating the matrix as a simple tensile load bearing component.

Double edge notched specimens were loaded elastically and the extension over a 4.25 in. gauge length was measured using the previously described compliance gauges. Values of EC/2 calculated using the results of the compliance terms are shown in Table 3.

Table 3. Calculations from Compliance Measurements Carried Out on Notched Specimens

W (in)	2a/w	EC/2* $(1/in)^2 \times 10^{-3}$	EC/2** $(1/in)^2 \times 10^{-3}$
.5	.1	7.97	8.71
	.2	9.45	–
	.3	8.49	7.65
	.4	–	11.05
	.5	9.47	9.06
	.6	12.55	12.40
	.7	11.34	12.24
	.8	16.01	16.00
	.9	19.76	–
1.0	.2	3.99	3.93
	.4	4.10	5.32
	.6	4.72	6.90
	.8	6.94	9.16
1.5	.2	2.43	2.17
	.4	2.58	3.38
	.6	3.47	3.04
	.8	6.14	7.65

* = 35 v/o; ** = 50 v/o; w = specimen width; 2a = notch length.

V% = 50

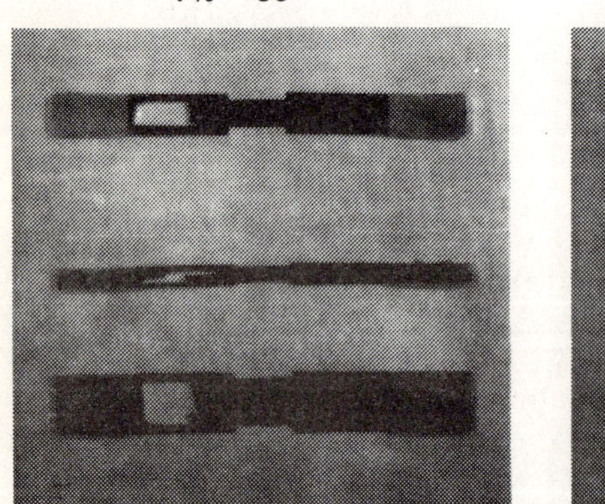

V% = 35

Figure 3. Photograph of Specimens after fracture. Notice the differences between 50 and 35 V%.

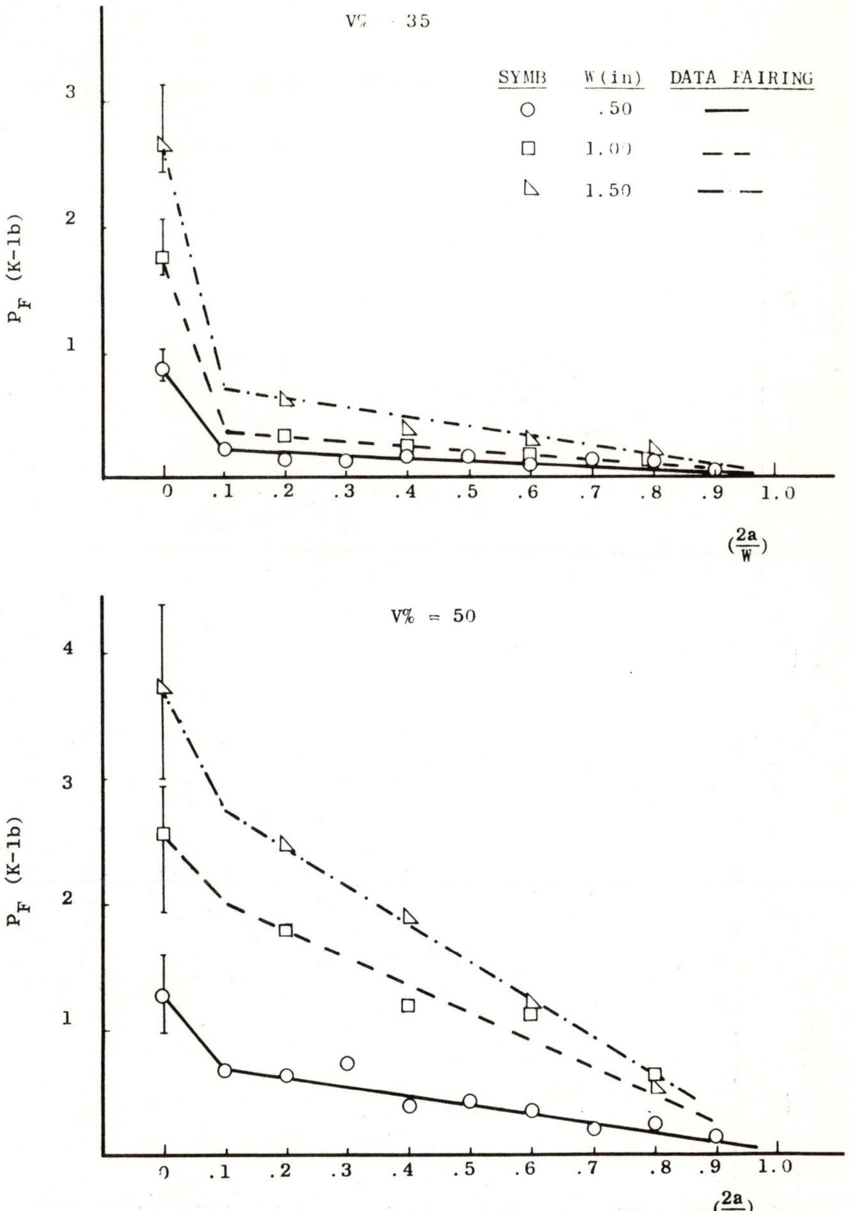

Figure 4. Fracture load versus crack length.

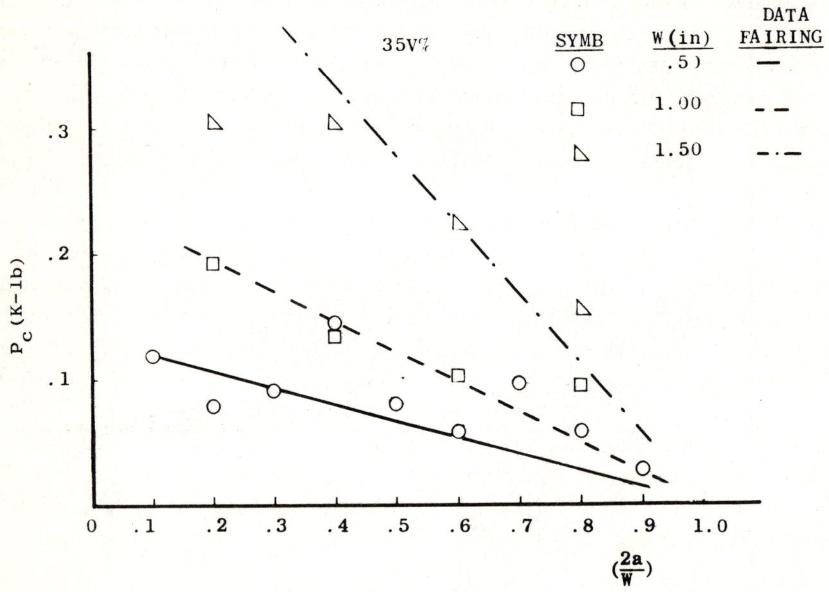

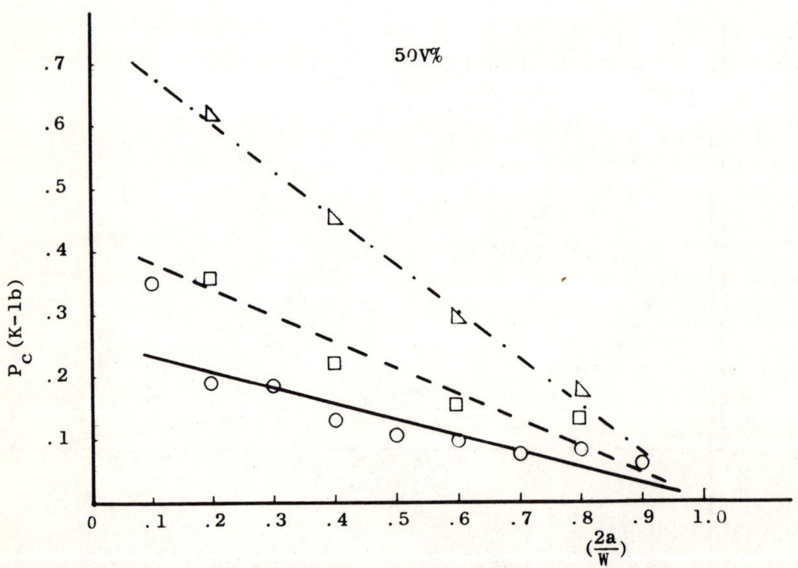

Figure 5. Fracture initiation load versus crack length.

The typical failure characteristics of the two groups of specimens, 35 v/o and 50 v/o are shown in Figure 3. It is to be noted that there appears to be two distinct failure modes. Specifically, both types of specimens initially split parallel to the fiber axis from a position at the tip of the precut slots. In the case of the 35 v/o material however, the split propagated only a small distance along the fiber axis before a transverse failure crack propagated through the fibers. In contrast, the split extended into the gripped portion of the 50 v/o specimen and failure occurred near to the grip.

The failure loads of double slotted specimens of carbon reinforced epoxy are shown in Figure 4. The presence of a small notch in the 35 v/o material produced a significant decrease in the load values obtained. In contrast, the failure loads of those specimens fabricated from 50 v/o material were reduced in direct proportion to the area of the cut slots. Failure was not the result of cracks propagating across the specimen from the premachined slots. Thus, in this respect at least, they were not crack sensitive. However, as discussed previously, appreciable damage resulted from cracks propagating from the cut slots parallel to the fiber axis. The loads at which this type of failure occurred are shown in Figure 5. The variation in the loads corresponds to that exhibited by the total loads to cause failure of the 35 v/o material. Thus, a similar crack sensitivity is exhibited.

DISCUSSION

i. Individual Fiber Tests

The strength data obtained from individual fibers was integrated numerically to produce the cumulative distribution curves shown in Figure 6. In these curves the

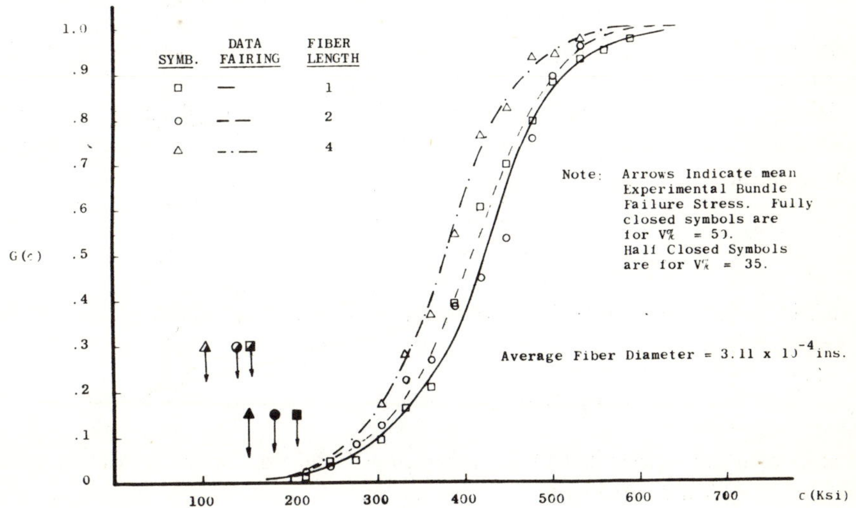

Figure 6. Cumulative frequency of experimental data for the strength of carbon fibers.

arrows indicate the failure stress of bundles of fibers. It is to be noted that failure occurred after a very small number of fibers, < 1%, would have failed. The shape of these curves is similar to those recently reported for boron fibers [9, 10]; thus, they can be described by the Weibull expression [11]:

$$G(\sigma) = 1 - \left[1 - \left(\frac{\sigma - \sigma^*}{\sigma_o}\right)^{\omega}\right]^{\alpha} \tag{1}$$

Where: α = $f(L/d)$, see Corten [12]

σ^* = Lower Limiting Strength, assumed here equal to 0 ksi

σ_o = Distribution Scale Factor

ω = Distribution Shape Factor, useful in describing the scatter of the data

This expression can be simplified using Poisson's approximation to give:

$$G(\sigma) = 1 - \exp\left[-\alpha\left(\frac{\sigma}{\sigma_o}\right)^{\omega}\right] \tag{2}$$

or by rearrangement and taking natural logarithms twice,

$$\ln\ln\left[\frac{1}{1 - G(\sigma)}\right] = \left[\ln \alpha - \omega \ln \sigma_o\right] + \omega \ln \sigma \tag{3}$$

both ω and the quantity $[\ln \alpha - \omega \ln \sigma_o]$ can then be determined using graphical methods.

Using the values of the parameters obtained from the individual fiber data, values of the mean strengths expected for each fiber length can be calculated using the expression:

$$\bar{\sigma} = \sigma_o\, \alpha^{-1/\omega}\, \Gamma\left(1 + \frac{1}{\omega}\right) \tag{4}$$

where Γ is the gamma function. The strength of a bundle of fibers, σ_B, can also be computed from:

$$\sigma_B = \sigma_o\, (\alpha\, \omega\, e)^{-1/\omega} \tag{5}$$

Values of $\bar{\sigma}$ and σ_B calculated using expressions 4 and 5 are included in Table 1 and shown graphically in Figure 7. The agreement between the calculated and

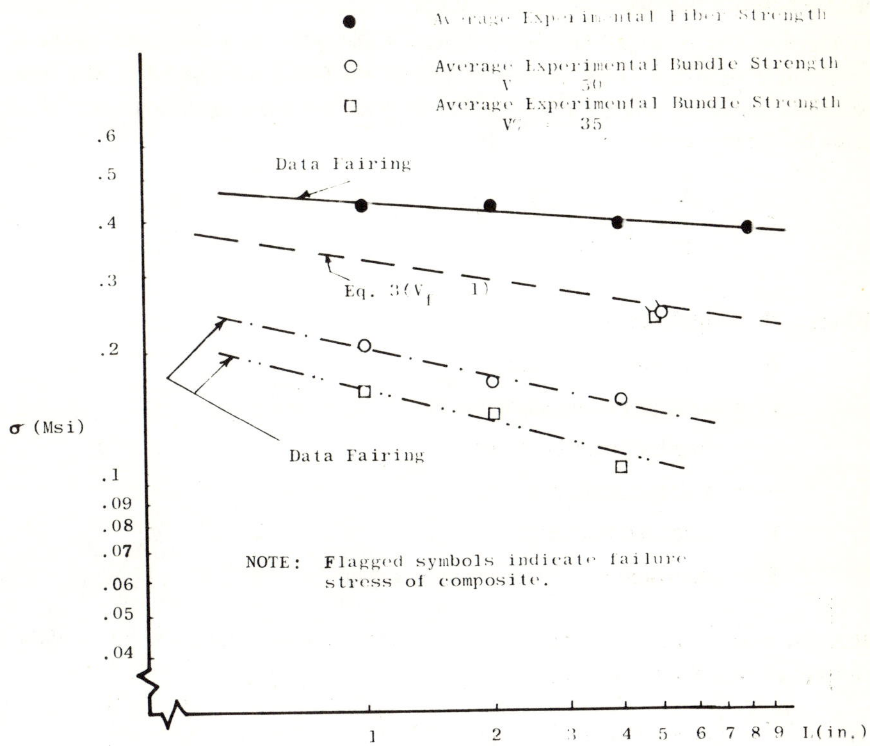

***Figure** 7. Theoretical and experimental fiber and bundle strength of carbon epoxy for various gauge lengths.*

theoretically obtained mean strength values is readily apparent. However, the experimental values of σ_B are relatively low. The reasons for this are not readily apparent. However, these low modulus fibers are known to be flaw sensitive and their strengths are affected by abrasion [13]. Despite careful technique, it was almost impossible to avoid abrading the individual fibers during routine handling of the bundle. Thus, a small strength loss would undoubtedly occur. By far, the biggest strength loss would occur during actual loading however, for the slight twist exhibited by the fibers in the bundle would cause them to be rubbed together. The addition of a matrix around the fibers would serve to protect the fibers during both handling and during the tensile test. Thus, the fibers would carry a larger load and a composite would exhibit higher strengths. Since the strength of a composite is no greater than that expected from a bundle, we intuitively believe that the matrix in these specimens functions to protect the fibers from abrasion and surface damage. The concept of "Ineffective Length" resulting from transfer of stress by shear through the matrix as discussed by Rosen [14] does not appear to apply.

ii. Compliance Results

Theoretical curves can be computed which describe the expected variation in compliance with the crack length of double slotted specimens fabricated from isotropic materials. Specifically, it is known that the crack extension force, G, is given by the expression:

$$G = \frac{2P^2}{EW} \left[\frac{d(EC/2)}{d(2a/w)}\right] \tag{6}$$

Where: P = applied load

E = modulus of elasticity

C = specimen compliance $\dot{\delta} = \text{Pt}$

a = length of each edge slot

w = width of specimen

t = specimen thickness

δ = extension of slotted specimen subjected to load, P.

It is also known that a suitable form for the stress intensity factor, K, for a double slotted tensile specimen is [15]:

$$K = \frac{P\sqrt{a}}{wt}\left[1.98 + 0.36\left(\frac{a}{w}\right) - 2.12\left(\frac{a}{w}\right)^2 + 3.42\left(\frac{a}{w}\right)^3\right] \tag{7}$$

Paris and Sih [16] have indicated that these expressions should be valid for composites providing certain orientation relationships are satisfied between the slots, the applied load and the reinforcement axis. The following relationship then holds:

$$C'K^2 = G \tag{8}$$

where,

$$C' = \left[\frac{A_{11}A_{22}}{2}\right]^{1/2}\left[\left(\frac{A_{22}}{A_{11}}\right)^{1/2} + \frac{(2A_{12} + A_{66})}{2A_{11}}\right]^{1/2}$$

where values of the individual parameters are shown in Table 4.

By substituting 7 and 8 into 6 it is found that:

$$\frac{P^2 a}{w^2 t^2} f^2\left(\frac{a}{w}\right) = \frac{2P^2}{C'EW}\left[\frac{d(EC/2)}{d(2a/w)}\right] \tag{9}$$

Table 4. Values of Constants Used to Calculate the Effective Modulus C′

Constant	*35 v/o*	*50 v/o*	*Reference*
$A_{11} = \frac{1}{E_{11}}$	1.34×10^{-7}	9.13×10^{-8}	Present
$A_{22} = \frac{1}{E_{22}}$	10^{-6}	10^{-6}	(22)
$A_{12} = -\frac{\nu_{12}}{E_{11}}$	-4.01×10^{-8}	-2.56×10^{-8}	(22)
$A_{66} = \frac{1}{G_{12}}$	2.00×10^{-6}	1.54×10^{-6}	(22)
C'	8.16×10^{-7}	7.23×10^{-7}	

Integration of expression 9 gives:

$$EC/2 = C'E/t^2 \left[1.96\left(\frac{a}{w}\right)^2 + 0.48\left(\frac{a}{w}\right)^3 - 2.05\left(\frac{a}{w}\right)^4 + 2.40\left(\frac{a}{w}\right)^5 - 1.16\left(\frac{a}{w}\right)^6 - 2.07\left(\frac{a}{w}\right)^7 + 1.46\left(\frac{a}{w}\right)^8\right] + A \quad (10)$$

The value of the integration constant, A, is obtained by applying the above expression to a parallel-sided specimen, i.e., by substituting $a = 0$ into Equation 10 consequently:

$$A = L/2wt^2 \quad (11)$$

Therefore, the compliance curve intercepts the vertical ordinate in the plot $EC/2$ vs $2a/w$ at a point which simply depends on the geometry of the specimen.

The compliance data previously presented in Table 3 is shown graphically in Figures 8 and 9. The theoretical curves given by the right-hand side of expression 10 are also shown for comparison. It is immediately apparent that the experimental results obtained for the 0.5 inch wide specimens compare favorably with the theoretical curves. However, the results obtained from both the 1.0 and 1.5 inch

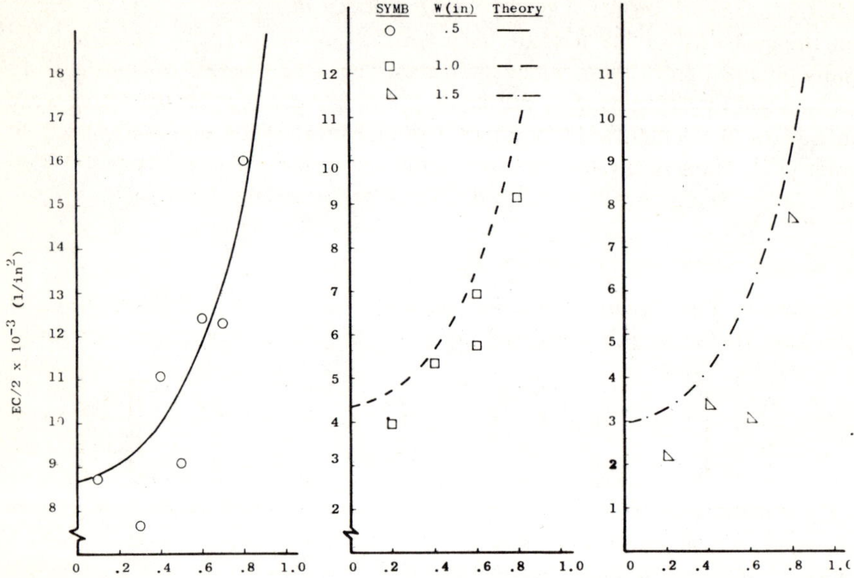

Figure 8. *Variation of compliance with crack length for carbon reinforced epoxy, V/O = .05.*

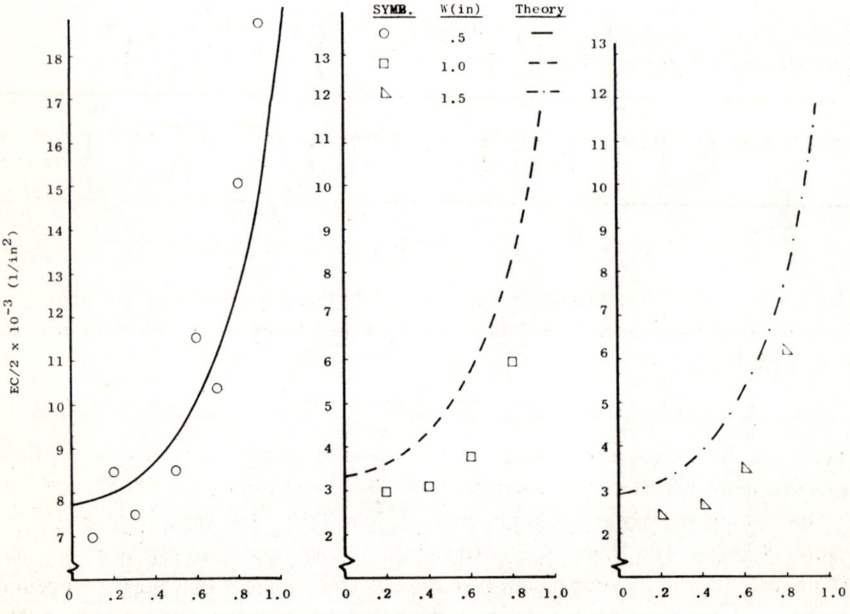

Figure 9. *Variation of compliance with crack length for carbon reinforced epoxy, V/O = .035.*

wide specimens are smaller than the theoretically expected values. It was assumed that the geometry of the test specimen and the testing procedure influenced the results obtained from these wider specimens. Thus, the left side of expression 10 was modified by multiplying by the factor, β. It was presumed that this factor was a function of the length:width ratio of the individual specimens as discussed by Bowie [17]. Numerical values of β were estimated from the figures to be 1.00, 1.40 and 1.74 for the 0.5 inch, 1.0 inch and 1.5 inch specimens respectively.

iii. Critical Stress Intensity Factors, K_c

Following the discussion of the previous section, the modified form of expression 10 can be rearranged and integrated to give an appropriate form for K:

$$K = \frac{P\sqrt{a}}{wt}\left[1.98 + 0.36\left(\frac{a}{w}\right) 2.12\left(\frac{a}{w}\right)^2 + 3.42\left(\frac{a}{w}\right)^3\right]\left[\frac{1}{\beta}\right]^{1/2} \qquad (12)$$

If the failure load, P_c, is substituted into expression 12 then values for K_c can be calculated. Actual values calculated for the 35 v/o material are shown in Table 5 compared to those obtained using expression 10 without the width correction factor, β. The difference between the two sets of data is quite striking for the unmodified expression exhibits a significant width effect that is not apparent in the modified data.

Table 5. Stress Concentration Factors Obtained for 35 V/O and 50 V/O Carbon Reinforced Specimens Edge Notched

W(in)	2a/w	K_F^{35} (ksi-in)	K_F^{35} (ksi-in) (corrected)	K_{sc}^{35} (ksi-in)	K_{sc}^{35} (ksi-in) (corrected)	K_{sc}^{50} (ksi-in)	K_{sc}^{50} (ksi-in) (corrected)
.5	.1	7.33		3.70		11.60	
	.2	6.24		3.57		8.47	
	.3	6.92		4.90		10.08	
	.4	10.14		8.87		8.24	
	.5	10.91		5.82		7.78	
	.6	6.73		5.72		7.90	
	.7	11.88		9.50		7.13	
	.8	12.74		7.14		9.50	
	.9	5.70		3.80		8.77	
1.0	.2	10.41	7.38	6.05	4.29	11.19	7.93
	.4	10.88	7.71	6.05	4.29	9.86	6.99
	.6	8.92	6.33	6.00	4.25	9.40	6.67
	.8	8.25	5.85	7.91	5.60	10.94	7.76
1.5	.2	15.76	9.05	7.85	4.51	15.96	9.17
	.4	14.45	8.30	11.16	6.41	16.46	9.46
	.6	14.18	8.15	10.78	6.19	14.08	8.09
	.8	13.75	7.90	10.58	6.08	12.03	6.91

It is important to recall that the double slotted specimens fabricated from 50 v/o material did not fail by simple crack propagation, for they split parallel to the fibers. Failure then occurred at loads that could be calculated using the remaining area and the failure stress of parallel sided specimens. However, if a specimen is considered to have failed when it splits, then a parameter K_{sc} can be calculated by substituting the splitting load into expression 12. The rationale behind this approach stems from the realization that K_c simply represents a combination of applied stress and crack length that is necessary to produce a critical tensile fracture stress, σ_y, at or near to the crack tip. Beaumont and Phillips [2] and others [18] have indicated that plastic matrix materials split when the shear stress, σ_{xy}, at the tip of the crack reaches a critical value. Also, Galliland, quoted by Beaumont et al, has deduced the magnitude of the relationships between σ_y and σ_{xy} at the crack tip. The ratio between the two stresses varies with the degree of orthotropy of the material for it is about 3:1 for an isotropic material and is about 11:1 for a 50 v/o carbon reinforced material. For specimens containing 50 v/o reinforcement therefore a value of the stress intensity factor necessary to cause splitting, K_{sc}, was calculated by substituting the appropriate value of the tensile stress to cause splitting into expression 12.

iv. Failure Mode

In the parallel sided specimens and in the bundles of individual fibers tested here, failure occurred by the cumulative or bundle failure mode. It has also been pointed out that failure occurred after very few of the fibers in each bundle had failed. Thus, the apparent stress necessary to cause failure of an individual fiber approaches the strength of the composite divided by the volume fraction of fibers in the specimen, i.e., $\sigma_f \approx 220$ ksi. Using the shear lag analysis of Cox [19] as quoted by Kelly [20]. The maximum shear stress generated adjacent to a fiber break in this type of material is about 0.19 σ_f or about 42 ksi. This is appreciably larger than the short beam shear strength of 12 ksi quoted by the specimen fabricator. Thus, it appears reasonable to conclude that the sudden application of this large shear load at a fiber break would almost certainly cause the matrix around the fiber to fracture. Thereafter stress transfer, if any, would be controlled by the frictional force acting across the crack. In the present case, the value of this frictional force would appear to be minimal and failure of the composite by a bundle mechanism is exhibited.

As shown in Figure 5, the stress necessary to cause splitting of a 50 v/o material containing a slot, $2a/w = 0.4$ is about 14 ksi. The mean stress per fiber is therefore 28 ksi and the maximum stress at the tip of the crack must be about equal to the maximum stress of about 220 ksi; thus, K_σ, is about 8. According to Galliland, the maximum shear stress concentration factor, K_τ, for this type of material is about 0.09 K_σ; thus the calculated shear stress at the crack tip is about 20 ksi. Similarly, the stress to cause splitting of a 35 v/o material is about 8 ksi. The mean stress per fiber is therefore 23 ksi and, K_σ, is probably slightly less than the 50 v/o material for fewer fibers are cut. However, by taking $K_\sigma = 8$ as the correct order of magnitude the tensile stress in the fibers at the crack tip is deduced to be 184 ksi. A linear interpolation between the stress concentrations caused by a slot in an isotropic material and that produced in a 50 v/o orthotropic material given by Galliland indicates that K_τ for 35 v/o material is about 0.11 K_σ. Thus, the maximum shear stress in this material is about 20 ksi or about the same as that required to cause splitting in the 50 v/o material.

Splitting of the matrix extends completely into the gripped portions of the reinforced 50 v/o material. However, for the case of the 35 v/o material the propagation of the crack stops not far from the crack tip. The reason for this is not immediately obvious; perhaps the local strength of the matrix-fiber bond increases sufficiently, or off-axis fibers span the shear crack to hinder its propagation. Most probably, the extent of the shear stress field is limited at the lower loads applied to the 35 v/o material. Thus, the propagating crack decelerates and stops as it moves out of the shear stress field. Notwithstanding, the longitudinal cracks in the 35 v/o material did not propagate into the grip portion of the specimen, the stress concentrating effect of the slot was not eliminated and transverse crack propagation was still possible. This occurred when the stress at the tip of the crack approaches 220 ksi.

CONCLUSIONS

We have shown that:

1. The strength of carbon fibers (Hy-E-13116) could be described by a Weibull expression.
2. The strength of carbon reinforced X-904 resin could be predicted in from a knowledge of the strength of individual carbon fibers.
3. The compliance of the unidirectional carbon reinforced specimen tested here could be described by the expression formalized for isotropic materials. However, slight modifications of the expressions were necessary to allow for differences in testing geometry.
4. The principles of linear elastic fracture mechanics can be used to describe the failure of carbon reinforced epoxy specimens irrespective of whether the specimens fail by transverse crack propagation or longitudinal splitting.

ACKNOWLEDGMENTS

The authors wish to acknowledge helpful discussions with Mr. H. T. Kulkarni and the financial support provided by the National Science Foundation.

REFERENCES

1. E. M. Wu, "Fracture Mechanics of Anisotropic Plates," *Composite Materials Workshop*, Technomic Publishing Co., Inc., (1968), p. 20.
2. P. W. R. Beaumont and D. C. Phillips, "Tensile Strength of Notched Composites," *J. Comp. Mat.*, Vol. 6 (1972), p. 32.

3. H. J. Konish, Jr., J. L. Swedlow and T. A. Cruse, "Experimental Investigation of Fracture in an Advanced Fiber Composite," *J. Comp. Mat.*, Vol. 6 (1972), p. 114.
4. M. E. Waddoups, J. R. Eisenmann and B. E. Kaminski, "Macroscopic Fracture Mechanics of Advanced Composite Materials," *J. Comp. Mat.*, Vol. 5 (1971), p. 446.
5. K. G. Kreider and L. Dardi "Fracture Toughness of Composites," United Aircraft Research Report L110865-1 (1972).
6. E. F. Olster and R. C. Jones, "Toughening Mechanisms in Continuous Filament Unidirectionally Reinforced Composites," Composite Materials: Testing and Design (second conference), ASTM STP 497, (1972), pp. 189–205.
7. P. W. R. Beaumont and A. S. Tetelman, "The Fracture Strength and Toughness of Fibrous Composites," UCLA-Eng-Report 7269 (1972).
8. D. M. Fisher, R. T. Busbey, and J. E. Srawley "Design and use of Displacement Gage for Crack Extension Measurements," NASA TN-D3724 (1966).
9. H. W. Herring, "Selected Mechanical and Physical Properties of Boron Filaments," NASA TN D-3202 (1966).
10. M. A. Wright and J. L. Wills, "The Tensile Failure Modes of Metal-Matrix Composite Materials," submitted to J. of Matls. Science (1973).
11. W. Weibull, "A Statistical Distribution Function with Wide Applicability," *J. Appl. Mech.* Vol. 18 (1951), p. 293.
12. H. T. Corten, "Micromechanics and Fracture Behavior of Composites," Modern Composite Materials, Addison Wesley Publishing Company (1967), p. 27.
13. J. Johnson, "Factors Affecting the Tensile Strength of Carbon Fibers," American Chem. Soc. Polymer Preprints, Vol. *9*, No. 2 (1968), p. 1316.
14. B. W. Rosen, "Tensile Failure of Fibrous Composites," *AIAA J.*, , Vol. 2 (1964), p. 1985.
15. Carl Osgood, "A Basic Course in Fracture Mechanics," Penton Education Division, Cleveland, Ohio, (1971).
16. P. C. Paris and G. C. Sih, "Stress Analysis of Cracks," Fracture Toughness Testing and Its Applications," ASTM STP 381 (1965), p. 60.
17. O. L. Bosie, "Rectangular Tensile Sheet with Symmetric Edge Cracks," *J. Appl. Mech.*, Vol 31 (1964), p. 208.
18. A. Kelly, "Interface Effects and the Work of Fracture of a Fibrous Composite," *Proc. Roy Soc., Sec. A.*, Vol. 319 (1970), p. 508.
19. H. L. Cox, "The Elasticity and Strength of Paper and Other Fibrous Materials," *Br. J. Appl. Phys.*, Vol. 3 (1952), p. 72.
20. A. Kelly, *Strong Solids*, Clarendon Press Oxford (1966).

Wave Surfaces Due to Impact on Anisotropic Plates

FRANCIS C. MOON

Aerospace and Mechanical Sciences Department
Princeton University, N.J. 08540

(Received December 4, 1971)

ABSTRACT

The stress waves induced in anisotropic plates by transverse, short-duration impact forces are examined in this report. The anisotropy is related to the layup angles of the fibers of a fiber composite laminated plate. Using a modification of Mindlin's approximate theory of plates, it is shown that both extensional and bending waves are generated by transverse impact. The magnitudes of the wave velocities in different directions are calculated for graphite fiber-epoxy matrix plates for various layup angles. Finally, the shapes of the wave fronts or wave surfaces due to point impact are also presented for the cases mentioned.

INTRODUCTION

THE SUCCESSFUL APPLICATION of advanced fiber composite materials to jet engine fan or compressor blades will depend in part on the ability of these materials to withstand the forces of impact due to foreign objects. Such impact can be the result of the ingestion of stones, nuts and bolts, hailstones, or birds into a jet engine. The relative velocity of the impacting body to the blade can be in the order of 450 meters per second (1500 ft/sec). The ingestion of objects of sizeable mass (e.g., birds) might involve the dynamics of the entire blade. The high speed impact of small objects will result in small impact times ($<50\ \mu$sec), and the initial transmission of impact energy into a local region of the blade. This initial energy will propagate into the rest of the blade in the form of stress waves. Although such high speed impact will involve local cratering or even complete penetration, long range damage away from the impact area can result from the reflection of stress waves (spalling) and focusing due to changes in geometry.

---

*This work was performed at the NASA Lewis Research Center, Cleveland, Ohio, while the author was a Summer Faculty Fellow, 1970.

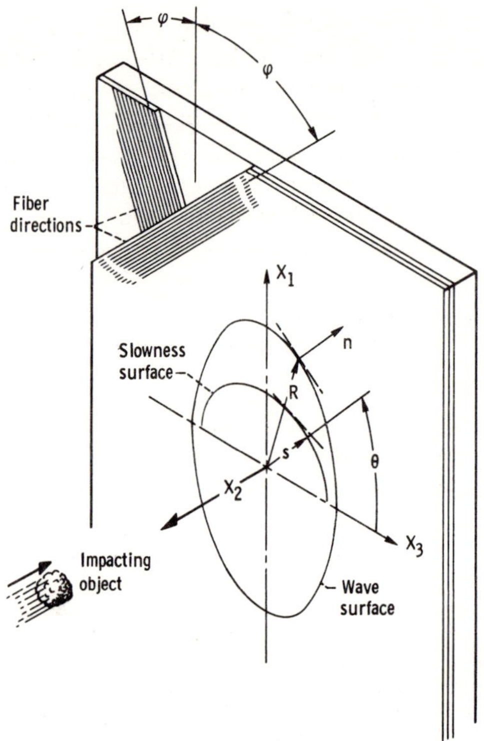

Figure 1. Diagram of plate element.

It is also observed that materials under high rates of strain exhibit an increased tensile strength and decreased ductility. Such evidence tends to validate the use of elastic wave analysis for the determination of the prefracture stresses induced by the impact forces. However, even if plastic waves do predominate, elastic precursor waves will bound the stressed impact zone.

In this paper, calculations of velocity and wave surfaces in anisotropic composite plates due to transverse impact forces are presented. These wave surfaces, for a given time after impact, bound the stressed region surrounding the impact point.

REVIEW OF BASIC EQUATIONS OF ANISOTROPIC ELASTIC PLATES

The composite plates under consideration are imagined to comprise a number of unidirectional plys (Figure 1). An equal number of plys lie at angles $\pm \phi$ from the symmetry axis in such a manner that bending-extensional coupling does not result. We also assume that the number of plys across the plate thickness is reasonably large, so that average properties across the plys can be used. This approximation will be valid for wavelengths greater that the ply thickness and certainly valid for wavelengths greater than the plate thickness.

The model used here is an effective modulous theory and does not constitute an effective stiffness model in the sense of Sun, Achenbach, and Herrman [10], in which the dynamics of fiber and matrix are separated. Such effects might become important when the wavelengths begin to approach the fiber diameter or fiber separation distances, which is not the case under study in this paper. Tauchert and Guzelsu [11] have measured dispersion in fiber composites at ultrasonic frequencies at which the wavelength is small enough to produce significant dispersion.

Thus, in place of the n-ply plate, the wave propagation in an equivalent anisotropic plate is being studied. The equivalent elastic constants are obtained from a static analysis of the n-ply composite plate.

The equations of motion for a linear anisotropic elastic body are [1]

$$t_{ij,j} = \rho \ddot{u}_i \tag{1}$$

(The double summation convention is assumed, the dots indicate time differentiation, and the notation $\phi,_j$ stands for $\partial\phi/\partial x_j$ where x_j is a Cartesian coordinate.) The vector u is the displacement, and body forces are assumed to be absent. In general, the stress tensor t_{ij} is related to the strains $e_{k\ell}$

$$t_{ij} = C_{ijk\ell} e_{k\ell} \tag{2}$$

There are only 21 independent elastic constants $C_{ijk\ell}$ in general. For orthotropic symmetry, which pertains to the composite plates under consideration, the stress-strain equations take the following matrix form:

$$\begin{bmatrix} t_{11} \\ t_{22} \\ t_{33} \\ t_{23} \\ t_{13} \\ t_{12} \end{bmatrix} = \begin{bmatrix} C_{11} & C_{12} & C_{13} & 0 & 0 & 0 \\ C_{12} & C_{22} & C_{23} & 0 & 0 & 0 \\ C_{13} & C_{23} & C_{33} & 0 & 0 & 0 \\ 0 & 0 & 0 & C_{44} & 0 & 0 \\ 0 & 0 & 0 & 0 & C_{55} & 0 \\ 0 & 0 & 0 & 0 & 0 & C_{66} \end{bmatrix} \begin{bmatrix} e_{11} \\ e_{22} \\ e_{33} \\ 2e_{23} \\ 2e_{13} \\ 2e_{12} \end{bmatrix} \tag{3}$$

The constants $C_{\alpha\beta}$ $(\alpha, \beta = 1, 2, \ldots, 6)$ are of course related to the $C_{ijk\ell}$. The strains are related to the displacements in the body by

$$e_{ij} = \frac{1}{2}(u_{i},_{j} + u_{j},_{i}) \tag{4}$$

Combining equations (1), (2), and (4) results in the following partial differential equations:

$$C_{ijk\ell}u_{k,\ell j} = \rho\ddot{u}_i \tag{5}$$

Wave propagation in anisotropic media has been studied for a long time [2 and 3]; however, few problems have been solved in which boundaries are present.

The approximate theory of anisotropic plates to be used in this study is due to Mindlin and coworkers [4 and 5]. In their theory the three-dimensional displacement is expanded in Legendre polynomials in the thickness direction.

$$u = \sum_n P_n(\eta)\, u^{(n)}(x_1, x_3, t) \tag{6}$$

where $\eta = x_2/b$ and b is half the plate thickness and where

$$P_0(\eta) = 1, \quad P_1(\eta) = \eta, \quad P_2(\eta) = \frac{3\eta^2 - 1}{2}$$

The functions $u^{(n)}$ have a physical significance (see Ref. 5, pp. 563-564): u_1^0, u_3^0 represent in-plane or extensional deformation; u_2^0 represents the transverse displacement of the plate; u_1^1 and u_3^1 are measures of the bending strains or $b\psi_1, b\psi_3$ where ψ_1 is the slope of the plate midsurface due to bending about the 3-axis; and u_2^1 is a measure of the thickness stretching.

To obtain the approximate equations of motion, a variational method is used [4]. Instead of solving equation (5) directly, the equations are integrated across the thickness:

$$\int_A \int_{-1}^{1} (t_{ij,i} - \rho\ddot{u}_j)\, \delta u_j\, b\, d\eta\, dA = 0 \tag{7}$$

where δu_j and t_{ij} are calculated using the series representation of the displacement (Eq. (6)). This leads to an infinite set of equations each involving higher modes of vibration of the plate:

$$bt^{(n)}_{\alpha j,\alpha} + \Big[P_n(\eta)\, t_{2j}\Big]_{-1}^{1} - t_{2j}^{(n)} = \rho b \frac{2}{2\eta + 1}\ddot{u}_j^{(n)} \tag{8}$$

where $\alpha = 1, 3$.

$$t_{\alpha j}^{(n)} = \int_{-1}^{1} P_n(\eta)\, t_{\alpha j}\, d\eta$$

$$t_{2j}^{(n)} = \int_{-1}^{1} \frac{dP_n}{d\eta} \; t_{2j} \, \eta$$

If impact forces are present on the upper surface of the plate, then the following boundary conditions are used to evaluate the second terms, $\left[P_n\,(\eta)t_{2j}\right]_{-1}^{1}$ in equation (8).

$$\begin{aligned} t_{22}\,(\eta = 1) &= q_2 \\ t_{22}\,(\eta = -1) &= 0 \\ t_{21}(\eta = \pm 1) &= 0 \\ t_{23}(\eta = \pm 1) &= 0 \end{aligned} \tag{9}$$

This scheme has been carried out for $n = 0, 1, 2$ for orthogonal symmetry. The equations for u_1^0 and u_3^0 are coupled to u_2^1, the thickness stretch. The flexural equations for u_2^0, u_1^1 and u_3^1 are found to depend on $u_2^{(2)}$. To truncate the infinite set of equations we drop higher order displacements $u_1^{(2)}$, $u_3^{(2)}$, etc. Next terms containing second derivatives in (8) for n=2 are dropped, keeping only the low frequency terms. This procedure results in explicit relations for u_2^1 and $u_2^{(2)}$ which are given below:

$$\left.\begin{aligned} q_2 - 2\left(C_{22}\frac{u_2^1}{b} + C_{12}\frac{\partial u_1^0}{\partial x_1} + C_{23}\frac{\partial u_3^0}{\partial x_3}\right) &= 0 \\ q_2 - 2\left(C_{22}\frac{3u_2^{(2)}}{b} + C_{12}\frac{\partial u_1^1}{\partial x_1} + C_{23}\frac{\partial u_3^1}{\partial x_3}\right) &= 0 \end{aligned}\right\} \tag{10}$$

Using these equations, the terms u_2^1 and $u_2^{(2)}$ can be eliminated from equations (8) (n=0,1). The resulting set of equations form the basis of our wave analysis:

$$\rho\frac{\partial^2 u_1^0}{\partial t^2} = \hat{C}_{11}\frac{\partial^2 u_1^0}{\partial x_1^2} + C_{55}\frac{\partial^2 u_1^0}{\partial x_3^2} + \left(C_{55} + \hat{C}_{13}\right)\frac{\partial^2 u_3^0}{\partial x_1\,\partial x_3} + \frac{C_{12}}{2C_{22}}\frac{\partial q_2}{\partial x_1} \tag{11}$$

$$\rho\frac{\partial^2 u_3^0}{\partial t^2} = \hat{C}_{33}\frac{\partial^2 u^0{}_3}{\partial x_3^2} + C_{55}\frac{\partial^2 u_3^0}{\partial x_1^2} + \left(C_{55} + \hat{C}_{13}\right)\frac{\partial^2 u_1^0}{\partial x_1\,\partial x_3} + \frac{C_{23}}{2C_{22}}\frac{\partial q_2}{\partial x_3} \tag{12}$$

$$\rho \frac{\partial^2 u_2^0}{\partial t^2} = C_{66} \frac{\partial^2 u_2^0}{\partial x_1^2} + C_{44} \frac{\partial^2 u_2^0}{\partial x_3^2} + C_{66} \frac{1}{b} \frac{\partial u_1^1}{\partial x_1} + C_{44} \frac{1}{b} \frac{\partial u_3^1}{\partial x_3} + \frac{1}{2b} q_2 \tag{13}$$

$$\rho \frac{\partial^2 u_1^1}{\partial t^2} = \hat{C}_{11} \frac{\partial^2 u_1^1}{\partial x_1^2} + C_{55} \frac{\partial^2 u_1^1}{\partial x_3^2} + \left(C_{55} + \hat{C}_{13}\right) \frac{\partial^2 u_3^1}{\partial x_1 \, \partial x_3} - \frac{3}{b} C_{66} \left(\frac{\partial u_2^0}{\partial x_1} + \frac{u_1^1}{b}\right) + \frac{C_{12}}{2C_{22}} \frac{\partial q_2}{\partial x_1} \tag{14}$$

$$\rho \frac{\partial^2 u_3^1}{\partial t^2} = \hat{C}_{33} \frac{\partial^2 u_3^1}{\partial x_3^2} + C_{55} \frac{\partial^2 u_3^1}{\partial x_1^2} + \left(C_{55} + \hat{C}_{13}\right) \frac{\partial^2 u_1^1}{\partial x_1 \, \partial x_3} - \frac{3}{b} C_{44} \left(\frac{\partial u_2^0}{\partial x_3} + \frac{u_3^1}{b}\right) + \frac{C_{23}}{2C_{22}} \frac{\partial q_2}{\partial x_3} \tag{15}$$

where

$$\hat{C}_{11} = C_{11} - \frac{C_{12}^2}{C_{22}}$$

$$\hat{C}_{33} = C_{33} - \frac{C_{23}^2}{C_{22}}$$

$$\hat{C}_{13} = C_{13} - \frac{C_{12}\, C_{23}}{C_{22}}$$

The first two equations (11), (12), determine the in-plane motion and the last three (13)-(15) govern the flexural waves. One can see that the impact pressure, $-q_2(x_1,x_3,t)$ can generate in-plane as well as flexural waves.

It should be noted that in the procedure used by Mindlin [4] the coefficients C_{44} and C_{66} in equations (13)-(15) were replaced by, $k_3 C_{44}$ and $k_1 C_{66}$, respectively. The correction constants k_1 and k_3 were adjusted in order to match the thickness shear vibration mode [4]. These terms will not enter the calculations presented herein.

WAVE PROPAGATION

Consider first the extensional motion which is governed by equations (11), (12). Assume a solution in the form below, which represents a plane wave traveling in the direction $\boldsymbol{n}$ called the wave normal.

$$\left.\begin{aligned} u_1^0 &= U_1\, f(\boldsymbol{n}\cdot\boldsymbol{x} - vt) \\ u_3^0 &= U_3\, f(\boldsymbol{n}\cdot\boldsymbol{x} - vt) \end{aligned}\right\} \tag{16}$$

Substituting these expressions into equations (11), (12) reveals that U_1, U_3, and v, the wave speed, must satisfy the following linear algebraic equations for a given $\boldsymbol{n}$:

$$\begin{bmatrix} A_{11} & A_{12} \\ A_{21} & A_{22} \end{bmatrix} \begin{bmatrix} U_1 \\ U_3 \end{bmatrix} = v^2 \begin{bmatrix} U_1 \\ U_3 \end{bmatrix} \tag{17}$$

where

$$A_{11} = \hat{C}_{11} \cos^2\phi + C_{55} \sin^2\phi$$

$$A_{22} = \hat{C}_{33} \sin^2\phi + C_{55} \cos^2\phi$$

$$A_{12} = A_{21} = \left(C_{55} + \hat{C}_{13}\right) \sin\phi \cos\phi$$

$$\boldsymbol{n} = (\cos\phi, \sin\phi)$$

Thus v^2 is a root of the equation

$$\Delta(\boldsymbol{n}, v^2) = \det\left(A_{ij} - \rho v^2 \delta_{ij}\right) = 0 \tag{18}$$

where δ_{ij} is the Kronecker delta ($\delta_{12} = \delta_{21} = 0$ and $\delta_{11} = \delta_{22} = 1$). The physically possible elastic constants $C_{\alpha\beta}$ will guarantee that A_{ij} is positive definite. This guarantees two positive real roots v_1^2 and v_2^2 for a given wave normal $\boldsymbol{n}$.

The ratio U_1/U_3 will be determined by substituting each root v^2 into the equations (18). Since $A_{ij} = A_{ji}$, the displacement vectors corresponding to the roots v_1^2 and v_2^2 will be orthogonal to each other. If the displacement direction, determined by U_1/U_3. is parallel to $\boldsymbol{n}$, the wave is called longitudinal; if the displacement corresponding to U_1/U_3 is normal to $\boldsymbol{n}$, the wave is called transverse. For isotropic materials it is known that the wave motion is longitudinal for the larger root, and transverse for the smaller root. For anisotropic materials, the velocity v depends on n and the motion is neither longitudinal nor transverse except for certain symmetry directions.

Consider next the bending equations (13)-(15). One can show that the only plane wave solutions of the form (17) that satisfy equations (13)-(15) are harmonic functions, that is,

$$\begin{bmatrix} u_1^1 \\ u_3^1 \\ u_2^0 \end{bmatrix} = \begin{bmatrix} -b\psi_3 \\ b\psi_1 \\ U_2 \end{bmatrix} e^{ik\,(n\cdot x\,-\,vt)} \tag{19}$$

For bending motion, the phase velocity v depends on the frequency $\omega = kv$ as well as the wave normal $\boldsymbol{n}$. Mindlin [4] has examined the dependence of v on ω for various material anisotropies.

Thus the behavior of the bending motion at the wave fronts cannot be determined in the same manner as was the extensional motion. Instead of finding a solution for the whole impact disturbed area of the plate, we consider the motion at the wave front only. Across this front, one imagines that certain quantities have discontinuities. The displacement and the stress are assumed to be continuous across the wave front but discontinuities in the second derivatives of U are assumed. Such waves are called acceleration waves ([6] Chapter 5).

Let $[\psi]$ denote the jump in the function $\psi\,(x_1,x_3)$ across the wave front. Then by assumption we have

$$\begin{aligned} \left[u_{2,1}^0\right] &= \left[u_{2,3}^0\right] = 0 \\ \left[u_{1,1}^1\right] &= \left[u_{1,3}^1\right] = 0 \\ \left[u_{3,1}^1\right] &= \left[u_{3,3}^1\right] = 0 \end{aligned} \tag{20}$$

The second derivatives are assumed to exist on both sides of the wave front; thus, we can write the equations of motion for bending (Eqs. (13)-(15)) on both sides of the wave front and subtract one from the other, from which results

$$\rho \left[\frac{\partial u_2^0}{\partial t^2}\right] = C_{66}\left[\frac{\partial^2 u_2^0}{\partial x_1^2}\right] + C_{44}\left[\frac{\partial^2 u_2^0}{\partial x_3^2}\right] \tag{21}$$

$$\rho \left[\frac{\partial^2 u_1^1}{\partial t^2}\right] = \hat{C}_{11} \left[\frac{\partial^2 u_1^1}{\partial x_1^2}\right] + C_{55} \left[\frac{\partial^2 u_1^1}{\partial x_3^2}\right] + \left(C_{55} + \hat{C}_{13}\right) \left[\frac{\partial^2 u_3^1}{\partial x_1\, \partial x_3}\right]$$

$$\rho \left[\frac{\partial^2 u_3^1}{\partial t^2}\right] = \hat{C}_{33} \left[\frac{\partial^2 u_3^1}{\partial x_3^2}\right] + \hat{C}_{55} \left[\frac{\partial^2 u_3^1}{\partial x_1^2}\right] + \left(C_{55} + \hat{C}_{13}\right) \left[\frac{\partial^2 u_1^1}{\partial x_1\, \partial x_3}\right]$$

The jump in acceleration, however, is not independent of the jump in the strain gradient. It can be shown that for a plane wave front with unit normal $\boldsymbol{n}$ the following relations hold:

$$\left[\frac{\partial^2 \psi}{\partial x_i\, \partial x_j}\right] = \frac{n_i n_j}{v^2} \left[\frac{\partial^2 \psi}{\partial t^2}\right] \tag{22}$$

The quantity v is called the wave front speed in the normal direction. This relation can then be used in the preceding equations to obtain linear algebraic relations between the discontinuities in acceleration across the front:

$$\rho v^2 = C_{66} \cos^2\phi + C_{44} \sin^2\phi$$

$$\begin{bmatrix} A_{11} & A_{12} \\ A_{21} & A_{22} \end{bmatrix} \begin{bmatrix} a_1 \\ a_2 \end{bmatrix} = v^2 \begin{bmatrix} a_1 \\ a_2 \end{bmatrix} \tag{23}$$

where

$$a_1 = \left[\frac{\partial^2 u_1^1}{\partial t^2}\right], a_2 = \left[\frac{\partial^2 u_3^1}{\partial t^2}\right]$$

$$n_1 = \cos\phi\,,\ n_3 = \sin\phi$$

and A_{ij} are exactly the same constants that occur in equation (18).

Thus the wave fronts associated with a jump in the bending accelerations $\partial^2 u_1^1/\partial t^2$ and $\partial^2 u_3^1/\partial t^2$ travel at the same speeds as the wave front associated with the extensional motion. There is another wave front corresponding to a jump in the quantity $\partial^2 u_2^0/\partial t^2$. For the case of a composite with symmetric ply orientation about the midplane,

$$C_{66} = C_{44}$$

The bending wave associated with the jump $[\partial^2 u_2^0/\partial t^2]$ is directionally isotropic.

If both extensional and bending motions are generated simultaneously by impact, the two extensional and two bending wave fronts will travel with the same wave speeds.

The analysis presented here is not unique. The same results can be obtained if one considers the equations of motion from the method of characteristics [6].

The velocity surfaces $\nu = \nu(\boldsymbol{n})$ have been computed for various fiber composites and ply configurations. These results are discussed in a later section.

WAVE SURFACES

In the preceding section we outlined how plane waves would travel in an anisotropic plate. The phase velocity of two of the modes was found to depend on the orientation of the wave normal to the symmetry axes of the plate. This angle we called ϕ. Suppose, then, that a plate receives a transverse impact at the origin of a coordinate system (r,θ). The disturbance can be thought of as a superposition of plane waves. To an observer at position (r_0,θ_0), the first signal to arrive may not be that corresponding to the wave normal $\phi = \theta_0$. If t is the arrival time, the first plane wave $\boldsymbol{n}(\phi)$ to arrive at the point $\boldsymbol{r}$ must satisfy (see Figure 1)

$$\boldsymbol{r} \cdot \boldsymbol{n}(\phi) = \nu(\phi)\, t$$

For a given time (say $t = 1$) the wave surface is defined as the locus of points $\boldsymbol{r}$ which satisfy (unpublished notes by Yih-Hsing Pao, Cornell University)

$$\boldsymbol{r} \cdot \boldsymbol{s} = 1 \tag{24}$$

where

$$\boldsymbol{s} = \frac{\boldsymbol{n}(\phi)}{\nu(\phi)}$$

The vector $\boldsymbol{s}$ is called the slowness vector, and the surface $1/\nu(\phi)$ is called the slowness surface. (A good discussion of the properties of velocity, slowness, and wave surfaces may be found in [2].

Instead of finding the first arrival wave $\boldsymbol{n}(\phi)$ for a given $\boldsymbol{r}$, we determine $\boldsymbol{r}$ for a given plane wave $\boldsymbol{n}$ such that equation (24) is satisfied, with $\boldsymbol{r}$ fixed. Then, the equation $\boldsymbol{s} \cdot \boldsymbol{r} = 1$ represents a line in the slowness plane and $\boldsymbol{r}$ the normal to that line. However, not all $\boldsymbol{s}$ are admissible; $\boldsymbol{s}$ has to be on the slowness surface. Thus, the line $\boldsymbol{s} \cdot \boldsymbol{r} = 1$ is tangent to the slowness surface and $\boldsymbol{r}$ is the normal vector to that surface. Suppose the slowness surface is given by the equation

$$g(\boldsymbol{s}) = 0 \tag{25}$$

Then

$$\boldsymbol{r} = \lambda \boldsymbol{N} \tag{26}$$

where

$$\boldsymbol{N} = \nabla g(\boldsymbol{s})$$

Substituting this expression for $\boldsymbol{r}$ into equation (24) yields

$$\lambda = \frac{\nu}{\boldsymbol{n} \cdot \nabla g(\boldsymbol{s})}$$

and

$$\boldsymbol{r} = \frac{\nu \nabla g(\boldsymbol{s})}{\boldsymbol{n} \cdot \nabla g(\boldsymbol{s})} \tag{27}$$

In our case ν and hence $1/\nu$ are roots of a quadratic equation (18), that is,

$$\nu^4 - a_1(\phi)\, \nu^2 + a_2(\phi) = 0$$

or

$$g(s) = a_2(\phi)\, s^4 - a_1(\phi)\, s^2 + 1 = 0$$

Thus,

$$\nabla g(s) = \frac{\partial g}{\partial s} e_s + \frac{1}{s} \frac{\partial g}{\partial \phi} e_\phi$$

$$\frac{\partial g}{\partial s} = 2s \left(2s^2 a_2 - a_1 \right)$$

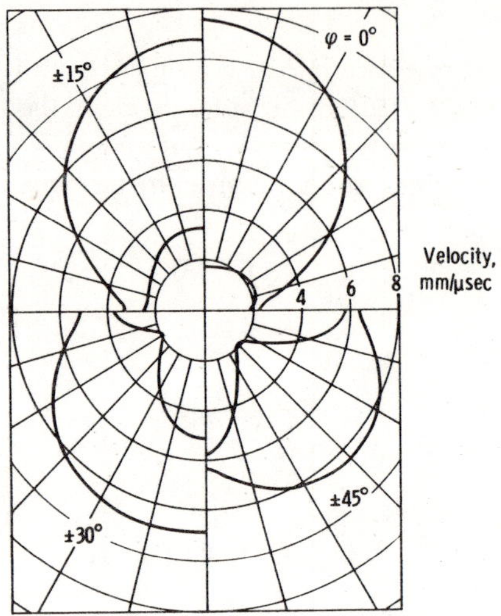

Figure 2. Velocity surfaces for 55% graphite fiber/epoxy matrix plates at layup angles 0°, ±15°, ±30°, ±45°.

$$\frac{1}{s}\frac{\partial g}{\partial \phi} = s\left(s^2\frac{\partial a_2}{\partial \phi} - \frac{\partial a_1}{\partial \phi}\right)$$

For each root ν there is a wave surface. It can be shown [3] that the outer surface, which is associated with the fastest velocity, is strictly convex. However, the slower velocity surface can result in a wave surface with cusp points.

The locus of $r(\phi)$ has been computed for various fiber composite systems and fiber layup angles. The results are discussed in the following section.

DISCUSSION OF NUMERICAL RESULTS

Velocity, slowness, and wave surfaces were calculated for a number of anisotropies corresponding to various fiber composite plates using a digital computer. The equivalent elastic constants for fiber-matrix systems at various layup angles were obtained by Chamis [7]. These constants, which are listed in Table 1, are based on a static analysis of an eight-ply plate using the known properties of each fiber-matrix ply.

The graphite-epoxy system contrasts with other composite systems because of its high stiffness ratio; $C_{11}/C_{33} = 24$ (zero layup angle). The velocity surfaces for layup angles of ±0°, ±15°, ±30°, and ±45° are shown in Figure 2. It is interesting to note that, as the fiber orientation approaches ±45°, the anisotropy in the larger wave velocity (quasi-longitudinal wave) diminishes, but that of the smaller-root (quasi-shear wave) increases.

Table 1. Stress-Strain Coefficients for 55 percent Graphite Fiber-Epoxy Matrix Composite

[All constants to be multiplied by 10^6 psi; data obtained from ref. 7]

0° Layup						±15° Layup					
27.95	0.3957	0.3957	0	0	0	24.56	0.4000	1.986	0	0	0
	1.170	0.4601	0	0	0		1.170	0.4558	0	0	0
		1.170	0	0	0			1.374	0	0	0
			0.3552	0	0				0.3552	0	0
				0.7197	0					2.310	0
					0.3552						0.3552
±30° Layup						**±45° Layup**					
16.48	0.4118	5.167	0	0	0	8.197	0.4279	6.758	0	0	0
	1.170	0.4400	0	0	0		1.170	0.4279	0	0	0
		3.093	0	0	0			8.179	0	0	0
			0.3552	0	0				0.3552	0	0
				5.491	0					7.082	0
					0.3552						0.3552

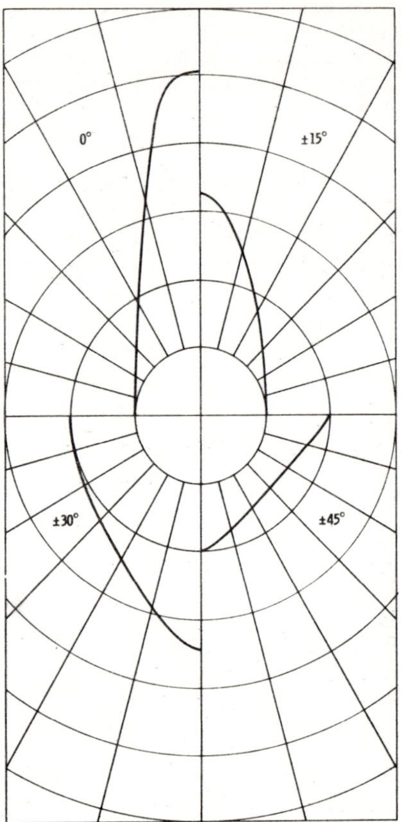

Figure 3. Slowness surfaces for 55% graphite fiber/epoxy matrix plates at layup angles 0°, ±15°, ±30°, ±45°.

The resulting wave surfaces for graphite-epoxy are shown in Figures 4,5. (The slowness surface is shown in Figure 3.) The inner surfaces show peculiar cusps and nonconvexity. This behavior is also characteristic of crystal systems such as zinc. Unlike the natural crystals, we can change the wave properties, without changing the material constituents, by varying the fiber layup angle. It becomes clear that, as the anisotropy in the outer wave is reduced (i.e., $\phi \rightarrow 45°$), the cusped behavior of the inner waves increases. This is due to the previously mentioned increase in shear wave anisotropy as $\phi \rightarrow 45°$ (Figure 2).

Another peculiar property of wave propagation in this composite system can be noted by examination of the ±45° fiber layup case (Figure 5). On the outer wave surface, the angle of the wave normal of the first arrival plane wave is listed. One can see that the distribution of plane wave normals is heavily concentrated at positions

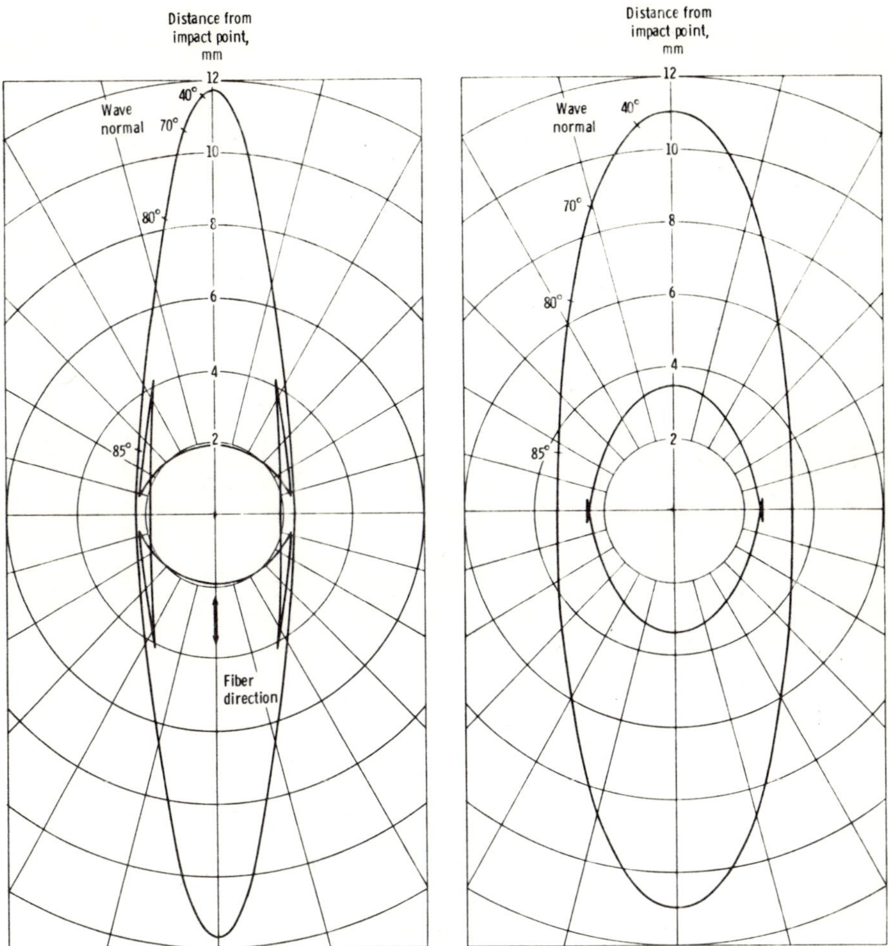

***Figure 4.** Wave surfaces for 55% graphite fiber/epoxy matrix plates at layup angles 0°, ±15°.*

on the wave surface close to the fiber directions. This might imply a focusing of waves along the fiber directions. For the other fiber orientations, the distribution of wave normals is also concentrated at those points on the wave surface close to the fiber directions but not as densely as in the ±45° layup case. The implications of this wave focusing along the fiber direction will not be made completely clear until the stress and displacement fields are found.

Similar results for the glass fiber-epoxy composite system have been calculated [9]. The ratio of stiffnesses for this case is $C_{11}/C_{33} = 3.1$ (zero layup angle). The wave surfaces for this system show features similar to the graphite-epoxy case. For a layup angle of ±15°, the quasi-shear wave velocity is almost isotropic. This results in a wave surface with no cusped behavior. Although not as marked as the graphite-

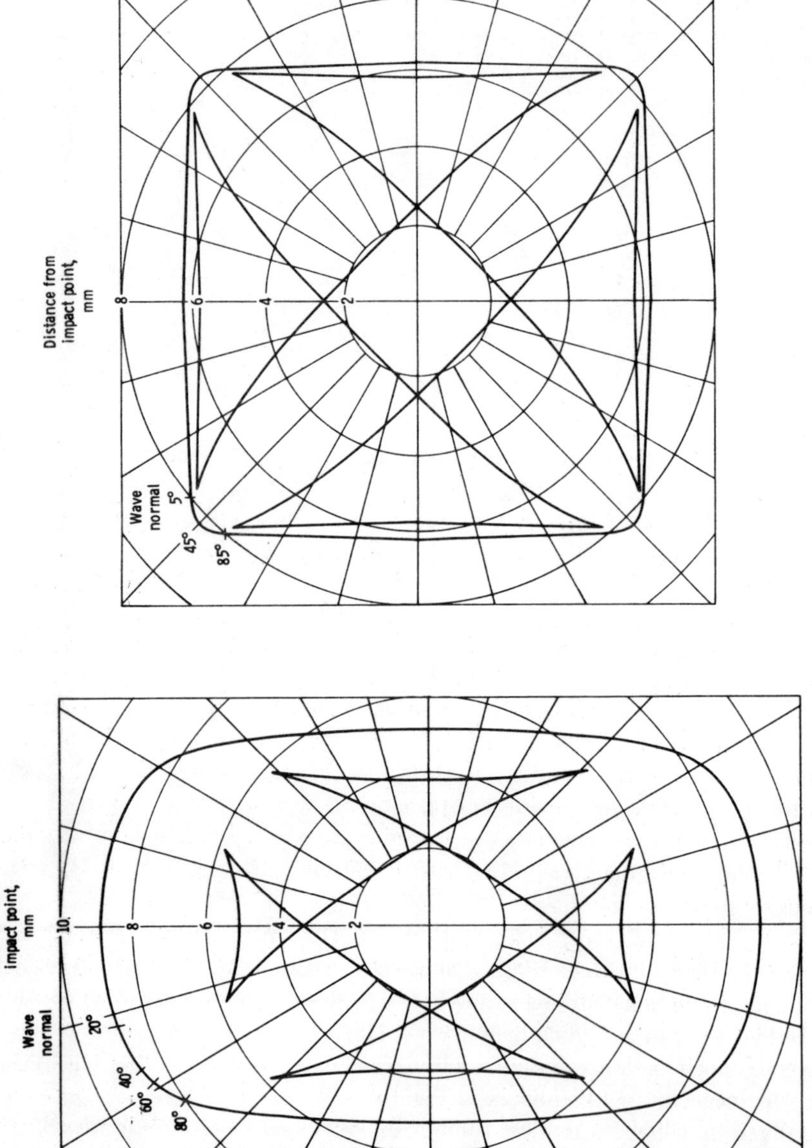

Figure 5. Wave surfaces for 55% graphite fiber/epoxy matrix plates at layup angles ±30°, ±45°.

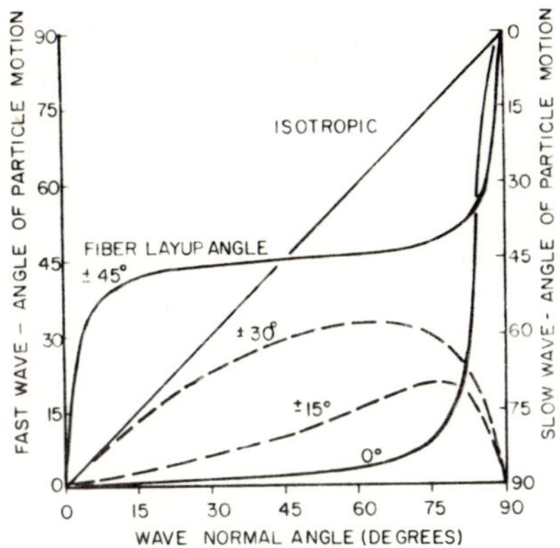

Figure 6. Direction of particle motion versus wave normal angle for 55% graphite fiber/epoxy matrix plates at layup angles 0°, ±15°, ±30°, ±45°.

epoxy case, this system also exhibits a wave normal focusing along the fiber directions.

The velocity and wave surfaces for a boron fiber/aluminum composite were also calculated [8]. However, the shear velocity is almost isotropic and no cusps appear on the wave surface.

Besides the velocity of the plane wave, the direction of particle motion relative to the wave normal has been calculated. For each wave normal there are two directions, which are the eigenvectors of the matrix A_{ij}. These two directions are orthogonal. In Figure 6 the direction of motion is plotted versus wave normal angle for various layup angles for graphite-epoxy and the isotropic case. For the latter, the motion is along the wave normal for the outer wave (longitudinal), and transverse to the wave normal for the slower wave. Such characterization as longitudinal or transverse cannot be made for anisotropic materials such as fiber composite systems. In fact for the 0°, ±45° layup angles, the direction of particle motion tends to lie close to the fiber directions for most wave normals.

The ±15° and ±30° layup angle cases present another departure from the isotropic case. In both cases $C_{55} > \hat{C}_{33}$. This means that for waves traveling in the X_3 direction, the faster wave becomes transverse and the slower wave becomes longitudinal. This behavior is also found in pine wood [1].

Recently a photoelastic study of anisotropic waves in a glass fiber reinforced plastic was made by Dally, Link, and Prabhakaran [9]. Their photographs show clearly the elliptic-like stress wave front patterns due to a small explosive charge produced in a small hole in the plate. These patterns are similar to the 0°, and ±15° layup angle wave patterns shown in this paper (Figure 4) and in [8].

Tsai and Pagano [12], have suggested the comparison of composites with an equivalent isotropic material defined by the constants

$$\hat{C}^{o}_{11} = (3\hat{C}_{11} + 3C_{33} + 2C_{13} + 4C_{55})/8$$

$$\hat{C}^{o}_{55} = (\hat{C}_{11} + \hat{C}_{33} - 2\hat{C}_{13} + 4C_{55})/8$$

The equivalent wave speeds would be

$$v^{o}_{2} = \left(\frac{\hat{C}^{o}_{11}}{\rho} \right)^{1/2} \qquad \text{(compressional)}$$

$$v^{o}_{2} = \left(\frac{\hat{C}^{o}_{33}}{\rho} \right)^{1/2} \qquad \text{(shear)}$$

For graphite-epoxy these calculations yield, $v^{o}_{1} = 7.32$ mm/μsec. and $v^{o}_{2} = 4.32$ mm/μsec. These may be compared with the anisotropic velocity surfaces in Figure 2.

ACKNOWLEDGMENT

The author wishes to thank Dr. C. C. Chamis and Mr. R. Johns of the NASA Lewis Research Center for suggestions and advise on this problem.

REFERENCES

1. S. G. Lekhnitskii, (P. Fern trans.), *Theory of Elasticity of an Anisotropic Elastic Body.* Holden-Day Publ., 1963.
2. Edgar A. Kraut, "Advances in the Theory of Anisotropic Elastic Wave Propagation," *Rev. Geophysics,* Vol. 1 (1963), p. 401.
3. M. J. P. Musgrave, "Elastic Waves in Anisotropic Media," *Progress in Solid Mechanics,* I. N. Sneddon and R. Hill, eds., Interscience Publ., Vol. 2 (1961), p.64.
4. R. D. Mindlin, "High Frequency Vibrations of Crystal Plates," *Quarterly of Applied Mathematics,* Vol. 19 (1961), p. 51.
5. R. D. Mindlin, and M. A. Medick, "Extensional Vibrations of Elastic Plates," *Journal of Applied Mechanics,* Vol. 26 (1959), p.561.
6. Richard Courant, and David Hilbert, *Methods of Mathematical Physics,* Interscience Publ., (1962).
7. C. C. Chamis, "Computer Code for the Analysis of Multilayered Fiber Composites – Users Manual," NASA TN D-7013, (1971).
8. F. C. Moon, "Wave Surfaces Due to Impact of Anisotropic Fiber Composite Plates," NASA TN D-6357, (1971).
9. J. W. Dally, J. A. Link, and R. Prabhakaran, "A Photoelastic Study of Stress Waves in Fiber Reinforced Composites," *Developments in Mechanics,* Vol. 6, Proceedings of the 12th Midwestern Mechanics Conference, (1971).
10. C. T. Sun, J. D. Achenbach, and G. Herrmann, "Continuum Theory for a Laminated Medium," *Journal of Applied Mechanics,* Vol. 35 (1968), p. 467.
11. T. R. Tauchert, and A. N. Guzelsu, "An Experimental Study of Dispersion Waves in a Fiber Reinforced Composite," Transactions ASME 71-APM-27, to appear in Journal of Applied Mechanics.
12. S. W. Tsai, and N. J. Pagano, "Invariant Properties of Composite Materials," *Composite Materials Workshop* ed. Tsai, Halpin, Pagano, (1968), p. 233.

The Vibration of Cantilever Beams of Fiber Reinforced Material

RENE B. ABARCAR

Department of Mechanical Engineering
University of the Philippines

AND

PATRICK F. CUNNIFF

Department of Mechanical Engineering
University of Maryland
College Park, Maryland 20742

(Received September 30, 1972)

ABSTRACT

Experimental results were obtained for the natural frequencies and the mode shapes of graphite-epoxy and boron-epoxy composite materials having fiber orientations of 0°, 15°, 30° and 90° with respect to cantilever beam axes. Certain elastic constants were experimentally determined and used in a programmed numerical solution in which rotary inertia, transverse shear, and coupled bending-torsion effects were included. The experimental results for the 15° and 30° beams gave a clear indication of the interaction between bending and twisting and good agreement was obtained with the numerical results.

INTRODUCTION

THE SUBJECT OF this paper is the effect of the fiber orientation on the mode of vibration of cantilever beams of unidirectional fiber reinforced composite materials. Bending of a beam where the reinforcing fibers are at an angle with the beam axis is accompanied by the twisting of the cross section around the beam axis. It is conceivable that even with small transverse displacements of the beam, resonance in the twisting mode could be present thus bringing into play a new spectrum of natural frequencies which is different from the beam which vibrates in a pure bending fashion.

COUPLED BENDING AND TORSION EFFECTS

Figure 1 shows a cantilever beam referred to a system of cartesian coordinates with the origin at the fixed end, the x and y-axes coincident with the principal axes of inertia of the cross section and the z-axis coincident with the beam axis. Using Lekhnitshii's results [1], under the simultaneous action of an applied bending moment M about the x-axis and an applied torque T about the z-axis, the deflection curve takes the form:

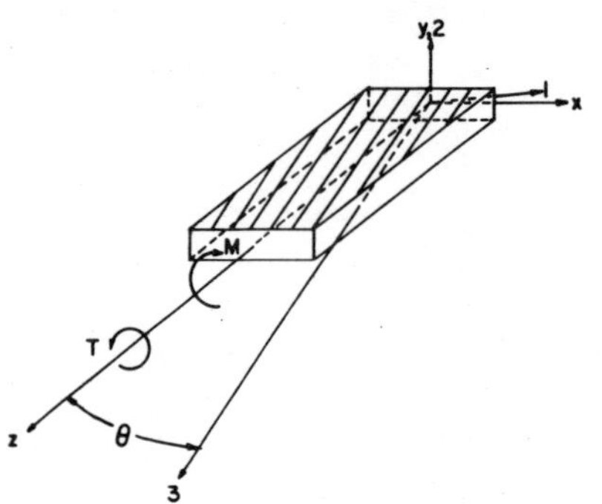

Figure 1. A beam of unidirectional fiber reinforced materials under the simultaneous action of a bending moment and a torque.

$$y = \frac{1}{2I_1}\left(M S_{33} - T\frac{S_{35}}{2}\right) z^2 \qquad (1)$$

with a relative angle of twist of:

$$\frac{d\Phi}{dz} = \frac{T}{C_t} - \frac{M}{2I_1} S_{35} \qquad (2)$$

In the case of unidirectional fiber reinforced composite materials, there are three mutually perpendicular planes of elastic symmetry. As such, and being homogeneous in the macroscopic scale, equations derived for homogeneous orthotropic materials are applicable to its analysis. Figure 1 also shows the relations between the material axes of elastic symmetry (1-1, 2-2, 3-3 coordinate system) and the beam principal axes (x, y, z coordinate system). The angle θ between the 3-3 axis and the z-axis represents the angle between the fiber direction and the beam axis. The coefficients of deformation for this case are given by:

$$S_{35} = 2\left(\frac{\sin^2\theta}{E_{11}} - \frac{\cos^2\theta}{E_{33}}\right)\sin\theta\cos\theta + \left(\frac{1}{G_{13}} - \frac{2\nu_{31}}{E_{33}}\right)(\cos^2\theta - \sin^2\theta)\sin\theta\cos\theta \qquad (3)$$

$$S_{33} = \frac{1}{E_{zz}} = \frac{\sin^4\theta}{E_{11}} + \frac{\cos^4\theta}{E_{33}} + \left(\frac{1}{G_{13}} - \frac{2\nu_{31}}{E_{33}}\right)\sin^2\theta\cos^2\theta \qquad (4)$$

VIBRATION THEORY

A. Governing Equations

At any point on the deflection curve of the beam, the slope is made up of two parts: (a) $\psi(z, t)$ is the angle of rotation made up of the contributions of direct bending due to an applied bending moment $M(z, t)$ and induced bending due to an applied torque $T(z, t)$, and (b) $\beta(z, t)$ is the angle of distortion due to direct shearing by the application of the shear force $Q(z, t)$.

The following summarizes the significant equations for the case of free vibrations of the beam:

$$\frac{dy(z)}{dz} = \psi(z) + \beta(z) = \psi(z) + \frac{Q(z)}{K' G_{zy} A} \tag{5}$$

$$\frac{d\psi(z)}{dz} = \frac{M(z)}{E_{zz} I_1} - \frac{T(z)}{C_{mt}} \tag{6}$$

$$\frac{dM(z)}{dz} + Q(z) + mk^2 \omega^2 \psi(z) = 0 \tag{7}$$

$$\frac{d\Phi(z)}{dz} = -\frac{M(z)}{C_{mt}} + \frac{T(z)}{C_t} \tag{8}$$

B. Numerical Solution

The numerical solution for the case of free vibrations has been developed following Myklestad's approach [2, 3]. The continuous cantilever beam is approximated by a system of concentrated masses and interconnecting fields. The beam is divided into n equal divisions of length Δz_i each and the mass of each segment is represented as a concentrated mass of point mass at the center of gravity of each segment.

From continuity and equilibrium considerations, we have the following relationship at the ith station:

$$\left\{ D \right\}_i^R = \left[T_s \right]_i \left\{ D \right\}_i^L \tag{9}$$

$$\left\{ D \right\}_i = \begin{Bmatrix} y \\ \psi \\ \Phi \\ M \\ Q \\ T \end{Bmatrix}_i \tag{10}$$

$$
\left[T_s\right]_i = \begin{bmatrix} 1 & 0 & 0 & 0 & 0 & 0 \\ 0 & 1 & 0 & 0 & 0 & 0 \\ 0 & 0 & 1 & 0 & 0 & 0 \\ 0 & -m_i k^2 \omega^2 & 0 & 1 & 0 & 0 \\ -m_i \omega^2 & 0 & 0 & 0 & 1 & 0 \\ 0 & 0 & -I_{pi}\omega^2 & 0 & 0 & 1 \end{bmatrix}
$$

Figure 2 shows the free body diagram of the field interconnecting station i to the station $i + 1$. Acting on the massless rod of length Δz_i are the shear force Q_i^R, bending moment M_i^R, and torque T_i^R at the left side and Q_{i+1}^L, M_{i+1}^L, T_{i+1}^L at the right side. From equilibrium considerations, we can obtain the following relationship between the right and left sides of the elastic element shown in Figure 2:

$$
\left\{D\right\}_{i+1}^{L} = \left[T_f\right]_i \left\{D\right\}_i^R \tag{12}
$$

where:

$$
\left[T_f\right]_i = \begin{bmatrix} 1 & \Delta z_i & 0 & \dfrac{(\Delta z_i)^2}{2E_{zz}I_1} & \left(\dfrac{\Delta z_i}{K'G_{zy}A} - \dfrac{(\Delta z_i)^3}{6E_{zz}I_1}\right) & -\dfrac{(\Delta z_i)^2}{2C_{mt}} \\ 0 & 1 & 0 & \dfrac{\Delta z_i}{E_{zz}I_1} & -\dfrac{(\Delta z_i)^2}{2E_{zz}I_1} & -\dfrac{\Delta z_i}{C_{mt}} \\ 0 & 0 & 1 & -\dfrac{\Delta z_i}{C_{mt}} & \dfrac{(\Delta z_i)^2}{2C_{mt}} & \dfrac{\Delta z_i}{C_t} \\ 0 & 0 & 0 & 1 & -\Delta z_i & 0 \\ 0 & 0 & 0 & 0 & 1 & 0 \\ 0 & 0 & 0 & 0 & 0 & 1 \end{bmatrix} \tag{13}
$$

The field and station transfer matrices were made dimensionless following the work of Pestel and Leckie [4]. The matrix operation also employed a shifted-matrix multiplication since the time required for this operation is much less than

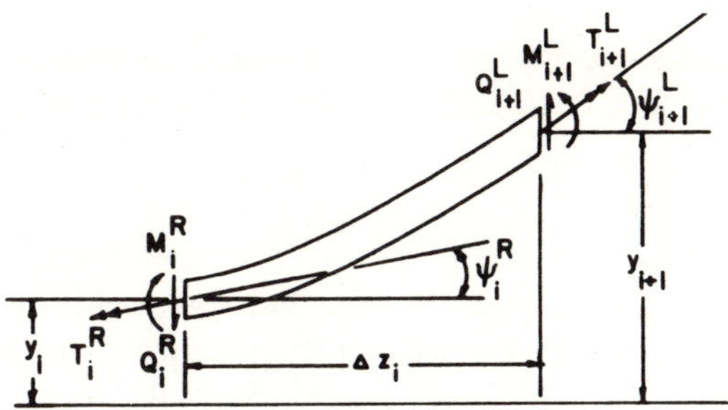

Figure 2. Free-body diagram of a field.

the direct multiplication of the station and field transfer matrices. The results from the computer program provided the natural frequencies and corresponding normal mode shapes.

C. Important Beam Parameters

The generalized torsional rigidity of a rectangular bar of orthotropic material of width b and thickness h (with $b/h > 3$) is [5]:

$$C_t = \frac{bh^3}{3} \frac{1 - 0.630 \dfrac{h}{b}}{\dfrac{1}{G_{xz}} - 0.630 \left(S_{35}^2 E_{zz} \dfrac{h}{b} \right)} \tag{14}$$

The coefficient K' for the angle of distortion is given by Langhaar [6] as $1/K' = 1.20$.

The elastic properties needed for predicting the dynamic response of the cantilever beams are the Young's moduli along two of the three principal axes of elastic symmetry, E_{11} and E_{33}, the shear moduli G_{12}, G_{23}, G_{13} and the Poisson's ratio ν_{31}.

EXPERIMENT

A. Specimens

The specimens used were graphite fiber and boron fiber reinforced epoxy beams with fiber orientations of 0°, 15°, 30° and 90° with respect to the beam axis. The graphite fiber reinforced beams were cut from Hercules designated 2002T sheets using Courtlauds HT-S graphite fibers. The boron fiber reinforced beams were from Hercules Prepreg type 2002B using an epoxy resin matrix No. BP-907 (Bloomingdale Department of American Cyanamid Company) reinforced with unidirectional boron filaments. The beams were of the approximate dimensions of 0.5 inches wide by 0.125 inches thick and a double cantilever length of 15 inches.

Thus, the length-to-thickness ratio is 60 for each cantilever beam. (The variation in thickness was as much as ± 0.002 inches and the variation in width was as much as ± 0.006 inches.)

B. Test Set-up

Figure 3 shows a diagram of the vibration test equipment which used an MB model PM-100 exciter driven by an oscillator through two model 2250 MB amplifiers. Each beam was supported between two fixed cylinders to minimize clamping effects. In order to monitor the beam response, the strain gages were placed on the beam as near as possible to the clamped support where the internal resisting moment and torque are maximum. The foil gages were the ED-DY-125AD-350 M-M strain gages which were glued to the beam surface by means of the Eastman 910 contact cement. The gages were used either singly or in combination to indicate longitudinal strains, lateral strain, or shear strains [7].

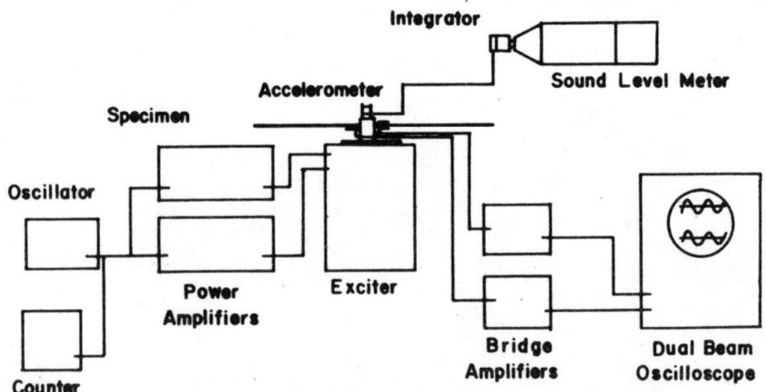

Figure 3. Instrument connection diagram.

C. Test Procedure

In the experiment, the beam specimen was made to undergo a harmonic motion by the application of a sinusoidal forcing function to the beam clamp. Peaking of the strain as observed in the oscilloscope indicated that a natural frequency had been reached. When this happened, the frequency was carefully adjusted in order to get maximum increase in the indicated strain.

For visualizing the mode shape, chalk powder (to contrast with the black beams) was sprinkled on top of the beam. The exciter table amplitude did not affect the value of the natural frequency and so it could be increased just enough to have a well defined mode shape at a certain natural frequency. The power amplifier was then turned off and the beam mode shape was photographed.

For the 15° and 30° beams, the bending modes had to be differentiated from the twisting modes. There were two sets of strain gages used. One indicated longitudinal strain and the other set indicated shear strain. Both these strains were viewed simultaneously with the dual beam oscilloscope. The bending modes indicated an almost constant ratio of the shear strain to the longitudinal strain. The twisting modes gave a different and increased ratio of the shear strain to the longitudinal strain indicating that resonance was in the twisting mode.

D. Experimental Results

Tables 1 and 2 show the predicted and measured natural frequencies and the ratios of the nth mode natural frequency to the 1st natural frequency for the beams. While two specimens were used for each test, the results for one specimen in each case is given for brevity. The number of modes excited for each particular specimen was limited by the capability of the equipment used. At frequencies higher than 5000 Hz., the displacement of the exciter tables was not enough to give a pronounced mode shape that could be visualized with the use of chalk powder.

Table 1. Comparison of Numerical and Experimental Results for the Graphite-Epoxy Beams

30° *Beam*

	Predicted				Experimental	
	G_{13} = 0.7479 × 10^6 psi		0.8465 × 10^6 psi			
Mode No.	f_n, *Hz.*	f_n/f_1	f_n, *Hz.*	f_n/f_1	f_n, *Hz.*	f_n/f_1
1	52.7	1.00	55.3	1.00	52.7	1.00
2	329.3	6.25	345.3	6.24	331.8	6.30
3	915.9	17.4	960.0	17.4	924.7	17.6
4	1767.0	33.5	1834.8	33.2	1766.9	33.5
4*	1896.5	36.0	1914.4	34.6	1827.4	34.7
5	2901.4	55.1	3036.6	54.9	2984.0	56.6
6	4263.9	80.9	4457.3	80.6	4432.4	84.1

15° *Beam*

Mode No.	f_n, *Hz.*	f_n/f_1	f_n, *Hz.*	f_n/f_1	f_n, *Hz.*	f_n/f_1
1	80.8	1.00	84.0	1.00	82.5	1.00
2	501.5	6.21	521.4	6.21	511.3	6.20
3	1376.0	17.0	1430.3	17.0	1423.4	17.2
3*	1579.3	19.5	1604.0	19.1	1526.9	18.5
4	2648.7	32.8	2754.8	32.8	2783.6	33.7
5	4189.0	51.8	4352.5	51.8	4364.6	52.9
5*	4772.5	59.1	4853.5	57.8	4731.6	57.4

The mode no. with * is a torsional mode.

Table 2. Comparison of the Theoretical and Experimental Results for the Boron-Epoxy Beams

30° Beam

	Predicted				Experimental	
G_{13} =	1.334 × 10^6 psi		1.551 × 10^6 psi			
Mode No.	f_n, Hz.	f_n/f_1	f_n, Hz.	f_n/f_1	f_n, Hz.	f_n/f_1
1	58.8	1.00	62.5	1.00	62.5	1.00
2	366.9	6.24	390.1	6.24	391.7	6.27
3	1020.4	17.4	1084.6	17.4	1090.5	17.4
4	1977.0	33.6	2097.5	33.6	2107.7	33.7
4*	2340.0	39.8	2343.1	37.5	2174.3	34.8
5	3235.5	55.0	3433.4	54.9	3542.4	56.7

15° Beam

Mode No.	f_n, Hz.	f_n/f_1	f_n, Hz.	f_n/f_1	f_n, Hz.	f_n/f_1
1	91.1	1.00	95.7	1.00	91.0	1.00
2	563.7	6.19	591.9	6.18	567.2	6.23
3	1543.4	16.9	1620.6	16.9	1575.5	17.3
3*	1865.3	20.5	1902.7	19.9	1767.4	19.4
4	2958.4	32.5	3106.4	32.5	3073.6	33.8
5	4681.1	51.4	4915.1	51.4	4926.7	54.1

The mode no. with * is a torsional mode.

Typical mode shapes for the beams are shown in Figures 4 and 5. In some cases the nodal lines are straight but skewed with respect to the beam axis, and in other cases the nodal lines are curved lines. These are indications of the interaction between bending and twisting.

One way by which the bending modes can be differentiated from the twisting modes is by comparing the directions of twisting of the free end. The bending modes experience the same direction of twist at the free end as shown by their having the same direction of the nodal line nearest that end. The twist of the free end for a torsional mode is in a direction opposite to that for bending.

DETERMINATION OF THE ELASTIC PROPERTIES

A. Principal Young's Moduli E_{11} and E_{33}

The Bernoulli-Euler theory applied to the transverse vibration of a cantilever

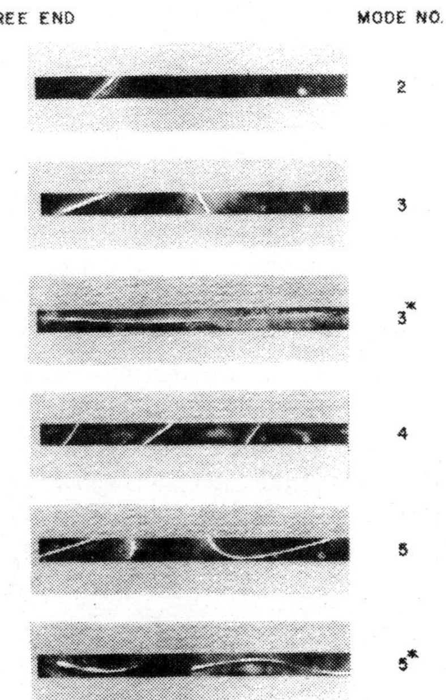

Figure 4. Mode shapes of the 15° graphite-epoxy beam.

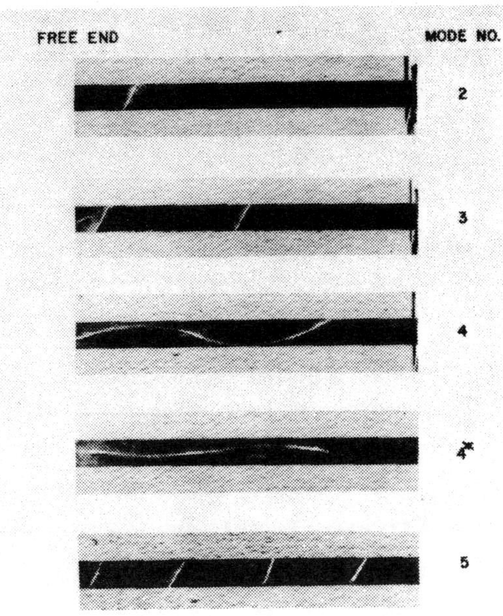

Figure 5. Mode shapes of the 30° boron-epoxy beam.

beam of isotropic material gives the fundamental frequency as:

$$\omega_1 = 3.5160\sqrt{E_{zz}I_1/m\ell^4} \qquad (15)$$

For the general case of bending without any torsional load, it is assumed that the fundamental frequency given by Equation (15) is not affected by any appreciable amount due to the rotatory inertia, transverse shear or coupled torsion effects. Hence, if the fundamental frequency is measured experimentally, [8], the value E_{zz} can be computed. It is also assumed that the dynamic modulus is frequency independent, and any departure of a measured n^{th} mode natural frequency from the corresponding natural frequency calculated using the Bernoulli-Euler theory is due to rotatory inertia, transverse shear and coupled torsion.

When the fibers are aligned with the beam axis (0° beam), the z–z direction is aligned with the 3-3 direction and the calculated dynamic Young's modulus is E_{33}. With the fibers perpendicular to the beam axis (90° beam) the z-z direction is aligned with the 1-1 axis and the corresponding dynamic Young's modulus is E_{11}.

Table 3 gives the value of E_{11} and E_{33} for the graphite-epoxy and the boron-epoxy beams, respectively, as computed from the experimentally measured fundamental frequencies.

Table 3. Principal Young's Moduli of the Beams Calculated from the Fundamental Frequency

Graphite-epoxy **$\rho = 0.056$ lb/in³**

	0° Beam		**90° Beam**	
Specimen	**f_1, *Hz.***	**$E_{33} \times 10^{-6}$, psi**	**f_1, *Hz.***	**$E_{11} \times 10^{-6}$, psi**
1	126.5	18.71	35.5	1.362
2	126.9	18.74	36.1	1.367
	Boron-epoxy **$\rho = 0.075$ lb/in³**			
1	148.2	36.66	46.7	3.851
2	147.7	37.42	46.7	3.838

B. Principal Longitudinal-Transverse Shear Moduli G_{12} and G_{23}

The natural frequencies of vibration of the Timoshenko beam of isotropic material are dependent both on the Young's modulus E and the ratio of the Young's modulus to the shear modulus, E/G. As applied to orthotropic materials

[9, 10], the solution is applicable using the longitudinal modulus E_{zz}, and the longitudinal-transverse shear modulus G_{zy}.

The value of E_{zz} computed from Equation (15) was used to calculate the natural frequencies of vibration of a Timoshenko beam with an assumed value of the ratio E_{zz}/G_{zy}. A value of E_{zz}/G_{zy} was then sought which would give a set of natural frequencies which match closely the set of natural frequencies obtained from experiment. The value of G_{zy} was then calculated.

The theoretical frequency ratios were calculated using the values of E_{zz} given in Table 3 and a ratio of E_{zz}/G_{zy} that best fits the set of experimental frequency ratios. Table 4 shows a typical set of results for one beam while Table 5 summarizes the results for G_{12} (using the 90° beams) and for G_{23} (using the 0° beam).

Table 4. Comparison for a Best Fit of the Experimental Frequency Ratios with the Theoretical Frequency Ratio for a 90° Graphite-epoxy Beam

$E_{11}/G_{12} = 3.7$

Frequency Ratio, f_n/f_1

Mode No.	*Theoretical*	*Experimental*
2	6.26	6.26
3	17.5	17.5
4	34.1	34.1
5	56.0	56.0
6	83.1	83.4
7	115.2	115.3
8	152.1	150.0
9	193.4	194.5

Table 5. Summary of G_{12} and G_{23} Values

Graphite-epoxy		*Boron-epoxy*	
$E_{11}/G_{12} = 3.7$	$G_{12} = 0.3686 \times 10^6$ psi	$E_{11}/G_{12} = 1.0$	$G_{12} = 3.845 \times 10^6$ psi
$E_{33}/G_{23} = 30.0$	$G_{23} = 0.6242 \times 10^6$ psi	$E_{33}/G_{23} = 50.0$	$G_{23} = 0.7408 \times 10^6$ psi

C. Poisson's Ratio ν_{31}

The Poisson's ratio ν_{31} was calculated as the average of the ratios of the lateral strains to the corresponding longitudinal strains measured in the tests of the 0° beam specimens. A 0° beam was vibrated with varying exciter table displacement at each natural frequency of vibration. The lateral strain and longitudinal strain readings were taken and the ratios were computed. The average Poisson's ratio obtained for the graphite-epoxy specimens was 0.30 which is in agreement with the value reported from static tests of the same type of composites.

The Poisson's ratio as computed for the boron-epoxy fibers was 0.1667 which is in error since it is below the values of the Poisson's ratios for either one of the constituents of the composite material. This discrepancy might be due to the fact that a smooth surface for gage placement is difficult to obtain when it comes to boron-epoxy composites because of the very hard boron fibers embedded in soft epoxy matrix. In view of the invalid experimental value, the static value of 0.267 reported from static tests [11] of the same type of boron-epoxy composites was used.

D. Shear Modulus G_{13}

The longitudinal modulus E_{zz} for each of the 15° and 30° beam specimens was calculated from the fundamental frequency using Equation (15) [11]. Returning to Equation (4), we now have the relationship to find G_{13} since we know E_{zz}, E_{11}, E_{33}, ν_{31} and θ. Table 6 gives the calculated values of G_{13}. Since the beams were cut from the same plate there should be only one value of G_{13}. The difference could be attributed to the difficulty in obtaining accurate alignment of fibers and accurate measuring off of the angles. In the vibration analysis, therefore, the two values are taken as the upper and lower values of the true G_{13}.

Table 6. Values of G_{13} Computed for Graphite-epoxy and Boron-epoxy Beams

Fiber Angle,	*Shear Modulus, $G_{13} \times 10^6$, psi*	
(degrees)	*Graphite-epoxy*	*Boron-epoxy*
15	0.8465	1.334
30	0.7479	1.551

COMPARISON OF THE THEORETICAL AND EXPERIMENTAL RESULTS

The values of the elastic properties obtained above were used in the computer program of the numerical solution of the vibration problem for the case of $n = 40$. Table 1 compares the computer results with the experimental results for the 30° and 15° graphite-epoxy beams while Table 2 compares the results for the 30° and 15° boron-epoxy beams. This comparison shows that there is very good agreement when the value of G_{13} computed for a particular fiber orientation is used in determining the natural frequencies of the beam of the same fiber orientation. Consequently, the natural frequencies of vibration could be satisfactorily determined with an accurate fiber alignment and precise measurement of the angle made by the fibers with the beam.

ACKNOWLEDGMENTS

This paper is part of a thesis submitted to the Department of Mechanical Engineering, University of Maryland, in partial fulfillment of the requirements for the Degree of Doctor of Philosophy.

The computer time for this project was supported by the National Aeronautical and Space Administration, Grant NsG-398 to the Computer Science Center of the University of Maryland.

The fiber reinforced beams were donated by Hercules, Inc.

NOMENCLATURE

A	cross-sectional area of the beam
S_{33}	coefficient of deformation
S_{35}	coefficient of deformation
C_{mt}	mutual stiffness coefficient $= 2 I_1/S_{35}$
C_t	Generalized torsional rigidity dependent on the cross-section and the appropriate shear moduli
E_{11}	Young's modulus in the 1-1 direction
E_{33}	Young's modulus in the 3-3 direction
E_{zz}	Young's modulus in the direction of the beam axis
G_{13}	shear modulus
G_{zy}	longitudinal-transverse shear modulus
I_1	moment of inertia of the beam cross-section with respect to the x-axis
I_p	polar moment of inertia of the beam element about the beam axis
k	radius of gyration of the beam differential element about an axis through its center normal to the y-z plane
K'	coefficient dependent on beam's cross-section
ℓ	beam length
M	bending moment
m	mass per unit length of beam
Q	shear force
T	torque
$[T_f]_i$	field transfer matrix
$[T_s]_i$	station transfer matrix
y	lateral displacement of the beam's neutral axis
β	angle of distortion due to direct shear
Φ	angle of twist of beam cross-section
ψ	angle of rotation due to the bending moment
ν_{31}	Poisson's ratio
θ	angle between the fiber direction and the beams longitudinal axis

REFERENCES

1. S. G. Lekhnitskii, *Theory of Elasticity of an Anisotropic Body*, Holden-Day, (1963), p. 184.
2. L. Meirovitch, *Analytical Methods in Vibrations*, MacMillan, (1967).
3. N. O. Myklestad, *Fundamentals of Vibration Analysis*, McGraw-Hill, (1956).
4. E. C. Pestel and F. A. Leckie, *Matrix Methods in Elastomechanics*, McGraw-Hill, (1963).
5. W. Voigt, *Lehrbuch der Kristallphysik*, Zeipzig-Berlin, (1928), p. 645.
6. H. L. Langhaar, *Energy Methods in Applied Mechanics*, John Wiley & Sons, (1962).
7. T. G. Beckwith and N. L. Buck, *Mechanical Measurements*, Addison-Wesley, (1961).
8. A. B. Schultz and S. W. Tsai, "Dynamics Moduli and Damping Ratios in Fiber-Reinforced Composites," *J. Composite Materials*, Vol. 2 (1968), p. 368.
9. T. J. Dudek, "Young's and Shear Moduli of Unidirectional Composites by a Resonant Beam Method," *J. Composite Materials,* Vol. 4 (1970), p. 232.
10. J. L. Nowinski, "On the Transverse Wave Propagation in Orthotropic Timoshenko Bars," *Int. J. Mechanical Science*, Vol. 11 (1969), p. 689.
11. J. M. Whitney and M. B. Riley, "Elastic Properties of Fiber Reinforced Composite Materials," *AIAA Journal*, Vol. 4 (1965), p. 1537.

Vibration and Buckling Analysis of Composite Plates and Shells*

J. A. McElman, *Lowell Technological Institute, Mechanical Engineering Department, Lowell, Massachusetts,* and A. C. Knoell, *Jet Propulsion Laboratory, Engineering Mechanics Division, Pasadena, California*

(Received June 12, 1971)

In the analysis of laminated composites it is known that a coupling exists between extension and bending if the plies are not balanced in number and fiber orientation. References 1 and 2 examined this effect for the bending, vibration and buckling of two, four and six ply laminates. The purpose of this note is to investigate the magnitude of this effect for buckling and vibration of doubly curved monocoque plates and shells of positive and negative Gaussian curvature. In addition, the effect of stacking sequence (the order in which individual plies are laid up) is examined. This effect is considered since it is analogous to that of eccentric stiffening of isotropic cylinders (References 3 and 4). Solutions are presented which provide a means of simply and economically assessing the magnitude of the coupling and stacking effects for various composite materials and geometric configurations.

The equations governing the prestressed vibration of doubly curved plates and shells (neglecting in-plane inertia) can be shown to be the following (Reference 5).

$$N_{x,x} + N_{xy,y} = 0$$
$$N_{y,y} + N_{xy,x} = 0 \qquad (1)$$
$$-M_{x,xx} - 2\,M_{xy,xy} - M_{y,yy} \pm \frac{N_x}{R_1} + \frac{N_y}{R_2} + N_x W_{,xx} + N_y W_{,yy} + M W_{,u} = 0$$

where the ($\pm$) refers to either positive or negative Gaussian curvature (Figure 1).

The solution to Equation (1) for the case of a simply supported shell of cross ply construction results in the following characteristic equation:

$$N_x\left(\frac{m\pi}{L}\right)^2 + N_y\left(\frac{n}{R_2}\right)^2 + M\omega^2 = K_{oo} - \left[\frac{K_{11}K_{23}^2 + K_{22}K_{13}^2 - 2\,K_{12}K_{13}K_{23}}{K_{11}K_{22} - K_{12}^2}\right] \qquad (2)$$

where

$$K_{oo} = D_{11}\left(\frac{m\pi}{L}\right)^4 + (2\,D_{12} + 4\,D_{66})\left(\frac{m\pi}{L}\right)^2\left(\frac{n}{R_2}\right)^2 + D_{22}\left(\frac{n}{R_2}\right)^4$$
$$+ 2\left[\pm\frac{B_{11}}{R_1}\left(\frac{m\pi}{L}\right)^2 + \frac{B_{22}}{R_2}\left(\frac{n}{R_2}\right)^2\right] + \left(\pm\frac{2\,A_{12}}{R_1R_2} + \frac{A_{11}}{R_1^2} + \frac{A_{22}}{R_2^2}\right)$$

*This paper presents the results of one phase of research carried out at the Jet Propulsion Laboratory, California Institute of Technology, under Contract No. NAS7-100, sponsored by the National Aeronautics and Space Administration.

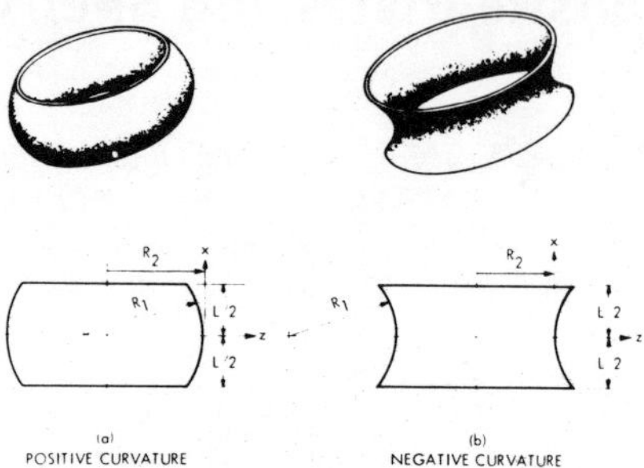

Figure 1. *Shell configurations indicating Gaussian curvature.*

and

$$K_{11} = A_{11}\left(\frac{m\pi}{L}\right)^2 + A_{66}\left(\frac{n}{R_2}\right)^2 \qquad K_{22} = A_{22}\left(\frac{n}{R_2}\right)^2 + A_{66}\left(\frac{m\pi}{L}\right)^2$$

$$K_{12} = (A_{12} + A_{66})\left(\frac{m\pi}{L}\right)\left(\frac{n}{R_2}\right) \qquad K_{23} = \left[\frac{A_{22}}{R_2} \pm \frac{A_{12}}{R_1}\right]\left(\frac{n}{R_2}\right) + B_{22}\left(\frac{n}{R_2}\right)^3$$

$$K_{13} = \left[\pm\frac{A_{11}}{R_1} + \frac{A_{12}}{R_2}\right]\left(\frac{m\pi}{L}\right) + B_{11}\left(\frac{m\pi}{L}\right)^3$$

where [A], [B] and [D] are stiffness matrices for a cross ply laminate and m is the number of axial half waves and n is the number of circumferential full waves. Equation (2) can also be used for the analysis of simply-supported, doubly-curved rectangular plates by changing (n/R_2) to $(n\pi/b)$.

The natural frequencies of prestressed plates can be determined from Equation (2) and buckling loads can also be determined by setting ω equal to zero and minimizing with respect to m and n.

Using Equation (2), numerical results were obtained for two materials: glass-epoxy and graphite-epoxy. The assumed properties of an individual ply are given as follows:

Material	E_{11} (psi)	E_{22} (psi)	G_{12}(psi)	ν_{12}	h (in)
Glass Epoxy	7.5×10^6	2.6×10^6	1.1×10^6	.25	.011
Graphite Epoxy	30.0×10^6	$.75 \times 10^6$	$.75 \times 10^6$	.25	.011

Using the ply properties given above, the material constants needed for Equation (2) were computed. Equation (2) was then solved numerically for the particular case under consideration.

Since it is impractical to perform a complete parametric study, data are presented for particular materials and shell geometries to demonstrate the effects of coupling and staking on shell response.

The shells considered had the following geometry:

$R_1 = \pm 64''$ (positive or negative Gaussian curvature); $R_2 = 16''$; $L = 40''$

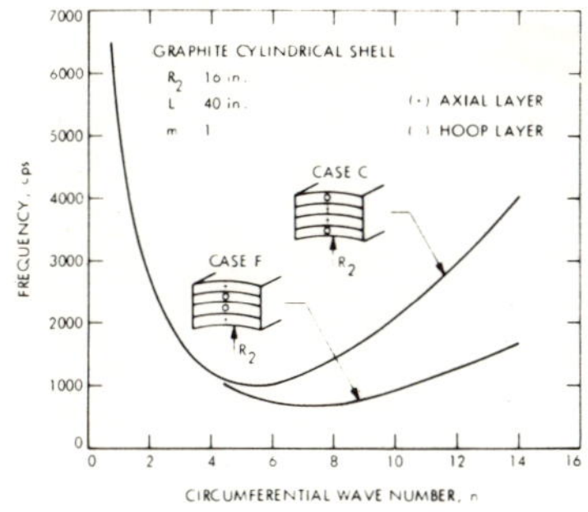

Figure 2. *Natural frequencies for four-ply graphite cylindrical shell showing effect of ply location.*

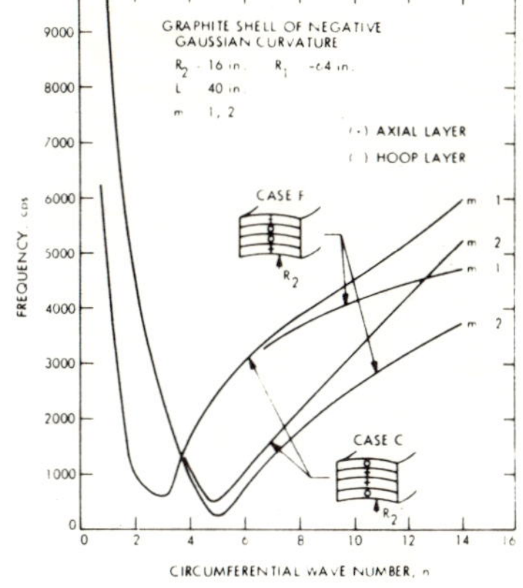

Figure 3. *Natural frequencies for four-ply shell of negative Gaussian curvature showing effect of ply location.*

Figures 2, 3, and 4 give the natural frequencies for cases C and F as a function of circumferential wave number for a cylinder and shells of both positive and negative Gaussian curvature. Since coupling effects can be neglected in these cases due to symmetry of the lay-up, the effect that is demonstrated is the effect of stacking sequence. Case C exhibits high circumferential stiffness as opposed to Case F, which has high axial stiffness. For the cylindrical shell (Figure 2) the lowest natural frequency with large circumferential stiffness (Case C) is 54% higher than that of the lowest frequency of the shell with high axial stiffness (Case F). At a circumferential wave number of 14 the difference is almost a factor of 2.5. Figure 3 demonstrates the effect of stacking sequence for a shell of negative Gaussian curvature where the minimum natural frequency occurs for $M = 2$ and $N = 5$. The lowest frequency for the shell with high circumferential stiffness (Case C) is twice that of the shell with high axial stiffness (Case F). For the shell with positive Gaussian curvature (Figure 4), large circumferential wave numbers must be reached before any significant difference is observed.

Figure 5 demonstrates the effect of coupling between extension and bending for a cylindrical shell. It shows that for cases B and E and cases A and D there are no significant differences in results. Both sets of cases have the same circumferential and axial stiffness. Cases B and E provide the largest value for the coupling matrix, [B], the smallest value of overall stiffness for all shell configurations considered, and hence produce the lowest natural frequencies. For the cylinder, cases A and D produce frequencies 23% higher than those calculated for cases B and E. Although not shown, the shells of negative and positive Gaussian curvature showed differences of the same order between the same case sets of 60% and 0%, respectively, for the lowest natural frequency. The shell of positive Gaussian curvature exhibited very little sensitivity to coupling effects until a high circumferential wave number was reached.

In an attempt to discover if the sensitivity to coupling effects is dependent on the stiffness of the material under consideration, natural frequencies were obtained for a

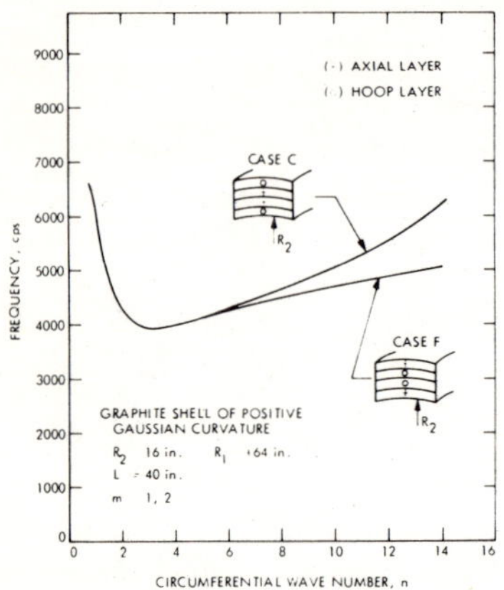

Figure 4. *Natural frequencies for four-ply shell of positive Gaussian curvature showing effect of ply location.*

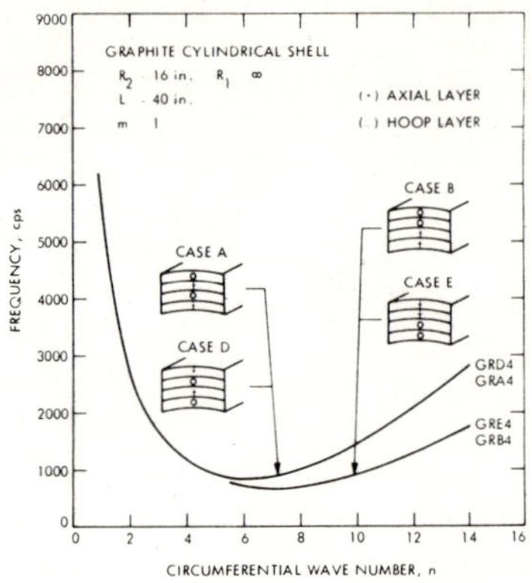

Figure 5. *Natural frequencies for four-ply graphite cylinder showing effect of coupling between extension and bending.*

two-ply cylindrical shell fabricated of glass epoxy. The results show that the low modulus material is slightly more sensitive than the high modulus material (e.g. 5% as compared to 1%). It must be emphasized, however, that the materials differed in characteristics other than just modulus of elasticity. For example, the graphite-epoxy considered has a shear modulus which is extremely low compared to the glass-epoxy composite. The results of this and countless other variables add greatly to the degree of difficulty in making overall conclusions and recommendations.

Buckling results were also obtained for a four-ply graphite cylinder loaded in axial compression. The results show the same overall effects as the vibration analysis with the difference in buckling load resultant varying by as much as 50%.

REFERENCES

1. J. M. Whitney, "Bending-Extensional Coupling in Laminated Plates Under Transverse Loading," *J. Comp. Matls*, Vol. 3 (1969), p. 20.
2. J. M. Whitney and A. W. Leissa; "Analysis of Heterogeneous Anisotropic Plates," *J. Appl. Mech.*, Vol. 36 (1969), p. 261.
3. M. F. Card, "Preliminary Results of Compression Tests on Cylinders With Eccentric Longitudinal Stiffness," NASA TM X-1004 (1964).
4. J. A. McElman, M. M. Mikulas and M. Stein, "Static and Dynamic Effects of Eccentric Stiffening of Plates and Cylindrical Shells," *AIAA Journal*, Vol. 4, No. 5 (1966).
5. J. A. McElman, "Eccentrically Stiffened Doubly Curved Shells," Thesis submitted as partial fulfillment of requirements for Ph.D. degree, Virginia Polytechnic Institute (1965).

Stress Gradients in Laminated Composite Cylinders

N. J. PAGANO, *Nonmetallic Materials Division, Air Force Materials Laboratory, Wright-Patterson AFB, Ohio 45433*

(Received February 3, 1971)

In a recent paper [1], (see also [2]) it was shown that extremely severe stress gradients can exist in the wall of a unidirectional, helical-wound cylinder under the common loadings applied in the laboratory, i.e., axial loading, internal pressurization, and torsion. In that study it was found that the stress gradients were drastically reduced in an orthotropic, symmetric laminated cylinder, such as an angle-ply. In fact, for the case of torsion, the shear stress distribution almost coincides with that which occurs in an isotropic cylinder. While an analogous smoothing of the stress field occurs under other loadings, the axial normal stress under axial loading and the hoop stress under internal pressure are somewhat different from the respective isotropic results. For example, refer to Figure 12 in [1]. In order to complete the treatment of the geometric design of tubular characterization specimens, the remaining stress components, e.g., the hoop and shear stresses induced in the axial loading experiment, should be considered. It is therefore the purpose of this note to present some results in this regard for laminated cylinders.

In order to discuss the stress field in a laminated cylinder, we shall consider the angle-ply configuration studied in [1], employing the same elastic layer properties used for the curves of Figure 12 [1]. Recall that these properties are representative of a highly anisotropic graphite-epoxy composite. We shall adopt the same notation utilized in [1], hence the definitions of the various symbols need not be repeated here.

The relationship of the maximum stresses in the gage section, σ_x and σ_θ, at the outer surface $r = R_1$ under pure torsion $\tau_o = 1$, to the helical angle α are shown in Figure 1. The results are indicated for $R/h = 10$, which is probably the minimum aspect ratio consistent with precise characterization. If the state of stress in each layer were uniform, the response would be given by classical lamination theory (LT) [3], in other words, the stresses in an angle-ply coupon under the membrane resultants $N_{xy} = \tau_o h$, $N_x = N_y = 0$. The fact that the cylinder and LT results of Figure 1 are almost coincident indicates nearly uniform layer stresses. This is more evident from Figure 2, where the variation of σ_θ across the wall thickness is shown for a $\pm 45°$ angle-ply for R/h values of 10 and 80. It is also interesting to note that the cylinder curves of Figure 1 are nearly reflections* of each other across the coordinate line $\alpha = 45°$.

Similar results for the axial loading $\sigma_o = 1$ are given in Figures 3-5, while

* The LT curves of fig. 1 are exact reflections of each other across the line $\alpha = 45°$.

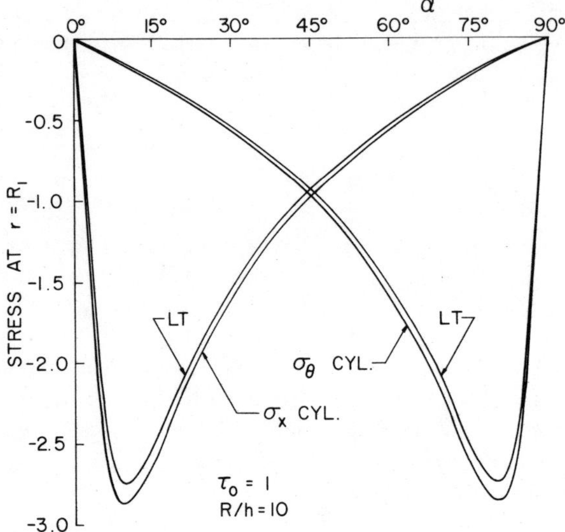

__Figure 1.__ Max. stresses in graphite-epoxy angle-ply composites under torsion.

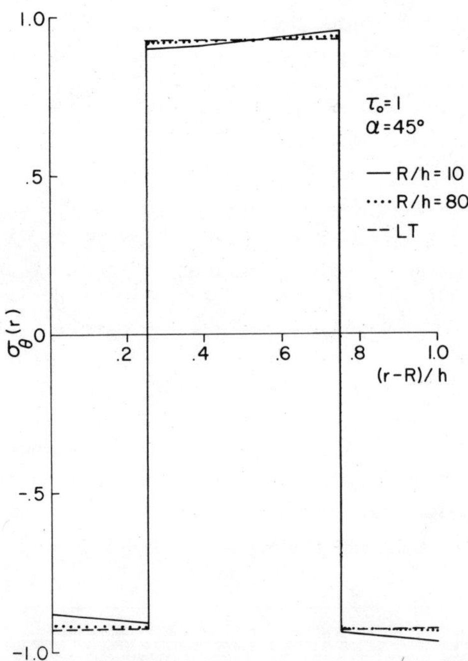

Figure 2. *Hoop stress distribution in ± 45° graphite-epoxy angle-ply under torsion.*

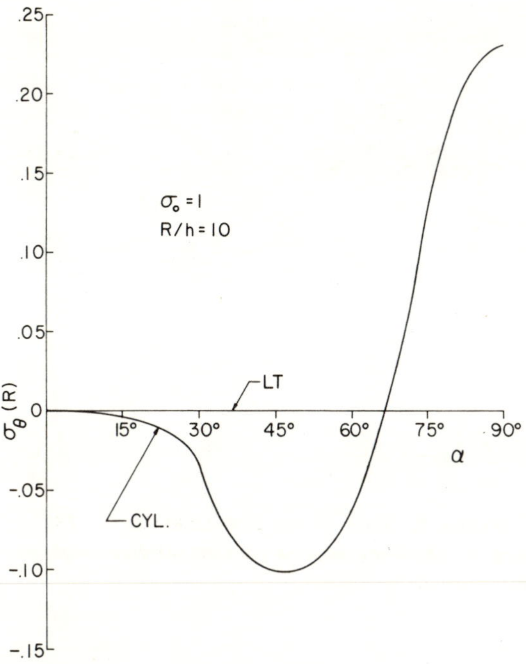

Figure 3. *Max. hoop stress in graphite-epoxy angle-ply composites under axial tension.*

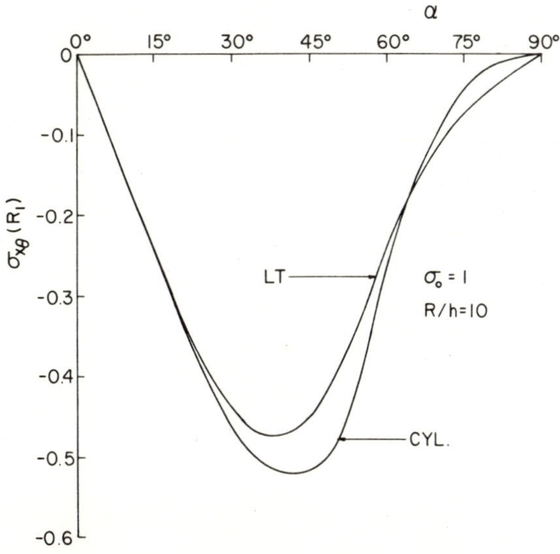

Figure 4. *Max. shear stress in graphite-epoxy angle-ply composites under axial tension.*

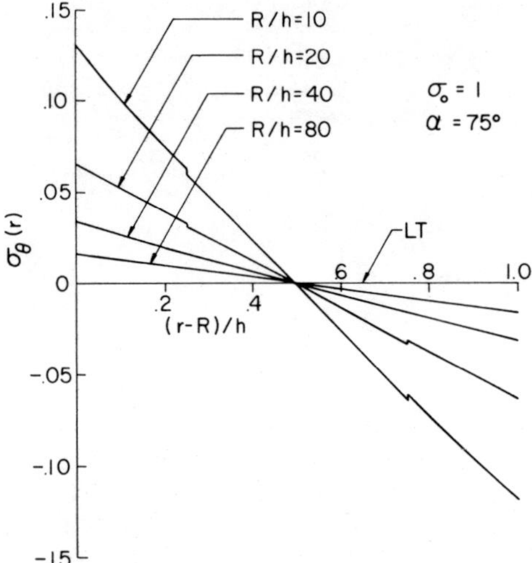

Figure 5. *Hoop stress distribution in ± 75° graphite-epoxy angle-ply under axial tension.*

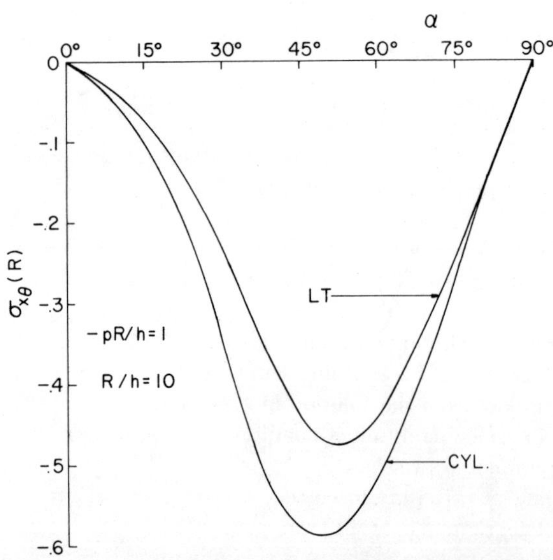

Figure 6. *Max. shear stress in graphite-epoxy angle-ply composites under internal pressure.*

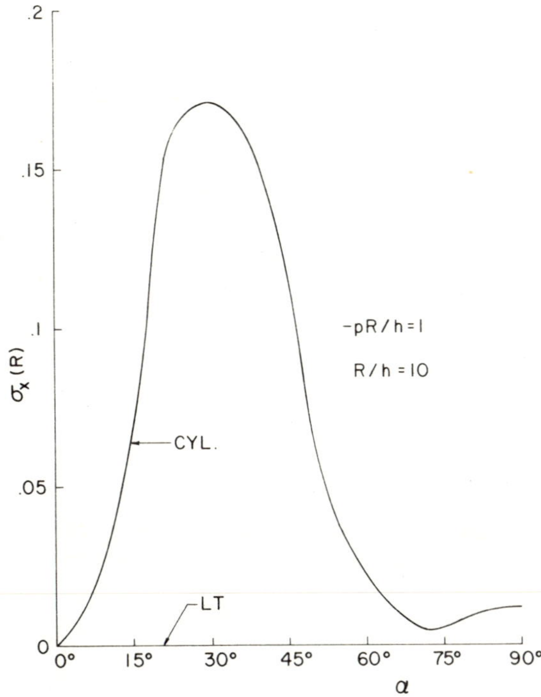

***Figure 7.** Max. axial stress in graphite-epoxy angle-ply composites under internal pressure.*

selected results for the internal pressure loading $-\dfrac{pR}{h} = 1$ are shown in Figures 6-8. For the same ratio $\dfrac{R}{h} = 10$, the respective cylinder and *LT* results are in much better agreement for the torsion problem than either axial or internal pressure loadings. Therefore, the stress gradients are generally steeper under the latter loadings.

These results have been presented because some of them might not have been anticipated from data given in [1], rather than to supply "design charts" which enable one to select an acceptable aspect ratio R/h for arbitrary loading conditions and laminate geometry. Although we have identified the important parameters and established some guidance in choosing an acceptable R/h value, this decision must, in general, follow from the solution of Equations (4)-(7) of [1], along with the appropriate interface continuity conditions, for the specific conditions employed in the experimental program.

It is appropriate at this time to correct a typographical error which appeared in the first complete sentence of [1] on page 376. This sentence should begin, "For example, although we do *not* have a rigorous proof". We might also add that a rigorous proof (by someone else) would certainly be welcomed.

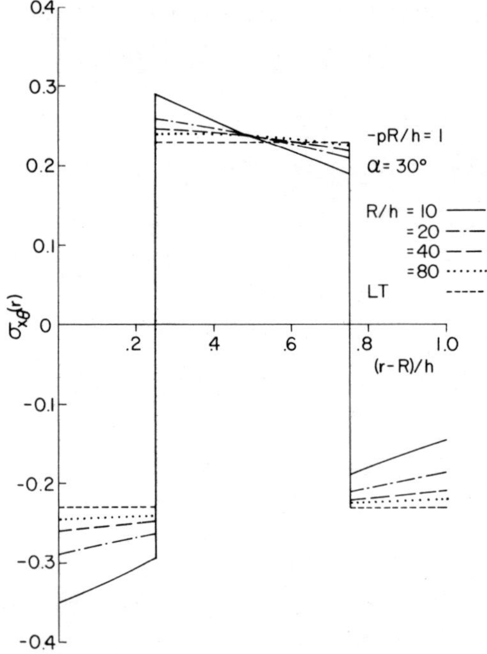

Figure 8. *Shear stress distribution in ± 30° graphite-epoxy angle-ply under internal pressure.*

ACKNOWLEDGMENTS

The author wishes to acknowledge the work of Mr. E. Guthrie for the computer analysis anl Mr. B. Maurer for his assistance in the numerical calculations.

REFERENCES

1. N. J. Pagano and J. M. Whitney, "Geometric Design of Composite Cylindrical Characterization Specimens," *J. Composite Materials,* Vol. 4 (1970), p. 360.
2. R. R. Rizzo and A. A. Vicario, "A Finite Element Analysis of Laminated Anisotropic Tubes (Part I—A Characterization of the Off-Axis Tensile Specimen)," *J. Composite Materials,* Vol. 4 (1970), p. 344.
3. Y. Stavsky, "Bending and Stretching of Laminated Aeolotropic Plates," *Proc. Am. Soc. Civil Engrs.; J. Engr. Mech. Div.,* Vol. 87 (1961), p. 31.

Bending of Laminated Anisotropic Composite Cylinders*

Hwa M. Zien, *The Babcock & Wilcox Company Alliance Research Center, Alliance, Ohio 44601*

(Received March 29, 1973)

Presented here are results relating to the effects of bending-shear coupling, layer interaction, and shear deflection of laminated anisotropic cylinders under bending. Cutler [1] obtained an elasticity solution to laminated composite cylinders under pure bending. Lehknitskii [2] and Adams, Maiti and Mark [3] solved for a case of cantilever cylinders where there was no coupling between bending and shear. Recently a three-dimensional elasticity solution was obtained for laminated anisotropic cantilever cylinders. This is summarized in the present note. An extension of Cutler's solution and the anisotropic cantilever solution provided solutions for statically determinate cylinders under applied concentrated forces and couples.

ELASTICITY SOLUTIONS

For a filament-wound fiber reinforced lamina, the constitutive relations are:

$$\begin{bmatrix} \sigma_Z \\ \sigma_\theta \\ \sigma_r \\ \tau_{Z\theta} \end{bmatrix} = \begin{bmatrix} C_{11} & C_{12} & C_{13} & C_{16} \\ C_{21} & C_{22} & C_{23} & C_{26} \\ C_{31} & C_{32} & C_{33} & C_{36} \\ C_{61} & C_{62} & C_{63} & C_{66} \end{bmatrix} \begin{bmatrix} e_Z \\ e_\theta \\ e_r \\ \gamma_{Z\theta} \end{bmatrix} \tag{1}$$

and

$$\begin{bmatrix} \tau_{r\theta} \\ \tau_{rZ} \end{bmatrix} = \begin{bmatrix} C_{44} & C_{45} \\ C_{54} & C_{55} \end{bmatrix} \begin{bmatrix} \gamma_{r\theta} \\ \gamma_{rZ} \end{bmatrix} \tag{2}$$

where C_{ij} are the elastic stiffnesses and Z, θ, and r are the axial, circumferential and radial coordinates, respectively.

Consider a laminated cylinder of length ℓ fixed at $Z = 0$. A force is applied at the free end ($Z = \ell$) in a direction perpendicular to the plane $\theta = 0$. By applying

*This work was performed under an internal research project supported by the Advanced Composites Department, Babcock & Wilcox Company.

the Saint-Venant principle, one obtained the elasticity solutions for each lamina in the following form

$$\begin{bmatrix} e_Z \\ e_\theta \\ e_r \\ \gamma_{Z\theta} \end{bmatrix} = \begin{bmatrix} ar \\ e_2(r) \\ e_3(r) \\ \dfrac{f_1(r)}{r} \end{bmatrix} (\ell - Z) \sin\theta + \begin{bmatrix} f_1(r) + br \\ f_2(r) \\ f_3(r) \\ e_6(r) \end{bmatrix} \cos\theta \tag{3}$$

and

$$\begin{bmatrix} \gamma_{r\theta} \\ \gamma_{rZ} \end{bmatrix} = \begin{bmatrix} e_4(r) \\ \dfrac{df_1(r)}{dr} \end{bmatrix} (\ell - Z) \cos\theta + \begin{bmatrix} f_4(r) \\ e_5(r) \end{bmatrix} \sin\theta \tag{4}$$

The solutions were obtained by the method of trial and error such that there resulted only three independent compatibility equations. The solutions also satisfied four of the six boundary conditions in stress resultants (similar to those given in [3]).

The three compatibility equations combined with three equilibrium equations and six constitutive relations resulted in a total of nine equations for nine functions (e_2, e_3, e_4, e_5, e_6, f_1, f_2, f_3, and f_4) of each lamina. Each of the constants a and b was found to have the same value in all the laminae. The solutions were uniquely determined by satisfying the remaining two boundary conditions, the conditions at the outer and inner surfaces of the cylinder, and the matching conditions between the adjacent laminae.

In Equation (3), terms with a factor $\cos\theta$ arise mainly from the transverse shear force. The primary effect of the bending-shear coupling is the induced shear strains $\gamma_{Z\theta}$ and γ_{rZ} which are linearly proportional to $(\ell - Z)$ or the applied moment. The deformed state of each lamina can have a curvature (due to $(f_1 + br)\cos\theta$ in e_Z) along the direction of the applied moment.

BENDING TESTS

On several selected cylindrical specimens (1/2 in. OD X 12 in. long) four-point bending tests were performed. The span between the supports was 10 in. Two forces of the same magnitude were applied at locations 2 in. from the middle of the span. The specimens used in the test were made of the following materials:

1. S-Glass/Epoxy $E_L = 6.5 \times 10^6$ psi, $E_T = 2.4 \times 10^6$ psi,

$G_{LT} = 0.98 \times 10^6$ psi

$\nu_{LT} = \nu_{TT} = 0.26$

2. GY-70/Epoxy $E_L = 33 \times 10^6$ psi, $E_T = 0.8 \times 10^6$ psi,

$G_{LT} = 0.7 \times 10^6$ psi

$\nu_{LT} = \nu_{TT} = 0.32$

3. PRD-III/Epoxy $E_L = 12.5 \times 10^6$ psi, $E_T = 0.9 \times 10^6$ psi,

$G_{LT} = 0.4 \times 10^6$ psi

$\nu_{LT} = \nu_{TT} = 0.32$

4. HMS/Epoxy $E_L = 33 \times 10^6$ psi, $E_T = 1 \times 10^6$ psi,

$G_{LT} = 0.65 \times 10^6$ psi

$\nu_{LT} = \nu_{TT} = 0.3$

The layup sequences of the specimens (starting from the outer layer) were:

Specimen 1 — 0° S-GL/(+30°, +45°, −45°, −45°, −25°, +20°) GY-70/ (0° S-GL)$_2$/0° GY-70/0° S-GL
(ply thickness: S-GL 0.0065 in., GY-70 0.0065 in.)

Specimen 2 — Same as Specimen 1 except that the second ply was +30° S-GL and that the ply thickness for GY-70 was 0.0075 in.

Specimen 3 — (33°, −19°) HMS/−23° PRD III/(35°, −45°, 25°) HMS/ 10° PRD III/−11° HMS/(−20°, 15°)PRD III
(ply thickness: HMS 0.0077 in., PRD III 0.0085 in.)

Specimen 4 — (±30° HMS, ±30° S-GL)$_3$/0° HMS
(ply thickness: 0.0077 in. each)

Specimen 5 — (0°)$_{11}$ GY-70
(total thickness: 0.075 in.)

RESULTS AND REMARKS

Results obtained from the elasticity solution and the bending tests are given in Table 1. To illustrate the effect of layer interaction, results from a strength of material approximation [4] are also given. There was good agreement between the test results and the elasticity solutions (columns 2, 3, 5, and 6 Table 1). However, for a unidirection cylinder (Specimen 5) the distortion in the cross-sectional shape due to concentrated loads was considerable and reliable readings in hoop strains were not obtained in the test.

Table 1. Strains (μin./in.) at Outer Surface of Laminated Composite Cylinders Under Four-Point Bending

Bending Moment = 10 lbs-in. and Shear Force = 10 lbs

Specimen	Axial Strain ($\theta = 90°$)			Hoop Strain ($\theta = 90°$)		Axial Strain ($\theta = 0°$)	Shear Strain ($\theta = 90°$)	Shear Strain $\gamma_{Z\theta}$ ($\theta = 0°$)
No.	Test	ES	SM	Test	ES	ES	ES	ES
(1)	(2)	(3)	(4)	(5)	(6)	(7)	(8)	(9)
1	−122	−113.2	−183	98	94.8	− 4.7	−16.0	− 50.8
2	−121	−118.0	−186	91	85.9	.4	.6	− 50.9
3	−103	− 99.3	−252	110	114.7	−14.5	−45.0	− 57.3
4	−115	−107.4	−230	113	117.0	− 1.0	− 3.7	− 43.2
5	− 29.9	− 32.5	− 32.5	21.6	8.8	0	0	−237.0

(Applied forces perpendicular to the plane $\theta = 0$; ES – elasticity solution; SM – strength of material approximation)

The results for laminated anisotropic cylinders are summarized as follows:

1. The interaction between layers is very significant. It increases the bending stiffness (columns 2, 3, and 4, Table 1). This effect was found much less significant in cylinders of transversely isotropic laminates [3].
2. The shear strain $\gamma_{Z\theta}$ and the hoop strain can be important factors in design analysis (columns 5, 6, and 9).
3. The bending-shear coupling can, in some cases, significantly affect the shear strain $\gamma_{Z\theta}$ (Specimens 1 and 3, column 8).
4. The axial strain due to a transverse shear force and bending-shear coupling, $(f_1(r) + br)\cos\theta$ in Equation (3), is relatively less significant (column 7).

A final remark is that for Specimen 5 which had very low effective shear modulus, the elasticity solution indicated an approximately 20 percent deflection due to the shear force.

REFERENCES

1. V. C. Cutler, "Bending Analysis of Directionally Reinforced Plastic Pipe," The Society of Plastic Industry, Inc., 16th Annual Technical and Management Conference, Reinforced Plastics Division (1961).
2. S. G. Lekhnitskii, *Theory of Elasticity of an Anisotropic Elastic Body*, Holden-Day, Inc., (1963).
3. S. F. Adams, M. Maiti and R. E. Mark, "Three-Dimensional Elasticity Solution of a Composite Beam," *J. Composite Materials*, Vol. 1 (1967), p. 122.
4. J. H. Faupel, *Engineering Design*, John Wiley and Sons, Inc., (1964).

Acoustic Emission Produced During Burst Tests of Filament-Wound Bottles*

MARVIN A. HAMSTAD AND T. T. CHIAO

University of California
Lawrence Livermore Laboratory
Livermore, California 94550

(Received May 12, 1973)

ABSTRACT

Acoustic emission was recorded during burst tests of filament-wound, composite pressure vessels. Organic and graphite fibers were tested, and two different epoxy resin systems were used: one with a low and another with a relatively high cure temperature. Acoustic emission was studied for the effects of different winding patterns, artificial flaws, winding-induced fiber fraying, different resins, and different fibers. Small effects produced in the vessels by changes in these variables were greatly magnified when they appeared as changes in acoustic emission. They would, in fact, be difficult or impossible to detect by other test means.

INTRODUCTION

LITTLE PROGRESS HAS been made in the development of nondestructive test methods for filament-wound pressure bottles. Acoustic emission is a promising technique to fulfill this need. But before the capabilities of acoustic emission can be assessed, it will be necessary to gain a more complete understanding of the acoustic emission produced during burst tests of filament-wound pressure bottles. It is toward this interim goal that the work described in this paper was directed. These results are presented with two primary purposes in mind. The first is to point out certain parameters and factors that influence the acoustic emission from filament-wound composites. Second, we hope to demonstrate the insights that can be gained

*This work was performed under the joint auspices of the National Aeronautics and Space Administration–Lewis Research Center–and the Atomic Energy Commission (NASA-AEC Purchase Request C-13980C).

by using acoustic-emission techniques during burst tests of filament-wound composites.

Recently it was reported [1] that for S-glass/epoxy filament-wound pressure bottles, standard acoustic-emission data is sensitive both to changes in the ratio of longitudinal wraps to hoop wraps and to changes in the order of wrapping. The author also concluded that acoustic-emission data could be used to detect a grossly flawed S-glass pressure bottle at about one-third its normal failure pressure. He further indicated that for the S-glass/epoxy system used in those tests, all or nearly all of the acoustic emission (at approximately 70 dB gain) was associated with fiber fracture. This latter result was based in part on the fact that a fracture-toughness, epoxy double-cantilever beam (*DCB*), strengthened with approximately 50 strands of single-end S-glass situated perpendicular to the cleavage plane, produced some 13,000 to 15,000 counts during testing to failure, while an epoxy *DCB* without strands produced approximately 400–600 counts during testing to failure.

SPECIMENS

Composite pressure bottles were wound using two basic wraps. Figure 1 shows a drawing of a typical bottle. The first basic wrap, called the longitudinal wrap, was

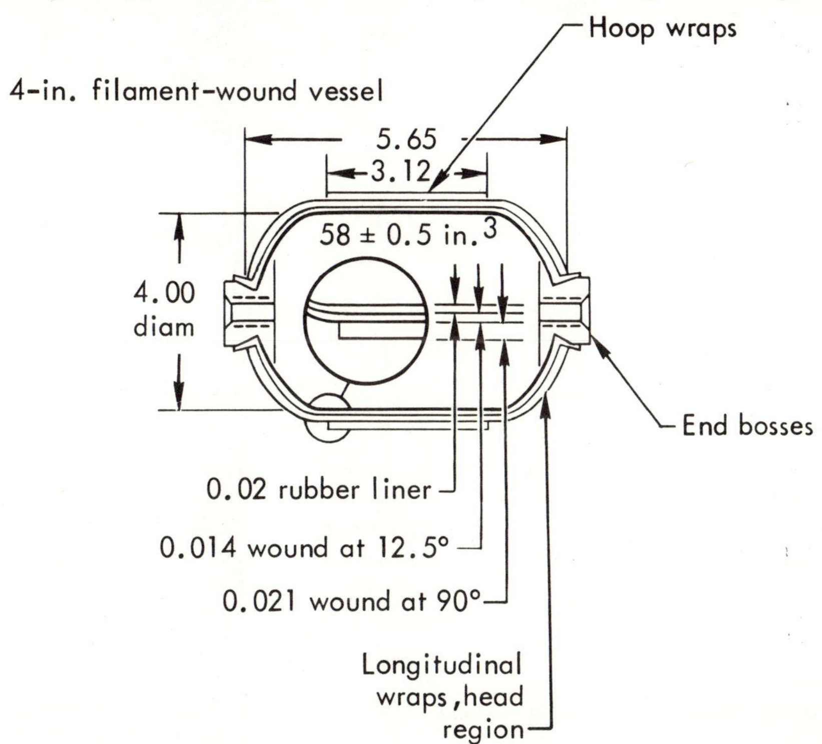

Figure 1. *Typical dimensions (in inches) of a filament-wound bottle.*

laid down at a fixed angle of between 7° to 13° from the longitudinal axis of the bottle. The second basic wrap, called the hoop wrap, was laid down at an angle of 90° to the longitudinal axis. The hoop wraps only appear on the cylindrical section of the bottle.

One of the primary winding variables is the ratio of longitudinal fiber stress σ_L to hoop fiber stress σ_H. This ratio, which we call the fiber stress ratio, was determined from a simple netting analysis applied to the cylindrical portion of the bottle; it gave an approximate measure of the ratio of the fiber stresses in the two different wraps at a given pressure. Hence, a bottle wound with a balanced design (based on the netting analysis) would have a fiber stress ratio $\sigma_L/\sigma_H = 1.0$. More detailed information on the winding conditions and strength characteristics of the bottles can be found in the papers by Chiao et al [2, 3, 4].

Four material systems were used to fabricate the bottles. These material systems, as well as the cure cycles, are summarized in Table 1.

Table 1. Material Systems

Fiber	*Epoxy Resin*	*Hardener*	*Curing Cycle*	
			Gell	*Cure*
PRD 49-III[a]: single end, 285 filaments per end	ERL 2258[b] 100 parts	ZZLB 0820[b] 30 parts	140°F for 16 hr	325°F for 2 hr
PRD 49-III[a]: single end, 285 filaments per end	DER 332[c] 100 parts	T403[d] 36 parts	Room temp. for 16–24 hr	165°F for 3 hr
Thornel[b] 400: single end, 1000 filaments per end	DER 332[c] 100 parts	T403[d] 36 parts	Room temp. for 16–24 hr	165°F for 3 hr
Thornel[b] 400: single end, 1000 filaments per end	ERL 2258 100 parts	ZZLB 0820 30 parts	140°F for 16 hr	325°F for 2 hr

[a]DuPont, Wilmington, Delaware

[b]Union Carbide, New York, New York

[c]Dow Chemical Co., Midland, Michigan

[d]Jefferson Chemical Co., Houston, Texas

TESTING AND ACOUSTIC EMISSION SYSTEMS

All the composite bottles were pressurized with fluid. A closed-loop system was used to control the rate of pressurization. All bottles were pressurized at a rate of 1000 psig/min.

A commercial piezoelectric acoustic-emission transducer was coupled to the cylindrical portion of each bottle with a viscous resin. In a few cases a transducer was also coupled to the "head" region of the bottles. The acoustic signal was processed through a preamplifier, a filter (usually set to pass 100–300 kHz), and a power amplifier. In total, the system provided some 55–75 dB of electronic gain as needed. The acoustic signal was presented visually in two ways. First, the signal was displayed on an oscilloscope as a function of time. In addition, a count was made of the number of times that the amplitude of the acoustic signal exceeded a voltage bias (nominally set at 1 V). This count was recorded on an x-y plotter as a function of pressure, both as a count rate (usually averaged over 1 or 2 sec) and as a cumulative total of counts.

ACOUSTIC EMISSION TESTS

PRD-49-III Fiber/Epoxy Vessels

Figures 2 and 3 illustrate the two broad types of acoustic-emission records of count rate and summation of counts as a function of internal pressure that can be

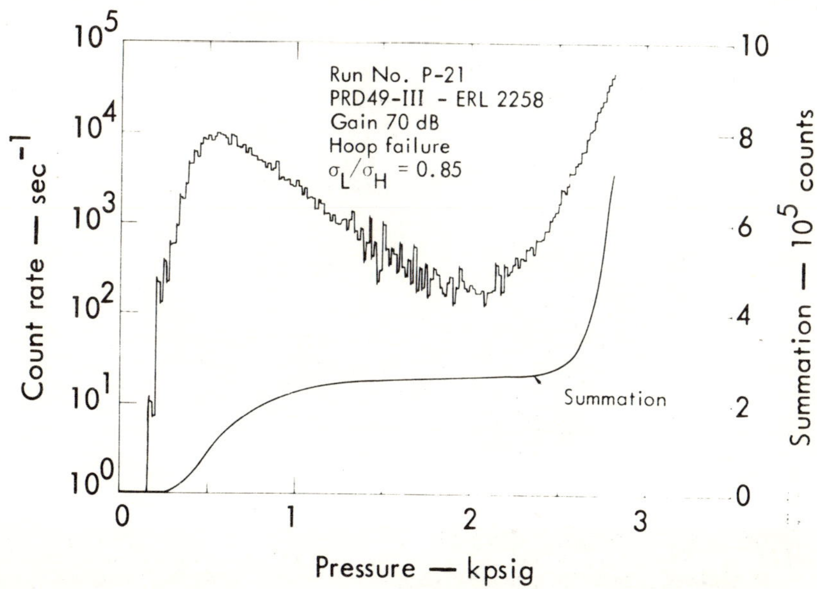

Figure 2. Acoustic emission from an organic-fiber bottle that experiences hoop failure.

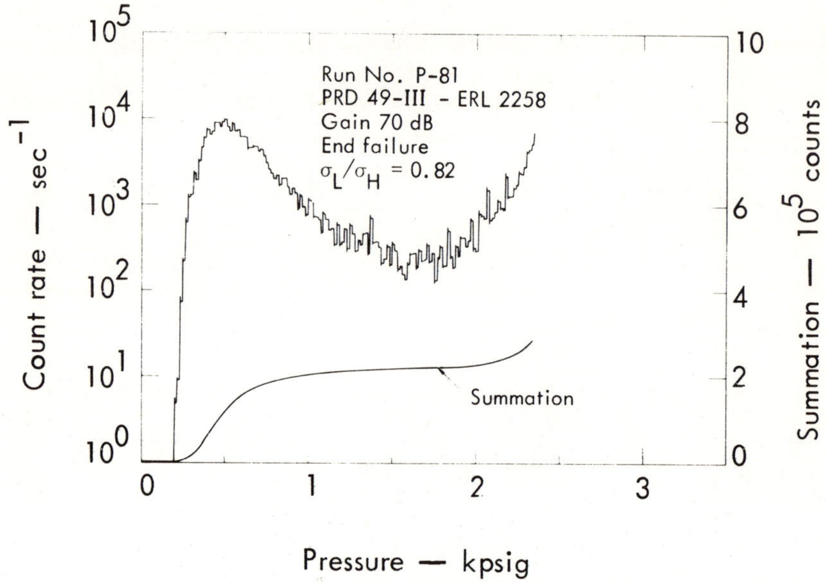

Figure 3. Acoustic emission from an organic-fiber bottle that experiences end failure.

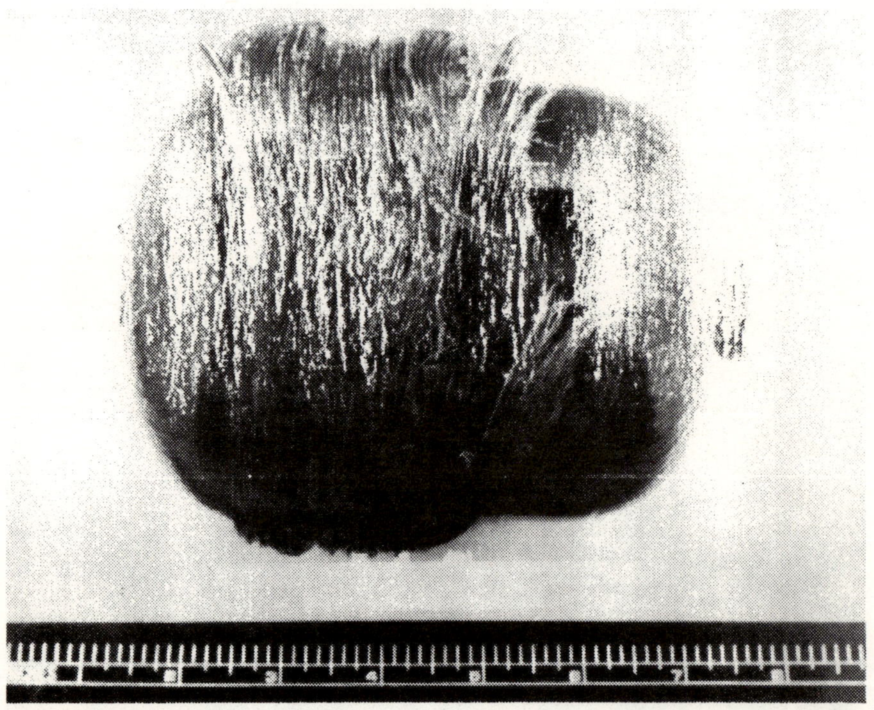

Figure 4. Typical hoop failure of an organic-fiber-reinforced pressure bottle.

obtained from testing of organic fiber composite bottles. The typical predominate failure modes that are associated with these two types of acoustic-emission responses are the so-called hoop failure and end failure. These are shown in Figures 4 and 5. The difference in acoustic emission and failure behavior between these two bottles is associated with how the bottle was wound. The behavior shown in Figures 3 and 5 results from one of the following: (a) a large stress ratio σ_L/σ_H (unbalanced design), (b) a weakness in the fitting region, or (c) failure to reinforce the ends of the hoop wraps. The behavior shown in Figures 2 and 4 results from a sufficiently low stress ratio σ_L/σ_H, adequately reinforced hoop wraps, and no weakness in the fitting region. The typical geometric patterns of acoustic emission shown in Figures 2 and 3 did not significantly change when the wrapping pattern was changed to the so-called interspersed pattern. The interspersed bottle was wound by the following wrapping sequence: two hoop layers, two longitudinal layers, two hoop layers, two longitudinal layers, and, finally, two hoop layers. The noninterspersed bottle was wound by first laying down four longitudinal layers and then six hoop layers. Unless otherwise stated, all bottles were wound with the noninterspersed pattern or with an intermediate pattern consisting of the sequence: two hoop layers, four longitudinal layers, and, finally, four hoop layers.

Figure 5. Typical end failure of an organic-fiber-reinforced pressure bottle.

Figure 6 shows the differences in the summation of acoustic emission versus internal pressure between unflawed bottles that typically experienced end failures

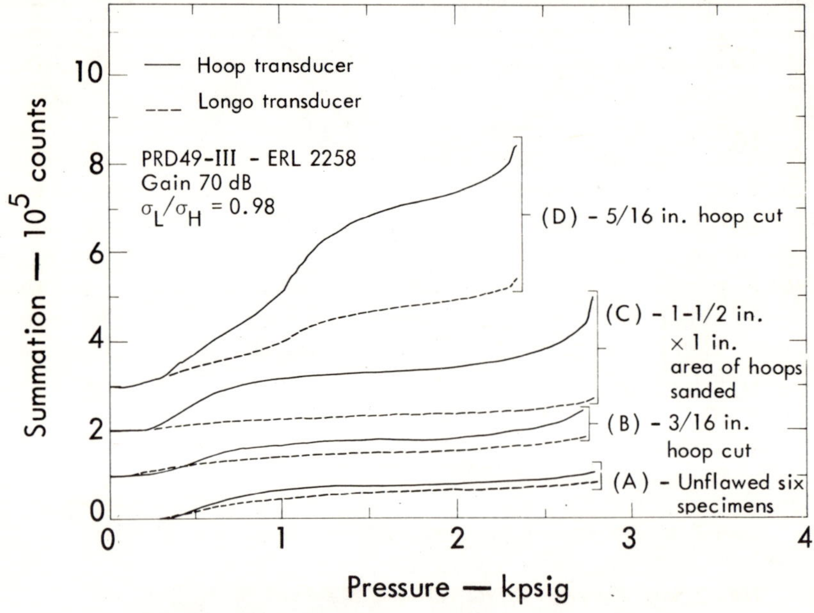

Figure 6. Acoustic emission from flawed organic-fiber bottles.

and bottles with various indicated flaws. In these tests two acoustic emission transducers were used. The transducer on the cylindrical portion of the bottle was called the hoop transducer, and the one on the "head" portion was called the longo transducer. The curves labeled (*A*) represent the typical summation of counts from both transducers for unflawed specimens. Curves (*B*), (*C*), and (*D*) show the typical emission from specimens with flaws: a 3/16 in. cut in the outer hoop wraps (*B*), a 1½ × 1 in. region of the outer hoop wraps sanded with emery paper (*C*), and a 5/16 in. cut in the outer hoop wraps (*D*). For clarity of presentation the zero summation line has been moved up one major division for curves (*B*), (*C*), and (*D*). In case (*B*) the flaw did not control the failure of the bottle. This bottle failed in the end or "head" region, as did the unflawed specimens. The failure location in the other two specimens, curves (*C*) and (*D*), was controlled by the artificial flaws.

Figure 7 compares the acoustic emission from two specimens that differ significantly only in the type of epoxy resin used. In one case the resin was the ERL 2258 system, while in the other case it was the lower-cure-temperature DER 332 system. Some of the properties of these two resin systems, when cured according to the curing cycles shown in Table 1, are summarized in Table 2.

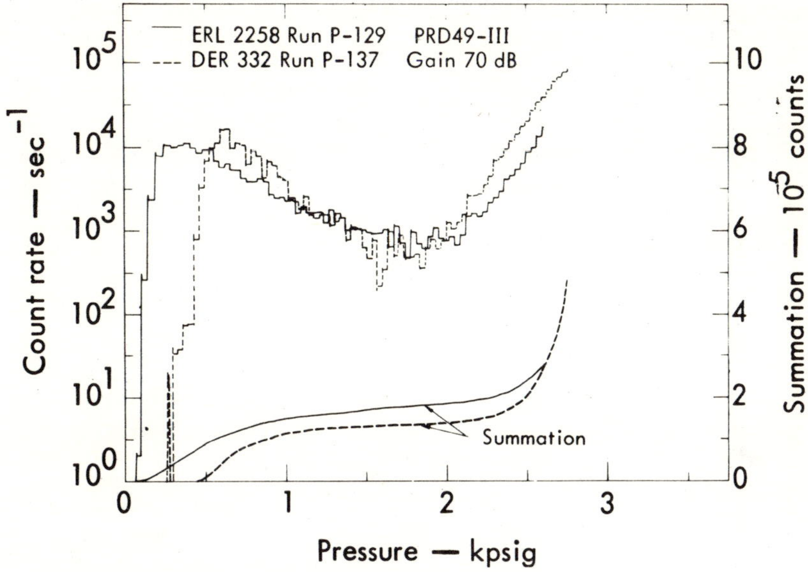

Figure 7. Acoustic emission from organic-fiber bottles with different epoxies.

Table 2. Some Properties of ERL 2258 and DER 332

Epoxy System	*Ultimate Tensile Strength (ksi)*	*Tensile Modulus (ksi)*	*Rupture Elongation (%)*
ERL 2258/ZZL 0820	16.5	557	6.2
DER 332/T-403	10.9	494	5.0

Thornel 400 – Fiber/Epoxy Vessels

Three typical acoustic-emission responses showing summation of counts versus internal pressure are shown in Figure 8 for filament-wound graphite bottles. These three curves represent the typical random variation between supposedly identical bottles. There is no significant deviation from the geometry of these curves when the ratio of hoop wraps to longitudinal wraps (σ_L/σ_H) is changed or when the bottle is wound using the interspersed pattern. The failure mode for the graphite bottles varies in such a fashion that even for identical specimens it is very difficult to find as many as two bottles that have failed in an identical fashion. A bottle that has been tested to failure is shown in Figure 9. Interspersed bottles or bottles with different stress ratios do not have characteristic failure modes either.

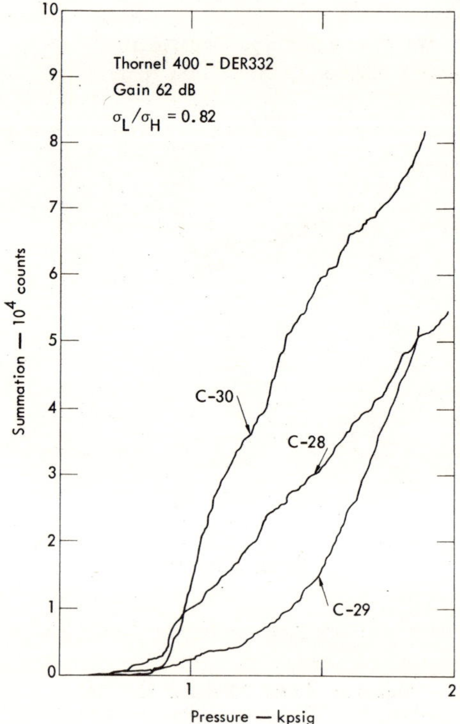

Figure 8. Acoustic emission from graphite-fiber bottles.

Figure 10 compares the count rate and summation of counts for a bottle wound with graphite yarn that tended to "fray" during the winding process with the count rate and summation of counts for a bottle wound with little or no "fraying" during winding.

The effect of the two different epoxy systems on the acoustic emission of otherwise nearly identical graphite bottles is shown in Figure 11. One bottle was wound using the ERL 2258 resin system, while the other was wound using the lower-cure-temperature DER 332 formulation.

DISCUSSION

The discussion and the interpretation of the above experimental results are based on the conclusion that the large majority of the acoustic-emission data gathered at the gain levels used in these tests comes from, or is associated with, fiber failure. The basis for this conclusion was alluded too briefly in the introduction and is discussed more completely in Reference [1]. Although these discussions primarily refer to S-glass-reinforced epoxies, the basic significant finding carries through to organic- and graphite-reinforced epoxies. Namely, the fiber failure stresses are much

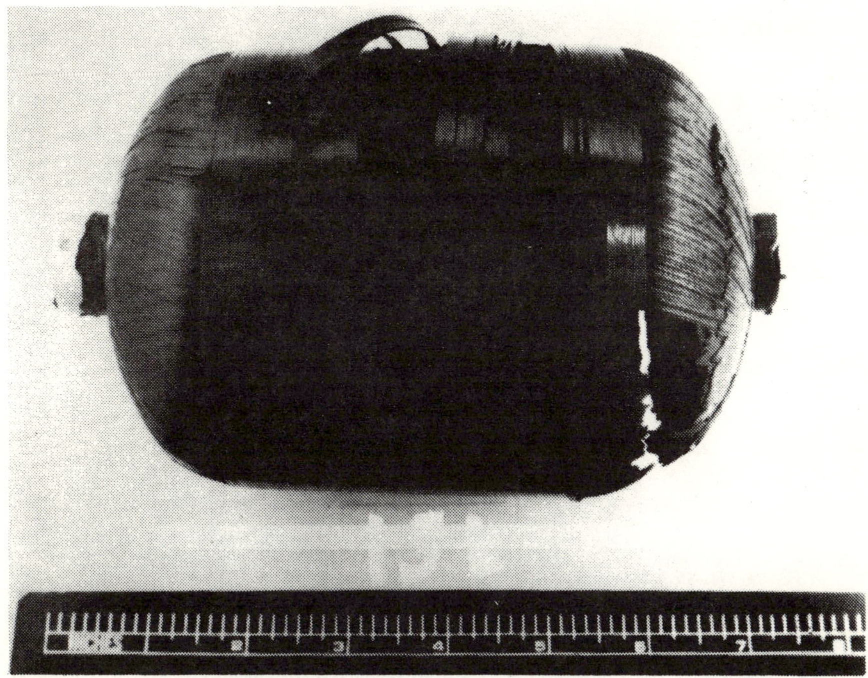

Figure 9. A failed graphite-fiber-reinforced bottle.

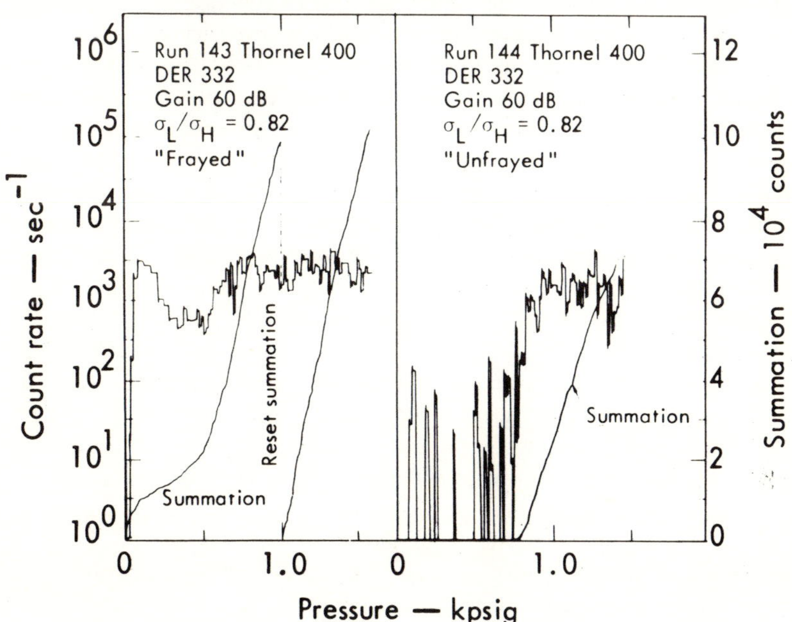

Figure 10. Acoustic emission from "frayed" and "unfrayed" graphite bottles.

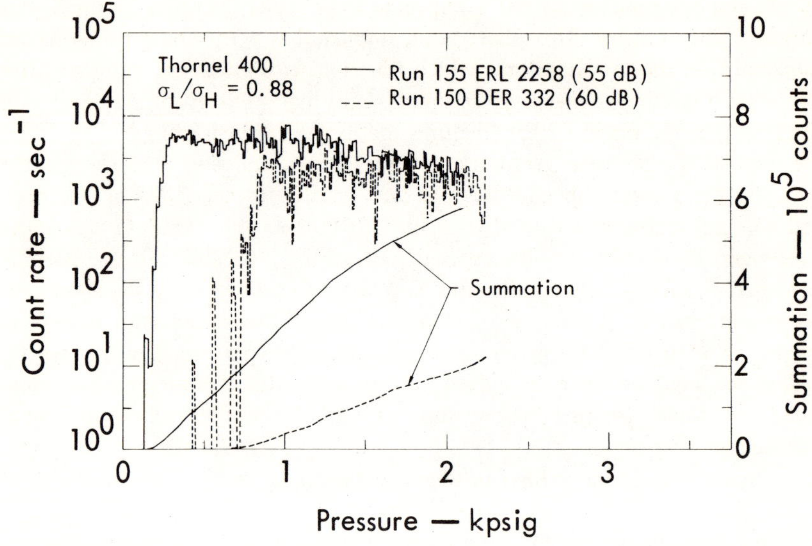

Figure 11. Acoustic emission from graphite fiber bottles with different epoxies.

higher than matrix failure or interfacial debonding stresses, and hence the energy released by these latter mechanisms is much less than that released by fiber failure.

The contrast between summation of counts and count rate shown in Figures 2 and 3 illustrates for organic fiber-reinforced epoxies one of the key applications of acoustic emission to composite pressure bottle design. As was pointed out in Reference [1] (for S-glass-reinforced epoxies), the rapid rise in summation of counts and count rate just prior to failure (shown in Figure 2) results from the many fiber breaks that occur in the relatively uniformly stressed hoop wraps. Thus, the acoustic-emission data can be used for organic and S-glass bottles to distinguish between bottle designs that efficiently load the hoop wraps prior to failure and those that do not. The acoustic-emission behavior shown in Figure 3 indicates that failure in this bottle occurred before the hoop wraps were highly loaded. The three basic causes for this behavior were pointed out in a previous section.

The rapid rise in count rate and summation counts just prior to failure has not been obtained for the graphite fiber-reinforced bottles in spite of several winding pattern changes. This result, coupled with the seemingly brittle failures that typically occur in these bottles (see Figure 9), leads us to conclude that the average number of fiber breaks per unit volume (i.e., the average density of fiber breaks) at which the graphite bottle fails is considerably smaller than the average density of fiber breaks for S-glass and organic bottles. Thus, the graphite bottle fails before the density of breaks undergoes the rapid rise that generates the rapid rise in the acountic-emission data just prior to failure of organic and S-glass bottles. Based on this result, the percentage of single-strand fiber strength that can be obtained in a composite graphite pressure bottle[1] is expected to be lower than that for the S-glass and organic fiber bottles. This observation implies that it may be possible to use acoustic-emission data to determine which fiber-matrix systems are more sensitive (in the sense that they experience complete failure at a lower average density of broken fibers) to the random fiber fractures that occur during loading.

The fact that the basic geometry of the acoustic-emission patterns for the organic- and the graphite-fiber bottles was not significantly altered when the interspersed winding sequence was used is an important observation. This result implies that for these two fibers the interspersed pattern will probably not degrade the performance of such vessels significantly, and may even offer a slight improvement. It was pointed out in Reference [1] that the use of the interspersed pattern in S-glass led to a significantly different acoustic-emission pattern. This pattern indicated a degraded bottle and, indeed, the experimental failure pressures were degraded by some 14–18% below those found for bottles having the non-interspersed pattern.

Both Figure 7 for organic-fiber bottles and Figure 11 for graphite-fiber bottles show that the use of a lower-cure-temperature epoxy system led to an increase in the internal pressure at which a significant amount of acoustic emission begins (defined as the pressure where the summation of counts rises above the zero line). The increase is on the order of two to four times the pressure at which acoustic emission begins in the high-cure-temperature resin bottle. At this point it is only possible to speculate on the cause for this change in acoustic-emission response. It may be that this phenomenon is related to the residual stresses or the fiber matrix configurations that are induced during the cure cycle. In the future a statistical study is planned to determine if the different matrix systems significantly change the average failure pressures of the filament-wound pressure bottles.

The result shown in Figure 10 demonstrates that acoustic emission could be used to check production processes if one desired to sort out graphite-reinforced bottles that had experienced significant "fraying" during winding. It would be necessary to pressurize each bottle to a pressure of only 400 or 500 psig to make this determination. The bottles that show the early acoustic emission could then be separated out. The actual mechanism through which the frayed yarn leads to early emission is not clear.

The results shown in Figure 6 for the flawed organic-fiber bottles indicate that it is possible to detect a flawed bottle when the flaw is controlling the bottle failure. Such flaws may be detected at approximately one-half the normal failure pressure. Note that in test (*B*), where the flaw did not control the bottle failure, that the summation of acoustic emission gave very little indication that the flaw was present. It is also clear from test (*C*) that the location of the acoustic-emission transducer with respect to the flaw location is important. In this case the longo transducer gave no visible indication that the flaw was present. In case (*D*) both

[1] Based on a large statistical sample of strands and bottles.

transducers gave adequate indication of the presence of the flaw. The reason for this difference can be partly explained by a difference in the mechanism of propagation of the flaws in cases (*C*) and (*D*). When the acoustic-emission signals from the hoop transducer in these two cases were viewed on an oscilloscope, the indication was that the flaw growth was proceeding in a different manner in each case. The acoustic-emission signals had very large amplitudes in case (*D*). Often the signals completely saturated the electronic equipment. This observation indicated that the flaw growth was proceeding by events involving large energy releases. In case (*C*) the acoustic-emission signals tended to be comparatively small (indicating growth by small energy releases). These rather small signals, which originated in the hoop wraps for case (*C*), were attenuated during propagation to the longo transducer and hence did not have enough amplitude to be counted by the longo acoustic-emission channel. One further point should be made with regard to these tests. In spite of relatively gross flaws, in case (*B*) the flaw did not control the failure, in case (*C*) the bottle failure pressure was not significantly degraded, and in case (*D*) the failure pressure was degraded only some 15%. Hence, it would seem that the organic-fiber bottle is relatively resistant to artificial flaws.

CONCLUSIONS

For S-glass and organic filament-wound pressure bottles, acoustic emission can be used to determine if the hoop wraps are highly loaded before bottle failure occurs.

Acoustic-emission techniques may offer a means to assess the average density of fiber breaks at bottle failure for different combinations of fibers and epoxy matrixes.

Acoustic emission indicates that bottles wound with Thornel-400 may fail at a much lower average density of fiber breaks than S-glass and PRD-49-III bottles.

Acoustic emission could be used to sort out graphite bottles that were frayed during the winding process.

Acoustic emission can sort out flawed organic fiber bottles if the flaw controls the bottle failure.

Use of a lower-cure-temperature epoxy matrix increased by a factor of between two and four the pressure at which graphite- and organic-fiber bottles first produce significant acoustic emission.

REFERENCES

1. M. A. Hamstad, "Acoustic Emission from Filament-Wound Pressure Bottles," Proc. 4th SAMPE Natl. Tech. Conf., Palo Alto, California (October, 1972).
2. T. T. Chiao and A. D. Commins, "Fiber Strength of S-Glass/Epoxy Composites Under Bi-Axial Loading," Proc. 4th SAMPE Natl. Tech. Conf., Palo Alto, California (October, 1972).
3. T. T. Chiao and M. A. Marcon, "Filament-Wound Vessel from An Organic Fiber/Epoxy System," Proc. 28th Annual Conf. on Reinforced Plastics/Composites Institute, SPI., Washington, D.C., February 6–10, 1973.
4. T. T. Chiao, R. L. Moore, and C. M. Walkup, "Graphite Epoxy Composites," submitted to SAMPE Quarterly.

Bulk Compressibility of Carbon Fibre Reinforced Plastics

A. SMITH *and* W. N. REYNOLDS, *Nondestructive Testing Centre;*
N. L. HANCOX, *Process Technology Division, Atomic Energy Research Establishment, Harwell, England*

(Received October 27, 1972)

A number of papers have recently been published reporting experimental verification of calculations of shear modulus of uniaxially reinforced composites, [1, 2, 3]. Reynolds and Hancox [4] have shown that for carbon fibre composites a nondestructive measurement of torsional shear modulus, combined with the volume fraction of fibres gives a good guide to the shear strength.

Calculations of composite elastic moduli in terms of those of the components yield theoretical values for the bulk compressibility. In this note we compare measured and theoretical values of bulk modulus for composites of different volume loadings made with surface treated and untreated types I and II fibres. We also measured the torsional moduli and strengths of the specimens and have used this information to relate the bulk modulus to the composite shear strength. When the bonding between fibres and resin is incomplete the stiffening effect of the fibres is reduced. Measurement of compressibility may therefore be used as a guide to the shear strength.

EXPERIMENTAL WORK

Uniaxial specimens 150mm X 6.5mm square were made from types I and II carbon fibre, some of which had been surface treated to give better adhesion and an anhydride cured epoxy resin cured for 3 hours at 120°C. Fibre volume loadings of 50, 60 and 70% were used. The centre 100mm of each specimen was ground down to a diameter and the torsional modulus obtained, see [5]. The bulk compressibility was then measured and the torsional strength and modulus remeasured.

The compressibility and hence bulk modulus were obtained as described by Smith [6]. A full range of measurements up to a pressure of 110MNm^{-2} was made at 25° and 60°C. The pattern of results was similar at both temperatures.

The results for bulk modulus, and shear modulus and strength are listed with other pertinent data in Table 1. The theoretical values of bulk modulus were computed by Heaton [7].

In Figures 1(a) and 1(b) the volume change measured at 110MNm^{-2} is shown as a function of fibre volume content. The theoretical values are also shown.

Table 1. Mechanical Properties of Carbon Fibre Reinforced Resins

Vol % Fibre	Fibre Type	Treated T or untreated U	Density g cm^{-3}	Bulk Modulus G Nm^{-2}		Theoretical Value	Shear Modulus G Nm^{-2}	Shear Strength M Nm^{-2}
				25°	60°			
50	1	U	1.574	7.20	6.93	8.72	3.85	16.5
		T	1.572	7.10	6.67		3.4	50.2
	2	U	1.502	8.22	7.63		4.3	48.2
		T	1.505	8.46	8.09		4.2	53.0
60	1	U	1.617	7.00	6.84	10.3	2.65	12.1
		T	1.638	7.53	7.15		4.5	51.0
	2	U	1.551	9.18	8.67		4.7	56.5
		T	1.545	9.24	8.82		4.7	62.6
70	1	U	1.684	7.58	7.20	11.9	4.2	10.4
		T	1.702	7.70	7.29		4.85	44.0
	2	U	1.610	10.28	9.93		6.05	55.7
		T	1.598	10.08	9.58		6.55	57.0

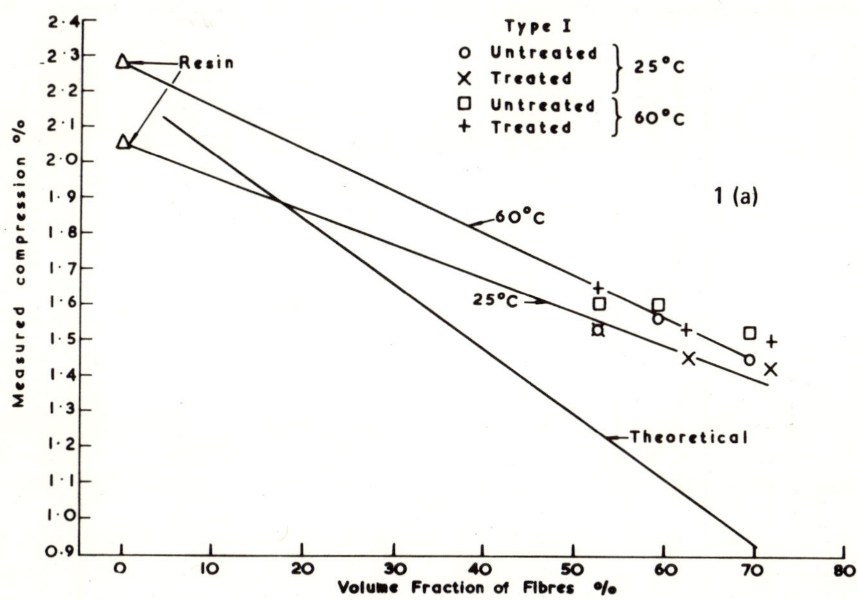

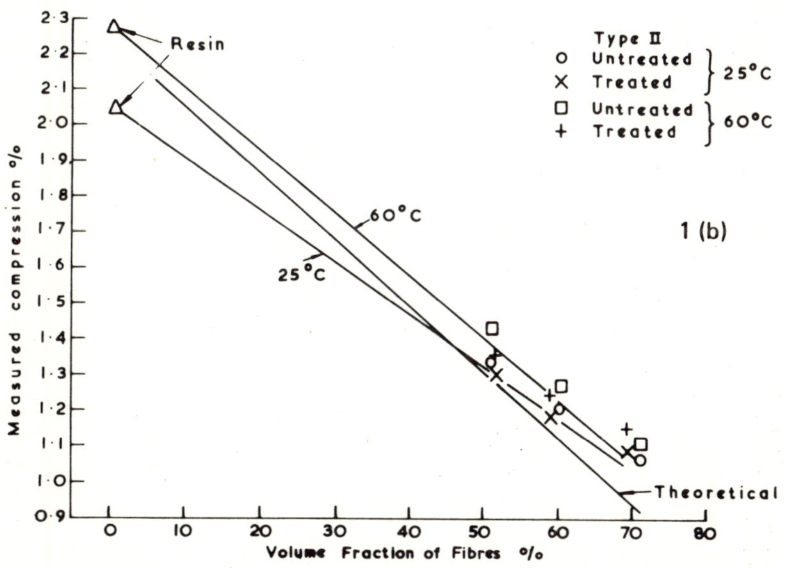

Figure 1. *Compressibility of specimens measured at 110 MNm⁻² (16000 psi) at 25 and 60° as a function of specimen density: (a) for Type I, (b) for Type II.*

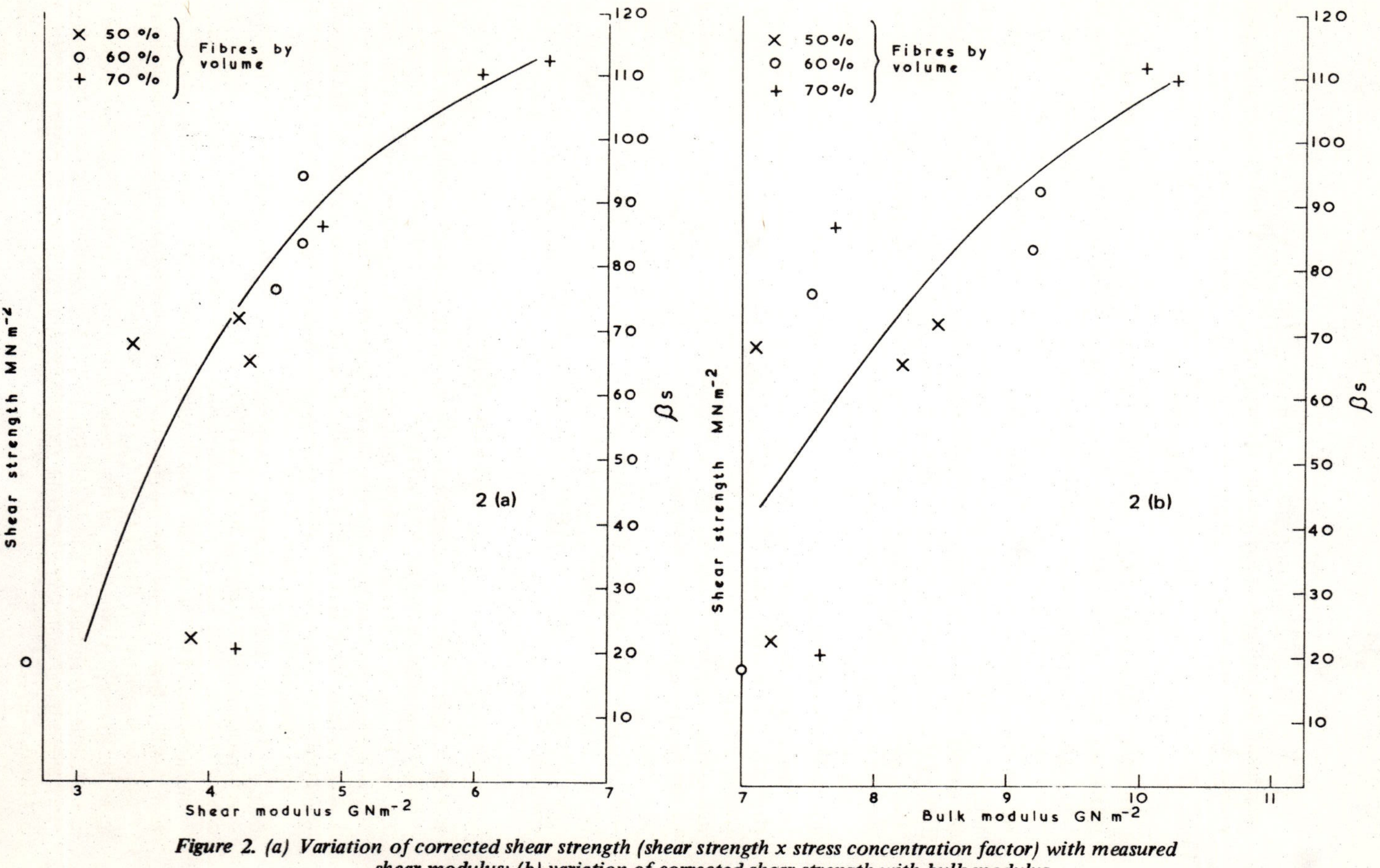

Figure 2. *(a) Variation of corrected shear strength (shear strength x stress concentration factor) with measured shear modulus; (b) variation of corrected shear strength with bulk modulus.*

A careful examination of the shear modulus and strength results for this batch of specimens in the light of the results reported in [4] shows that though individual results deviate the overall behaviour does not differ markedly from average.

DISCUSSION

Several interesting points are shown by Figures 1(a) and 1(b). As expected increasing fibre content decreases the amount of compression noted, but it is at first sight surprising that the lower modulus type II fibre should have a greater effect pro rata than type I. As fibre transverse and shear moduli are believed to be similar the explanation must be in the weaker bonding of the type I fibre even after surface treatment. Surface treatment causes a measureable decrease in compressibility in all cases except 50% type I and 70% type II. However it may be seen from Table 1 that the shear modulus result for a 50% type I specimen is anomalous. Thus bulk and shear moduli measurements appear comparable in consistency. Results taken at 60°C are very similar to those made at the lower temperature.

Since one of the purposes of this work was to find nondestructive tests for shear strength it is interesting to see how the bulk modulus can be used for this purpose. In Figure 2(a) measured shear moduli have been plotted against measured shear strength, S x a stress concentration factor β which allows for the stress concentrating effects of introducing fibres into the matrix. The values used are those determined experimentally by [4]. In Figure 2(b) the bulk moduli corresponding to the shear moduli of the specimens used here are plotted against the same values of corrected shear strength (S x β). Although there is considerable scatter in both cases it should be recalled that each point refers to a strength measurement on a single specimen. Further work with much better statistics would be required to establish whether there is a good correlation between bulk modulus and shear strength.

REFERENCES

1. P. R. Goggin, "The elastic constants of carbon fibre composites," Atomic Energy Research Establishment Report, AERE R 6838 (1972).
2. C. G. Brown, N. L. Hancox and W. N. Reynolds, "NDT of carbon fibre reinforced plastics – a study of an ultrasonic resonance technique," *Journal Physics D*, Vol. 5 (1972), p. 782.
3. R. D. Adams, M. A. O. Fox, R. J. L. Flood, R. J. French and R. L. Hewett, "The dynamic properties of unidirectional carbon and glass reinforced plastics in torsion and flexure," *J. Comp. Materials*, Vol. 3 (1969), p. 594.
4. W. N. Reynolds and N. L. Hancox, "Shear strength of the carbon resin bond in carbon fibre reinforced epoxies," *J. Physics D.*, Vol. 4 (1971), p. 1747.
5. N. L. Hancox, "The use of a torsion machine to measure shear strength and modulus of unidirectional CFRP," *J. Mat. Sci.*, Vol. 7 (1972), p. 1030.
6. A. Smith, "High pressure bulk modulus test rig – a nondestructive test for specimens of composite material," *J. Physics E.*, Vol. 5 (1972), p. 276.
7. M. D. Heaton, "A calculation of the elastic constants of a unidirectional fibre reinforced composite," *J. Physics D.*, Vol. 1 (1968), p. 1039. "A calculation of the elastic constants of a unidirectional composite containing transversely isotropic fibres," *J. Physics D.*, Vol. 3 (1970), p. 672.